# Mineral-Based Composite Catalytic Materials

# Mineral-Based Composite Catalytic Materials

Guest Editors

**Pengwei Huo**
**Xin Liu**
**Zhi Zhu**
**Yunlei Zhang**

MDPI

Basel • Beijing • Wuhan • Barcelona • Belgrade • Novi Sad • Cluj • Manchester

*Guest Editors*
Pengwei Huo
School of Chemistry and Chemical Engineering
Jiangsu University
Zhenjiang
China

Xin Liu
School of Chemistry and Chemical Engineering
Jiangsu University
Zhenjiang
China

Zhi Zhu
School of Chemistry and Chemical Engineering
Jiangsu University
Zhenjiang
China

Yunlei Zhang
School of Chemistry and Chemical Engineering
Jiangsu University
Zhenjiang
China

*Editorial Office*
MDPI AG
Grosspeteranlage 5
4052 Basel, Switzerland

This is a reprint of the Special Issue, published open access by the journal *Catalysts* (ISSN 2073-4344), freely accessible at: https://www.mdpi.com/journal/catalysts/special_issues/M9XW5N2P6L.

For citation purposes, cite each article independently as indicated on the article page online and as indicated below:

| Lastname, A.A.; Lastname, B.B. Article Title. *Journal Name* **Year**, *Volume Number*, Page Range. |
|---|

**ISBN 978-3-7258-5587-2 (Hbk)**
**ISBN 978-3-7258-5588-9 (PDF)**
**https://doi.org/10.3390/books978-3-7258-5588-9**

# Contents

catalysts

MDPI

Article

# PAN/$TiO_2$ Ultrafiltration Membrane for Enhanced BSA Removal and Antifouling Performance

**Yinshan Xie [1], Xinning Wang [1], Hulin Li [1], Tao Wang [2], Wei Feng [2,*] and Jian Li [1,*]**

[1] Laboratory of Environmental Biotechnology, School of Environmental and Civil Engineering, Jiangnan University, Wuxi 214122, China; jameson618@163.com (Y.X.); sanni.wxn@gmail.com (X.W.); lihulin03@163.com (H.L.)

[2] School of Food Science and Technology, Jiangnan University, Wuxi 214122, China; twang3813@gmail.com

* Correspondence: jjfengwei@jiangnan.edu.cn (W.F.); jian.li@jiangnan.edu.cn (J.L.)

**Abstract:** Membrane separation has been widely utilized to eliminate pollutants from wastewater. Among them, a polyacrylonitrile (PAN) ultrafiltration (UF) membrane has presented outstanding stability, and distinguished chemical and thermal properties. However, UF membranes inevitably incur fouling issues during their operation procedure caused by contaminant adhesion on the membrane surface, which would restrict the operational efficiency and increase the maintenance cost. The conventional physical and chemical cleaning is not an effective technique to reduce the fouling due to the additional chemical addition and inevitable structure damage. Recently, UF membranes combined with photocatalytic materials are suggested to be a useful approach to conquer the membrane fouling issues. Herein, $TiO_2$ nanoparticles were utilized to blend with a PAN casting solution for fabricating a composite UF membrane via a phase inversion method. With a certain $TiO_2$ addition, the obtained membranes presented an enhancement of hydrophilicity, which could promote the water permeability and antifouling performance. The optimized M3 membrane prepared with 15.0 wt% PAN and 0.6 wt% $TiO_2$ exhibited an excellent water permeability up to 207.0 L $m^{-2}$ $h^{-1}$ $bar^{-1}$ with an outstanding 99.0% BSA rejection and superior antifouling property. In addition, the photocatalytic $TiO_2$ nanoparticles endowed the M3 membrane with a remarkable self-cleaning ability under the UV irradiation. This facile construction method offered new insight to enhance the UF membrane separation performance with an enhanced antifouling ability.

**Keywords:** polyacrylonitrile; ultrafiltration; membrane fouling; $TiO_2$; photocatalysis

## 1. Introduction

Much attention has been paid to water quality and available water resources due to the speedy development of industrialization, populations and increased water pollution [1]. Various separation methods, e.g., ion exchange, membrane separation and adsorption, have been deployed to handle the pollutants from wastewater [2]. Membrane separation technologies, based on microfiltration (MF), ultrafiltration (UF) and nanofiltration (NF) membranes, have exhibited excellent advantages with high separation efficiency and low energy consumption, which outperform other traditional separation techniques [3,4]. Among them, UF membranes with high water permeation, a small footprint and modularity have been extensively deployed in water purification and wastewater treatment [5,6]. Generally, UF membranes are mainly fabricated with polymer materials such as polyethersulfone (PES), polyvinylidene fluoride (PVDF) and polyacrylonitrile (PAN) [7,8]. Compared to the other polymer materials, PAN possesses outstanding solubility in organic solvents, and a prepared PAN UF membrane exhibits distinguished chemical and thermal properties [9,10]. However, during wastewater treatment, especially for those systems containing organic compounds, the separated pollutants can be concentrated on the membrane surface and inside the pores via hydrophobic and electrostatic interactions. As a result, the gathered contaminants need to be further removed or it could reduce the membrane separation efficiency and service life [11,12]. Hence, it remains challenging to explore an effective way for membrane fouling elimination.

Efficient methods such as adjusting membrane hydrophilicity, roughness and surface charge and optimizing the hydrodynamic condition have been utilized to improve the membrane fouling resistance [13,14]. Surface hydrophilicity majorization is an efficient method that contributes to reduce the membrane fouling. Thus, multiple approaches, such as surface grafting of antifouling materials, incorporating a hydrophilic substance during the membrane fabrication process and introducing inorganic particles into the polymer materials, have been employed to prepare a fouling-resistant UF membrane [15]. However, these modified UF membranes are still prone to sustain the fouling issues after a long-time filtration via the conventional physical and chemical cleaning, which is derived from the irreversible fouling [16]. Currently, advanced oxidation processes (AOPs), such as photocatalysis, Fenton and ozonation, have proven to be effective technologies to degrade the organic pollutants [17]. Among these AOPs, photocatalysis can effectively handle the contaminants with less chemical reagent usage and energy consumption [18,19]. Thus, filtration membranes combined with photocatalytic oxidation technology can be an efficient method to alleviate the membrane fouling issues [20,21]. Specifically, an electron of the semiconductor photocatalyst inside the membrane can be excited from the valence band (VB) to the conduction band (CB) facilitated with visible or UV light [22,23]. Thus, a positive charge and electron–hole pair are formed in the VB band during the process, which can generate the hydroxyl radicals (·OH) and eventuate the degradation of contaminants on the membrane surface [24,25]. For example, a nano-ZnO particle was employed in the fabrication of a membrane matrix to obtain a photocatalysis PVDF UF membrane with a phase inversion approach, which exhibited enhanced fouling resistance and excellent self-cleaning under UV irradiation [26]. Meng et al. prepared a g-$C_3N_4/Bi_2MoO_6$-blended polysulfone UF membrane to improve the bovine serum albumin (BSA) rejection and antifouling performance with a high photocatalytic self-cleaning property under UV light irradiation [27]. A polysulfone (PSF) UF membrane combined with resorcinol–formaldehyde (RF) and a β-FeOOH nanoparticle exhibited an excellent water permeability (376.3 L $m^{-2}$ $h^{-1}$ $bar^{-1}$) with BSA retention (97.5%), which could also effectively degrade the organic fouling under visible light illumination based on the cooperation of $H_2O_2$ [28].

Based on the highly photocatalytic, hydrophilic and inexpensive properties, titanium dioxide ($TiO_2$) has been broadly deployed in water treatment [29,30]. Previous studies have proved that membranes decorated with $TiO_2$ particles could improve the membrane separation performance, porosity and tensile strength [31]. With the desired hydrophilic properties, immobilization of $TiO_2$ particles endows the obtained membrane with enhanced hydrophilicity, inducing a better water permeability, and fouling resistance [32]. In addition, $TiO_2$ can be utilized to modify the membrane structure. A highly interconnected pore structure of a PVDF membrane was obtained with the introduction of $TiO_2$ into the matrix, resulting in a promoted pure water flux [33]. Moreover, immobilization of $TiO_2$ particles with UF membranes can effectively remove organic foulants in the presence of UV illumination. Accordingly, a $TiO_2$-based UF membrane could absorb incident photons and generate ·OH for decomposition of the contaminants existing on the membrane surface [34]. Damodar et al. fabricated composite PVDF membranes via mixing $TiO_2$ nanoparticles in a casting solution and the derived UF membrane presented a significantly improved antifouling property of the membrane under UV irradiation [35]. Thus, incorporation of $TiO_2$ particles into the UF membrane can improve the membrane antifouling performance and extend the membrane reusability.

In this work, we decorated a PAN UF membrane with $TiO_2$ particles through a phase inversion method for wastewater purification with improved membrane performance, including the pure water permeability, BSA rejection, flux recovery rate and self-cleaning property. Moreover, surface morphologies and chemical composition of the obtained membranes were explored. Meanwhile, the effects of PAN and $TiO_2$ particle concentration for the membrane separation performance were detected. In addition, the self-cleaning ability of fabricated membranes and the mechanism were investigated under UV irradiation.

## 2. Results and Discussion

### 2.1. Morphologies of Membranes

The surface and cross-section morphologies of the obtained UF membranes are presented in Figure 1. According to Figure 1a, the M0 membrane possessed a microporous and smooth structure, which was generated with the phase inversion of the casting solution during the phase-conversion process [36]. With the introduction of $TiO_2$, crystal nuclei around 50 nm were uniformly exposed on the M1 membrane surface (Figure 1b). After an adequate concentration of $TiO_2$ was introduced to the membrane (Figure 1c,d), the M3 membrane presented a uniform distribution of particles with a small size on the membrane surface. However, continuously increased $TiO_2$ particle loading leads to the aggregation of the particles (Figure 1e,f), which may clog the membrane pores. Meanwhile, accumulation of $TiO_2$ particles would render the membrane surface rougher, resulting in a serious membrane fouling and a decline in water permeability. In addition, the PAN-based membranes displayed a typical asymmetric cross-sectional porous structure with finger-shaped macrovoids (Figure 1g–i), which was consistent with the non-solvent during the phase inversion process [37]. Meanwhile, the deposition of $TiO_2$ particles inside the membranes did not significantly change the cross-section of the PAN membrane, indicating an excellent accommodation of the PAN matrix and $TiO_2$ particles.

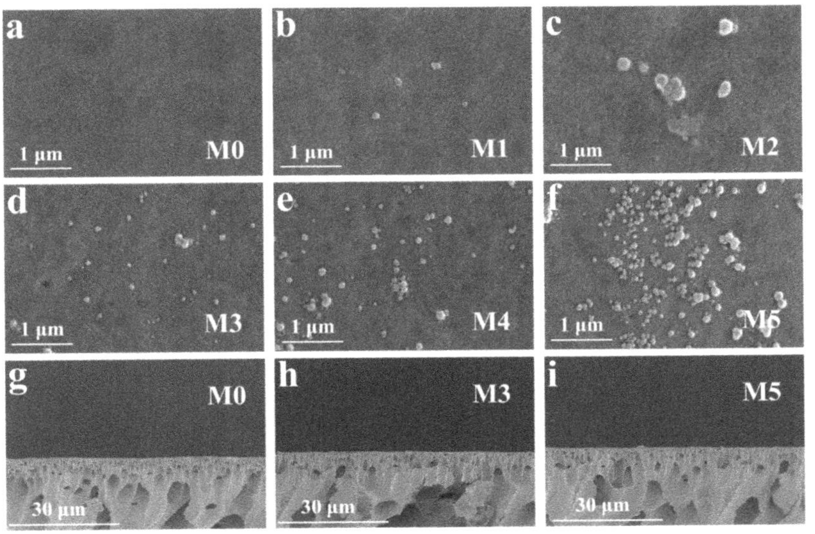


**Figure 1.** Surface SEM morphologies of (**a–f**) M0-M5 membranes; cross-sectional SEM morphologies of (**g**) M0, (**h**) M3 and (**i**) M5 membranes.

The XRD patterns of the obtained membranes are displayed in Figure 2. According to the XRD spectrum of M0, a broad diffraction peak was found around $2\theta = 17.0°$, corresponding to the (100) crystallographic plane of the PAN hexagonal structure [38,39]. As mentioned, the shape of the M0 membrane represented that the PAN membrane mainly possessed the amorphous structure with a negligible crystal feature [29]. Regarding the introduction of $TiO_2$, a new diffraction peak appeared at around $2\theta = 26.5°$ for M1–M5 membranes, which was attributed to the (101) crystal planes of $TiO_2$ nanoparticles [40], indicating the successful loading of $TiO_2$ particles on PAN UF membranes.

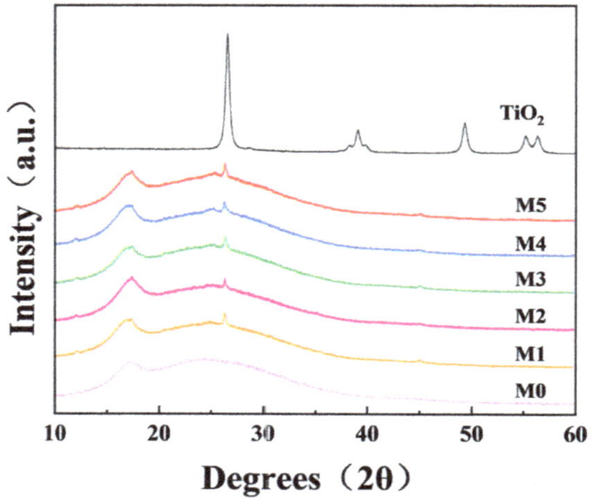


**Figure 2.** XRD spectra of prepared UF membranes.

### 2.2. Chemical Properties of the Membranes

FTIR was employed to detect the chemical compositions of the synthetic membrane surface (Figure 3). As exhibited in Figure 3a, four distinctive absorption peaks of PAN membranes at 2940 $cm^{-1}$, 2243 $cm^{-1}$, 1729 $cm^{-1}$ and 1453 $cm^{-1}$ corresponded to the stretching vibrations of C-H, -C ≡ N-, -C=O and -$CH_2$, respectively [41,42]. The peaks around 3500 $cm^{-1}$ were assigned to the bending vibration of -OH bonds of the $TiO_2$ particles [43]. In addition, the spectrum of $TiO_2$ particles exhibited a steep declined intensity of 1000 $cm^{-1}$ to 500 $cm^{-1}$, manifesting the existence of the O-Ti-O bond [29]. As a result, the addition of $TiO_2$ nanoparticles into the PAN membrane rendered an increased peak around 900 $cm^{-1}$ to 700 $cm^{-1}$ in Figure 3b, indicating the successful inclusion of $TiO_2$ nanoparticles into the PAN membrane matrix.

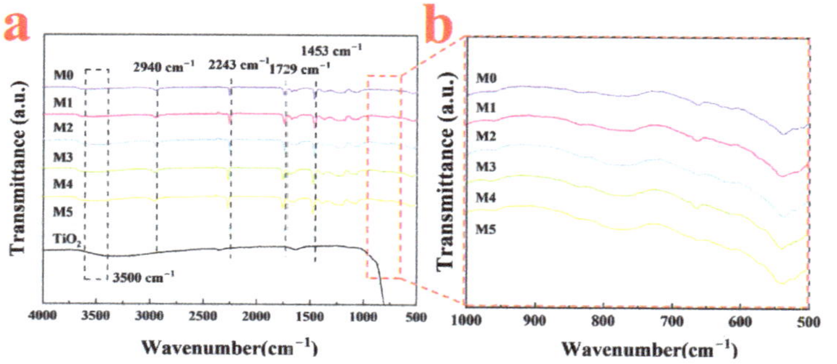


**Figure 3.** (**a**) Complete and (**b**) partial FTIR spectra of various membranes.

The hydrophilicity of the $TiO_2$-decorated membranes was revealed (Figure 4a) with a sessile drop technique. Compared to the M0 membrane, contact angles for M1 to M5 membranes declined, which was caused by the intrinsically hydrophilic $TiO_2$ particles improving the hydrophilicity of the membranes [44]. Meanwhile, the addition of $TiO_2$ nanoparticles improved the PAN membrane surface, which could promote the relevant wettability based on the Wenzel equation [45]. With an enhanced hydrophilicity, composite UF membranes could promote the water permeability and antifouling performance [46]. However, minor changes of the contact angle were detected for M4 and M5 membranes compared with the M3 membrane. It was mainly attributed to the accumulation of $TiO_2$ particles on the M3 membrane, inducing a rougher surface with decreased hydrophilicity [37]. In addition, the incorporation of hydrophilic $TiO_2$ particles endowed the PAN polymer with a faster solvent and non-solvent exchange speed over the phase inversion of the casting solution [47]. Thus, a continuous improvement in membrane porosity was detected for the obtained M0 to M5 membranes (Figure 4b).

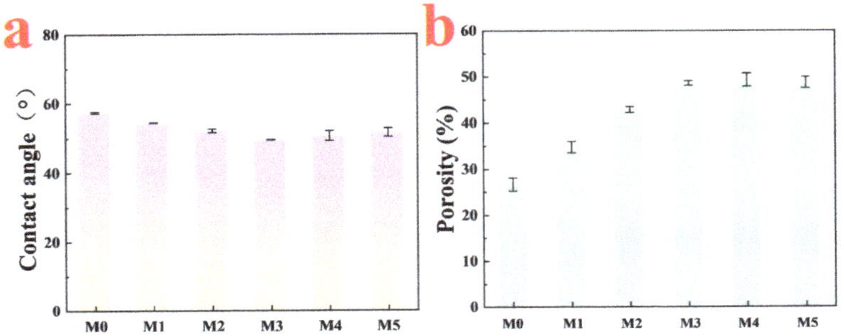


**Figure 4.** (**a**) Contact angle and (**b**) porosity results of prepared membranes.

### 2.3. Separation Performance of the Membranes

It was demonstrated that the structures and properties of the PAN/$TiO_2$ UF membranes were influenced using the concentrations of PAN power and $TiO_2$ loading, which could affect the resultant membrane performance. The pure water permeability and BSA rejection of obtained membranes are exhibited in Figure 5. As presented in Figure 5a, the water permeability of the $PAN_{13}$ membrane was measured up to 300.5 L $m^{-2}$ $h^{-1}$ $bar^{-1}$ with limited BSA rejection of 48.6%. With an increased viscosity of the PAN casting solution, a gradually declined water permeation and improved BSA rejection were detected, which caused the value of the $PAN_{15}$ membrane to reach 120.2 L $m^{-2}$ $h^{-1}$ $bar^{-1}$ and 99.1%. In addition, the water permeability suffered a decline with an increased PAN polymer concentration, while the BSA rejection remained almost constant. Considering the water permeability and BSA rejection, $PAN_{15}$ was regarded as the optimized membrane for fabricating the composite membranes with different additions of $TiO_2$ particles.

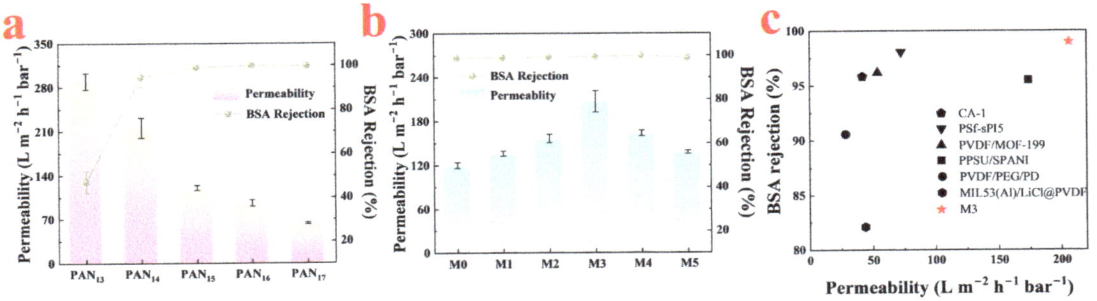


**Figure 5.** Water permeability and BSA rejection for (**a**) PAN; (**b**) PAN/$TiO_2$ membranes; (**c**) performance comparison of M3 membrane with other UF membranes.

As presented in Figure 5b, PAN UF membranes decorated with $TiO_2$ presented an improved water permeability from M0 to M3 (207.0 L $m^{-2}$ $h^{-1}$ $bar^{-1}$), which was explained with the higher hydrophilic membrane surface and porosity. However, the water permeability suffered a decline with the excess $TiO_2$ particle loading, which was consistent with the result of surface morphologies. The excess $TiO_2$ particles could gather on the

PAN membrane surface and blocked the pores of the membrane, inducing an enhanced resistance for transferring water molecules [32]. Furthermore, $TiO_2$ nanoparticles migrated into the PAN membrane during the phase inversion process and formed the agglomerates, increasing the membrane thickness and roughness, which led to a flux decline [48]. Simultaneously, the BSA rejection for the $TiO_2$-decorated PAN membranes remained above 99.0%. Figure 5c and Table 1 compare the separation performance of the M3 membrane with other reported UF membranes, indicating a superiority in water permeability and BSA rejection for the M3 membrane.

**Table 1.** Performance comparison of M3 membrane with other UF membranes.

| Membrane | Operating Pressure (bar) | Permeability (L m$^{-2}$ h$^{-1}$ bar$^{-1}$) | BSA Concentrations (g/L) | BSA Rejection (%) | Ref. |
|---|---|---|---|---|---|
| CA-1 | 3.5 | 41.6 ± 0.4 | 1.0 | 95.9 ± 0.6 | [49] |
| PSf-sPI5 | 5.0 | 72.1 ± 0.8 | 0.8 | 98.0 ± 0.5 | [50] |
| PVDF/MOF-199 | 3.5 | 54.2 ± 0.4 | 1.0 | 96.2 ± 0.2 | [51] |
| PPSU/SPANI | 1.5 | 173.3 ± 2.6 | 1.0 | 95.5 ± 0.6 | [52] |
| PVDF/PEG/PD | 4.1 | 28.8 ± 0.6 | 1.0 | 90.5 ± 0.5 | [53] |
| MIL53(Al)/LiCl@PVDF | 1.0 | 43.6 ± 1.0 | 1.0 | 82.1 ± 0.9 | [7] |
| M3 | 2.0 | 207.0 ± 10.4 | 1.0 | 99.0 ± 0.3 | This work |

However, the membrane would suffer from membrane fouling during the BSA filtration. The membrane antifouling properties were investigated through a filtration of the BSA solution (Figure 6). As shown in Figure 6a, the M0 membrane possessed the value of 16.30%, 90.22% and 83.71% for *FRR*, $R_t$ and $R_{ir}$, respectively. This result revealed that the M0 membrane suffered a serious membrane fouling formed with the adhesion of BSA molecules on the membrane surface. With the addition of $TiO_2$, the value of *FRR* for obtained membranes continuously improved to 37.13% for the M3 membrane. Meanwhile, the $R_t$ and $R_{ir}$ declined to 74.65% and 62.87%, respectively. The increased *FRR* and decreased $R_t$ and $R_{ir}$ value for the M3 membrane manifested the enhanced antifouling ability, which was attributed to the improved hydrophilicity of the membrane surface decorated with $TiO_2$ particles [54]. However, the excess $TiO_2$ addition endowed the M4 and M5 membranes with inferior *FRR* properties. This was caused by multiple-$TiO_2$-particle incorporation rendering a rougher membrane surface, resulting in enhanced adsorption to BSA molecules and a reduced membrane flux recovery ratio. Overall, by comprehensively considering the water permeability, BSA rejection and antifouling property, the M3 membrane was deemed as the optimized membrane.

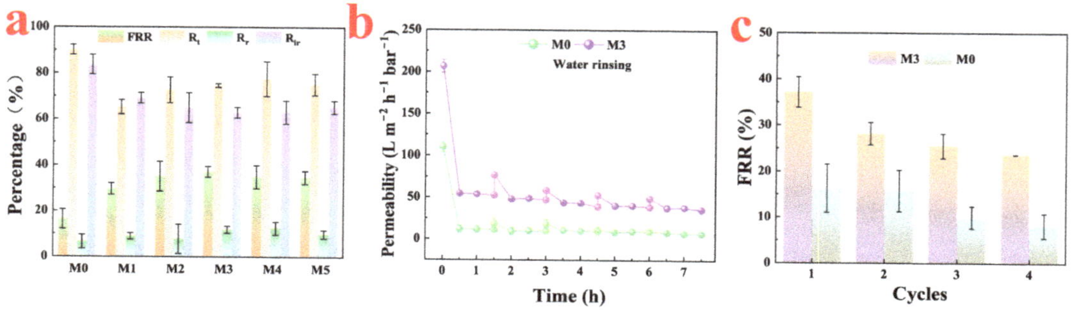


**Figure 6.** (**a**) The *FRR*, $R_t$, $R_r$ and $R_{ir}$ of membranes during the BSA ultrafiltration; (**b**) membrane performance and (**c**) *FRR* value in the process of water rinsing cycles.

A cyclic four-step ultrafiltration test for the BSA rejection solution was adopted to further measure the membrane antifouling performance in Figure 6b. It could be found that M0 and M3 membranes possessed stable BSA rejection above 99.0%, while for M0 and M3 membranes, the BSA fouling endowed the membranes with a continuous water permeability decline in the 1.5 h BSA filtration, which dropped to 10.8 and 52.3 L $m^{-2}$ $h^{-1}$ $bar^{-1}$, respectively. After that, the fouled membranes were washed in a pure water filtration, and obtained a certain flux recovery under the hydraulic shear force. Figure 6c summarizes the *FRR* of M0 and M3 membranes during the period. According to the data, the water permeability of the M0 and M3 membranes recovered to 18.07 and 76.55 L $m^{-2}$ $h^{-1}$ $bar^{-1}$ after the first water rinsing, which possessed the *FRR* of 16.27% and 37.13%, respectively. After that, the membranes were continuously filtrated in the BSA solution and water. It could be found that water permeability of the membranes dropped constantly in the BSA solution with the decreased *FRR* value, and that of the M3 membrane was higher than that of the M0 membrane in the process, while the water permeability had merely 23.58% *FRR* caused by the multiple contaminants adhered to the M3 membrane. Thus, the resultant M3 membrane incorporated with $TiO_2$ particles possessed a higher flux recovery rate compared to the pristine M0 membrane, but the extreme membrane fouling still existed after four BSA ultrafiltration tests.

### 2.4. Self-Cleaning Performance of the Membranes

For pursuing a better membrane antifouling performance, the designed membranes were tested in BSA ultrafiltration combined with UV illumination. Figure 7 presents the self-cleaning performance of the obtained membranes in a four-step filtration–irradiation process. Similarly, the water permeability of M0 and M3 membranes suffered a quick decrease in the 1.5 h BSA filtration as was discussed above (Figure 7a). After exposure to UV light irradiation for another 1.0 h, the M3 membrane decorated with photocatalytic $TiO_2$ particles derived an obvious permeation recovery, which reached 113.2 L $m^{-2}$ $h^{-1}$ $bar^{-1}$ with 52.41% *FRR* as depicted in Figure 7b. Compared with the value of water rinsing, the photocatalytic irradiation process endowed the M3 membrane with a great *FRR* promotion. However, the water permeability of the M0 membrane merely recovered to 20.2 L $m^{-2}$ $h^{-1}$ $bar^{-1}$, which was consistent with the value of water rinsing, revealing the useless effect of UV irradiation to the M0 membrane without $TiO_2$ particles. After that, the membranes were transferred to the same ultrafiltration process as before for four cycles. It demonstrated that the water permeability and *FRR* value of the M3 membrane finally declined to 63.85 L $m^{-2}$ $h^{-1}$ $bar^{-1}$ and 30.0% in the following cycles, respectively. It was explained with the not-fully-covered photocatalytic $TiO_2$ nanoparticles on the membrane surface, which could not degrade the whole of the BSA contaminants. Meanwhile, the M0 membrane also possessed a dropped water permeability and *FRR* value, which was consistent with Figure 6c. Thus, the M3 membrane cooperated with UV illumination and derived a great antifouling improvement compared to both the M0 membrane and M3 membrane without irradiation. In addition, M0 and M3 membranes maintained an excellent rejection above 99.0% towards BSA, indicating the high UV-irradiated stability of the PAN-based membrane. These results suggested that M3 incorporated with the $TiO_2$ membrane possessed an outstanding self-cleaning property and reusability.

The photoluminescence (PL) spectral was utilized to detect the photocatalytic property of the M3 membrane under the UV irradiation (Figure 8). As mentioned, terephthalic acid (TA) could be adopted to investigate the existence of ·OH radicals [55]. It could be found that a distinctive peak at $\lambda$ = 425 nm was presented, which was associated with hydroxyterephthalic acid (HTA) [56]. With the addition of the M0 membrane, the corresponding solution displayed no difference with the pristine TA solution. However, a continual improvement in the fluorescence intensities was measured with the membranes incorporated with $TiO_2$ particles, indicating the production of ·OH radicals in the UV irradiation process. In addition, the increased $TiO_2$ particle concentration rendered the PL spectral with a continuous enhancement, which represented the generation of more ·OH radicals. Thus, with the existence of ·OH radicals, the fouled M3 membrane could present a remarkable self-cleaning performance.

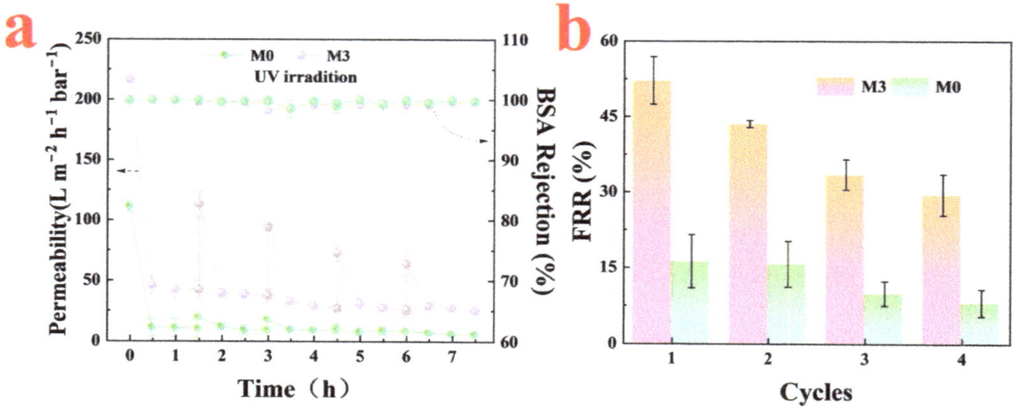


**Figure 7.** (**a**) Membrane performance and (**b**) *FRR* value in the process of UV irradiation cycles.

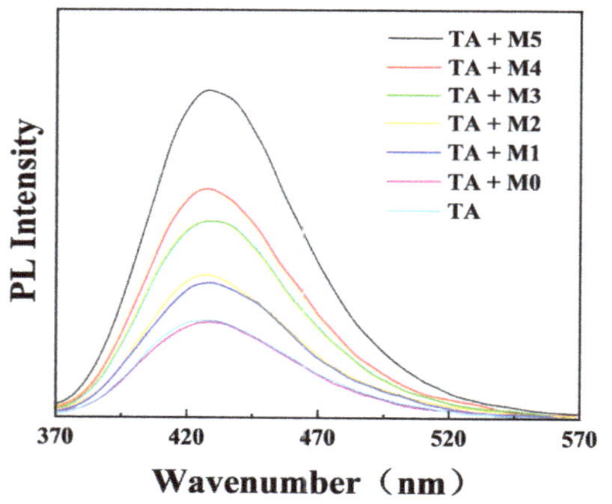


**Figure 8.** PL spectra of the solution including TA with different membranes.

## 3. Materials and Methods

### 3.1. Materials

N,N-Dimethylformamide (DMF, 99.7% purity) and terephthalic acid ($H_2BDC$, 99.0% purity) were provided by Sigma-Aldrich (Shanghai, China). Polyacrylonitrile (PAN) (MW = 100 kDa) was purchased from Solvey (Shanghai, China) Co., Ltd. Bovine serum albumin (BSA, MW = 66 kDa) and titanium dioxide ($TiO_2$, 99.7% purity) were supplied by Shanghai Aladdin Biochemical Technology Co., Ltd (Shanghai, China).

### 3.2. Preparation of PAN/$TiO_2$ Membranes

The fabrication process of PAN membranes is shown in Figure 9. Firstly, the PAN powders were dried in an oven for 12 h at 60 °C. Then, various qualities of PAN powders were dispersed in DMF (13 wt% to 17 wt%) and stirred for 12 h at 70 °C to obtain a homogeneous casting solution. After defoaming for 4 h, the prepared solution was poured onto a glass plate and scraped using a casting knife with the gap of 200 μm; then, the liquid films were immersed into the deionized (DI) water and soaked for 24 h to obtain the PAN membranes.

**Figure 9.** Schematic illustration of the preparation process for PAN/$TiO_2$ membranes.

For PAN/$TiO_2$ composite UF membranes, different amounts of $TiO_2$ powders were firstly dispersed in DMF and sonicated for 30 min. After that, a certain concentration of PAN powders was added into the prepared mixture solution and the above membrane fabrication operation was repeated to obtain the PAN/$TiO_2$ composite membranes. The corresponding composition of the membranes is listed in Table 2.

**Table 2.** Casting solution composition of the membranes.

| Membrane ID | Compositions | | |
|---|---|---|---|
| | DMF (mL) | PAN (wt%) | $TiO_2$ (wt%) |
| $PAN_{13}$ | 20.0 | 13.0 | - |
| $PAN_{14}$ | 20.0 | 14.0 | - |
| $PAN_{15}$/M0 | 20.0 | 15.0 | - |
| $PAN_{16}$ | 20.0 | 16.0 | - |
| $PAN_{17}$ | 20.0 | 17.0 | - |
| M1 | 20.0 | 15.0 | 0.2 |
| M2 | 20.0 | 15.0 | 0.4 |
| M3 | 20.0 | 15.0 | 0.6 |
| M4 | 20.0 | 15.0 | 0.8 |
| M5 | 20.0 | 15.0 | 1.0 |

### 3.3. Membrane Characterization

A scanning electron microscope (SEM, Hitachi SU8010, Japan) was deployed to inspect the morphologies of dried UF membranes. The chemical composition of the membranes was revealed with Fourier Transform Infrared (FTIR, Bruker Vertex 70, Switzerland). X-ray diffraction (XRD, Rigku Ultima IV, Japan) was used to investigate the crystal phase of the membrane. The hydrophilicity of the membranes was detected with a water contact angle device (OCA20, Dataphysics Instruments, Germany).

Membrane porosity was investigated with a reported approach [57]. Specifically, a cut membrane (2 cm × 2 cm) was immersed in DI water for 24 h; then, the excess water on the surface was carefully swiped and the weight of the obtained membrane was noted as $W_1$ (g). After that, the wet membrane was placed into an oven at 60 °C for 24 h and weighed as $W_2$ (g). The membrane porosity ($\varepsilon$) was measured as follows:

$$\varepsilon = \frac{\frac{W_1 - W_2}{D_w}}{\frac{W_1 - W_2}{D_w} + \frac{W_2}{D_p}} \times 100\% \quad (1)$$

where $D_w$ and $D_p$ are the water density (0.998 g/cm$^3$) and polymer density (1.25 g/cm$^3$), respectively.

### 3.4. Ultrafiltration and Antifouling Performance of Membranes

The separation performances of prepared UF membranes, concerning water permeability and BSA rejection, were evaluated with a crossflow cell setup (Hangzhou Saifei Membrane Separation Technology Co., Ltd., Hangzhou, China) with an effective area of 7.07 $cm^2$ as shown in Figure 10.

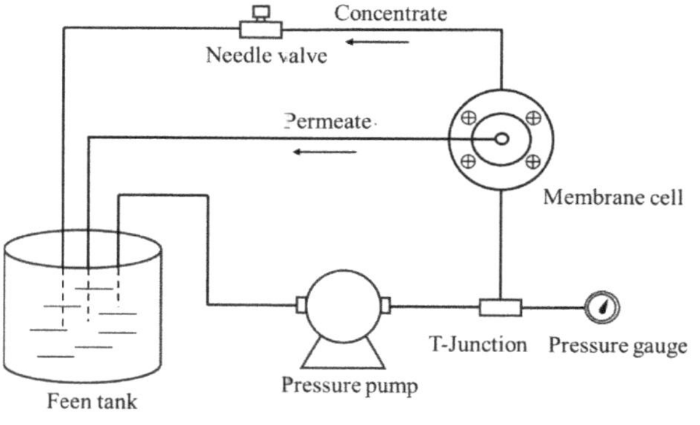


**Figure 10.** The schematic diagram of the crossflow cell setup.

After the pre-pressure process (0.2 MPa, 30 min), the pure water permeation ($J$) and BSA rejection ($R$) were calculated as follows:

$$J = \frac{V}{A \times t \times P} \tag{2}$$

$$R = \left(1 - \frac{c_p}{c_f}\right) \times 100\% \tag{3}$$

where $V$, $A$, $t$ and $P$ are permeated volume (L), filtration area ($m^2$), time (h) and operation pressure (bar). $c_p$ and $c_f$ represent the BSA concentration of the permeate and feed solutions, respectively. The feed BSA concentration was 1.0 g/L and a UV-vis absorption spectrometer (Shimadzu UV1800, Japan) was applied to measure the concentration of BSA.

The antifouling performance of membranes was detected with the flux recovery ratio ($FRR$), total fouling ($R_t$), the reversible fouling ratio ($R_r$) and the irreversible fouling ratio ($R_{ir}$) as follows [36,58]:

$$FRR = \frac{J_{w2}}{J_{w1}} \times 100\% \tag{4}$$

$$R_t = \left(1 - \frac{J_P}{J_{w1}}\right) \times 100\% \tag{5}$$

$$R_r = \left(\frac{J_{w2} - J_P}{J_{w1}}\right) \times 100\% \tag{6}$$

$$R_{ir} = \left(\frac{J_{w1} - J_{w2}}{J_{w1}}\right) \times 100\% = R_t - R_r \tag{7}$$

Specifically, the pure water permeability ($J_{w1}$) was detected at 0.2 MPa for 30 min. After that, a 1 g/L BSA solution was utilized as a feed solution and the permeability ($J_P$) was recorded every 10 min in a 30 min period. Then, the membranes were washed with pure water for 30 min and the water permeability ($J_{w2}$) of the refreshed membrane was obtained.

### 3.5. Self-Cleaning Performance of Membranes

The self-cleaning property of the PAN/$TiO_2$ membranes was tested with a photocatalytic degradation experiment of BSA using a 300 W xenon lamp (Perfect Light, PLS-SXE 300). Firstly, the membrane was pre-compacted in a 1 g/L BSA solution for 1.5 h at 2.0 bar, and the permeation and BSA rejections were recorded. After the filtration, a fouled membrane was transferred into 50 mL of DI water to be irradiated under UV light for 1 h. Meanwhile, a fluorescent probe test was deployed to investigate the mechanism during the self-cleaning process under UV irradiation, which was reported in the published literature [55].

## 4. Conclusions

In this work, a photocatalytic self-cleaning PAN UF composite membrane with $TiO_2$ decoration was synthesized via a facile phase inversion method. The characterization of membranes exhibited that the $TiO_2$ particles loaded on the membrane successfully. The incorporation of $TiO_2$ into the PAN polymer material endowed the UF membrane with a more hydrophilic, rough and porous surface. With the optimized fabrication condition of 15.0 wt% PAN and 0.6 wt% $TiO_2$, the M3 membrane presented the highest water permeation of 207.0 L $m^{-2}$ $h^{-1}$ $bar^{-1}$ and outstanding 99.0% BSA rejection. In addition, the incorporation of $TiO_2$ endowed the M3 membrane with a superior *FRR* of 37.13%, which was 2.28 times higher than that of the pristine M0 membrane, indicating that the PAN membrane decorated with $TiO_2$ enhanced the antifouling properties under water rinsing. Furthermore, remarkable self-cleaning ability was exhibited with the photocatalytic activity of $TiO_2$, and the value of *FRR* suffered a promotion to 52.41% and the BSA rejection remained constant. Consequently, the UF membrane incorporated with $TiO_2$ provided an effective method to conquer the membrane fouling problems in wastewater treatment.

**Author Contributions:** Conceptualization, J.L.; methodology, T.W.; resources, W.F.; data curation, X.W.; writing—original draft preparation, Y.X.; writing—review and editing, J.L.; visualization, H.L.; funding acquisition, W.F. All authors have read and agreed to the published version of the manuscript.

**Funding:** This research was funded by the Natural Science Foundation of Jiangsu Province, grant number: BK20211237.

**Data Availability Statement:** All data generated or analyzed during this study are included in this published article.

**Acknowledgments:** The financial support from the Natural Science Foundation of Jiangsu Province (Grant No. BK20211237).

**Conflicts of Interest:** There are no conflicting interest declared by the authors.

## References

1. Ullah, R.; Khitab, F.; Gul, H.; Khattak, R.; Ihsan, J.; Khan, M.; Khan, A.; Vincevica-Gaile, Z.; Aouissi, H.A. Superparamagnetic zinc ferrite nanoparticles as visible-light active photocatalyst for efficient degradation of selected textile dye in Water. *Catalysts* **2023**, *13*, 1061. [CrossRef]
2. Lu, F.; Astruc, D. Nanomaterials for removal of toxic elements from water. *Coord. Chem. Rev.* **2018**, *356*, 147–164. [CrossRef]
3. Wang, M.; Xu, Z.; Hou, Y.; Li, P.; Sun, H.; Niu, Q.J. Photo-Fenton assisted self-cleaning hybrid ultrafiltration membranes with high-efficient flux recovery for wastewater remediation. *Sep. Purif. Technol.* **2020**, *249*, 117159. [CrossRef]
4. Yabalak, E.; Ozay, Y.; Gizir, A.M.; Dizge, N. Water recovery from textile bath wastewater using combined subcritical water oxidation and nanofiltration. *J. Clean. Prod.* **2021**, *290*, 125207. [CrossRef]
5. Wang, X.; Sun, M.; Zhao, Y.; Wang, C.; Ma, W.; Wong, M.S.; Elimelech, M. In situ electrochemical generation of reactive chlorine species for efficient ultrafiltration membrane self-cleaning. *Environ. Sci. Technol.* **2020**, *54*, 6997–7007. [CrossRef] [PubMed]
6. Yue, R.; Raisi, B.; Rahmatinejad, J.; Ye, Z.; Barbeau, B.; Rahaman, M.S. A photo-Fenton nanocomposite ultrafiltration membrane for enhanced dye removal with self-cleaning properties. *J. Colloid Interface Sci.* **2021**, *604*, 458–468. [CrossRef]
7. Cheng, L.; Zhou, Z.; Li, L.; Xiao, P.; Ma, Y.; Liu, F.; Li, J. PVDF/MOFs mixed matrix ultrafiltration membrane for efficient water treatment. *Front. Chem.* **2022**, *10*, 985750. [CrossRef]

8. Orudzhev, F.; Alikhanov, N.; Amirov, A.; Rabadanova, A.; Selimov, D.; Shuaibov, A.; Gulakhmedov, R.; Abdurakhmanov, M.; Magomedova, A.; Ramazanov, S.; et al. Porous hybrid PVDF/$BiFeO_3$ smart composite with magnetic, piezophotocatalytic, and light-emission properties. *Catalysts* **2023**, *13*, 874. [CrossRef]
9. Xu, Z.; Wei, C.; Jin, J.; Xu, W.; Wu, Q.; Gu, J.; Ou, M.; Xu, X. Development of a novel mixed titanium, silver oxide polyacrylonitrile nanofiber as a superior adsorbent and its application for MB removal in wastewater treatment. *J. Braz. Chem. Soc.* **2018**, *29*, 560–571. [CrossRef]
10. Swaminathan, S.; Muthumanickkam, A.; Imayathamizhan, N. An effective removal of methylene blue dye using polyacrylonitrile yarn waste/graphene oxide nanofibrous composite. *Int. J. Environ. Sci. Technol.* **2015**, *12*, 3499–3508. [CrossRef]
11. Fernando, T.L.D.; Ray, S.; Simpson, C.M.; Gommans, L.; Morrison, S. Remediation of fouling on painted steel roofing via solar energy assisted photocatalytic self-cleaning technology: Recent developments and future perspectives. *Adv. Eng. Mater.* **2022**, *24*, 2101486. [CrossRef]
12. Ren, G.; Li, R.; Zhao, M.; Hou, Q.; Rao, T.; Zhou, M.; Ma, X. Membrane electrodes for electrochemical advanced oxidation processes: Preparation, self-cleaning mechanisms and prospects. *Chem. Eng. J.* **2023**, *451*, 138907. [CrossRef]
13. Kamali, M.; Suhas, D.P.; Costa, M.E.; Capela, I.; Aminabhavi, T.M. Sustainability considerations in membrane-based technologies for industrial effluents treatment. *Chem. Eng. J.* **2019**, *368*, 474–494. [CrossRef]
14. Cao, X.-L.; Zhou, F.-Y.; Cai, J.; Zhao, Y.; Liu, M.-L.; Xu, L.; Sun, S.-P. High-permeability and anti-fouling nanofiltration membranes decorated by asymmetric organic phosphate. *J. Membr. Sci.* **2021**, *617*, 118667. [CrossRef]
15. Geng, Z.; Yang, X.; Boo, C.; Zhu, S.; Lu, Y.; Fan, W.; Huo, M.; Elimelech, M.; Yang, X. Self-cleaning anti-fouling hybrid ultrafiltration membranes via side chain grafting of poly(aryl ether sulfone) and titanium dioxide. *J. Membr. Sci.* **2017**, *529*, 1–10. [CrossRef]
16. Huang, S.; Ras, R.H.A.; Tian, X. Antifouling membranes for oily wastewater treatment: Interplay between wetting and membrane fouling. *Curr. Opin. Colloid Interface Sci.* **2018**, *36*, 90–109. [CrossRef]
17. Yang, Q.; Ma, Y.; Chen, F.; Yao, F.; Sun, J.; Wang, S.; Yi, K.; Hou, L.; Li, X.; Wang, D. Recent advances in photo-activated sulfate radical-advanced oxidation process (SR-AOP) for refractory organic pollutants removal in water. *Chem. Eng. J.* **2019**, *378*, 122149. [CrossRef]
18. Rosman, N.; Wan Salleh, W.N.; Jaafar, J.; Harun, Z.; Aziz, F.; Ismail, A.F. Photocatalytic filtration of zinc oxide-based membrane with enhanced visible light responsiveness for ibuprofen removal. *Catalysts* **2022**, *12*, 209. [CrossRef]
19. Li, Y.; Ma, Y.; Li, K.; Chen, S.; Yue, D. Photocatalytic reactor as a bridge to Link the commercialization of photocatalyst in water and air purification. *Catalysts* **2022**, *12*, 724. [CrossRef]
20. Lee, J.-P.; Choi, S.; Cho, S.; Song, W.-J.; Park, S. Fabrication of carbon nanofibers decorated with various kinds of metal oxides for battery applications. *Energies* **2021**, *14*, 1353. [CrossRef]
21. Zhou, M.; Chen, J.; Yu, S.; Chen, B.; Chen, C.; Shen, L.; Li, B.; Lin, H. The coupling of persulfate activation and membrane separation for the effective pollutant degradation and membrane fouling alleviation. *Chem. Eng. J.* **2023**, *451*, 139009. [CrossRef]
22. Wenderich, K.; Mul, G. Methods, mechanism, and applications of photodeposition in photocatalysis: A review. *Chem. Rev.* **2016**, *116*, 14587–14619. [CrossRef] [PubMed]
23. Sisay, E.J.; Kertész, S.; Fazekas, Á.; Jákói, Z.; Kedves, E.Z.; Gyulavári, T.; Ágoston, Á.; Veréb, G.; László, Z. Application of $BiVO_4$/$TiO_2$/CNT composite photocatalysts for membrane fouling control and photocatalytic membrane regeneration during dairy wastewater treatment. *Catalysts* **2023**, *13*, 315. [CrossRef]
24. Zhang, H.; Mane, A.U.; Yang, X.; Xia, Z.; Barry, E.F.; Luo, J.; Wan, Y.; Elam, J.W.; Darling, S.B. Visible-light-activated photocatalytic films toward self-cleaning membranes. *Adv. Funct. Mater.* **2020**, *30*, 2002847. [CrossRef]
25. Deemter, D.; Coelho, F.E.B.; Oller, I.; Malato, S.; Amat, A.M. Assessment of a Novel Photocatalytic $TiO_2$-Zirconia Ultrafiltration Membrane and Combination with Solar Photo-Fenton Tertiary Treatment of Urban Wastewater. *Catalysts* **2022**, *12*, 552. [CrossRef]
26. Hong, J.; He, Y. Polyvinylidene fluoride ultrafiltration membrane blended with nano-ZnO particle for photo-catalysis self-cleaning. *Desalination* **2014**, *332*, 67–75. [CrossRef]
27. Meng, M.; Li, B.; Zhu, Y.; Yan, Y.; Feng, Y. A novel mixed matrix polysulfone membrane for enhanced ultrafiltration and photocatalytic self-cleaning performance. *J. Colloid Interface Sci.* **2021**, *599*, 178–189. [CrossRef]
28. Bao, X.; Liu, Q.; Yang, J.; Wang, F.; Yu, F.; Yu, J.; Yang, Y. Cascading in-situ generation of $H_2O_2$ and Fenton-like reaction in photocatalytic composite ultrafiltration membrane for high self-cleaning performance in wastewater treatment. *J. Membr. Sci.* **2022**, *660*, 120866. [CrossRef]
29. Wu, J.; Yi, S.; Wang, Y.; Yao, J.; Gao, W. Polymer-based $TiO_2$ nanocomposite membrane: Synthesis and organic pollutant removal. *Int. J. Smart Nano Mater.* **2021**, *12*, 129–145. [CrossRef]
30. Zhu, C.; Liu, G.; Han, K.; Ye, H.; Wei, S.; Zhou, Y. One-step facile synthesis of graphene oxide/$TiO_2$ composite as efficient photocatalytic membrane for water treatment: Crossflow filtration operation and membrane fouling analysis. *Chem. Eng. Process.-Process Intensif.* **2017**, *120*, 20–26. [CrossRef]
31. Pan, Z.; Cao, S.; Li, J.; Du, Z.; Cheng, F. Anti-fouling $TiO_2$ nanowires membrane for oil/water separation: Synergetic effects of wettability and pore size. *J. Membr. Sci.* **2019**, *572*, 596–606. [CrossRef]
32. Shi, F.; Ma, Y.; Ma, J.; Wang, P.; Sun, W. Preparation and characterization of PVDF/$TiO_2$ hybrid membranes with ionic liquid modified nano-$TiO_2$ particles. *J. Membr. Sci.* **2013**, *427*, 259–269. [CrossRef]
33. Kim, B.S.; Lee, J. Macroporous PVDF/$TiO_2$ membranes with three-dimensionally interconnected pore structures produced by directional melt crystallization. *Chem. Eng. J.* **2016**, *301*, 158–165. [CrossRef]

34. Sun, T.; Liu, Y.; Shen, L.; Xu, Y.; Li, R.; Huang, L.; Lin, H. Magnetic field assisted arrangement of photocatalytic $TiO_2$ particles on membrane surface to enhance membrane antifouling performance for water treatment. *J. Colloid Interface Sci.* **2020**, *570*, 273–285. [CrossRef] [PubMed]
35. Damodar, R.A.; You, S.-J.; Chou, H.-H. Study the self cleaning, antibacterial and photocatalytic properties of $TiO_2$ entrapped PVDF membranes. *J. Hazard. Mater.* **2009**, *172*, 1321–1328. [CrossRef]
36. Ma, W.; Pan, J.; Ren, W.; Chen, L.; Huang, L.; Xu, S.; Jiang, Z. Fabrication of antibacterial and self-cleaning CuxP@g-$C_3N_4$/PVDF-CTFE mixed matrix membranes with enhanced properties for efficient ultrafiltration. *J. Membr. Sci.* **2022**, *659*, 120792. [CrossRef]
37. Dmitrenko, M.; Kuzminova, A.; Zolotarev, A.; Markelov, D.; Komolkin, A.; Loginova, E.; Plisko, T.; Burts, K.; Bildyukevich, A.; Penkova, A. Modification strategies of polyacrylonitrile ultrafiltration membrane using $TiO_2$ for enhanced antifouling performance in water treatment. *Sep. Purif. Technol.* **2022**, *286*, 120500. [CrossRef]
38. Wang, W.; Wang, C.; Gao, Q.; Chen, M.; Wang, Y.; Yao, Z. A new perspective on the internal structure of polyacrylonitrile-based preoxidized fibers through ultrathin sections. *Polym. Degrad. Stab.* **2019**, *167*, 269–276. [CrossRef]
39. Zhang, Y.; Wang, L.; Li, L.; Wang, H.; Dong, X.; Pan, Y.; Wang, T. Insight into the influences of thermal crosslinking on the transition from polyacrylonitrile based ultrafiltration membrane to organic solvent nanofiltration membrane. *J. Membr. Sci.* **2023**, *679*, 121694. [CrossRef]
40. Akhter, P.; Nawaz, S.; Shafiq, I.; Nazir, A.; Shafique, S.; Jamil, F.; Park, Y.-K.; Hussain, M. Efficient visible light assisted photocatalysis using $ZnO/TiO_2$ nanocomposites. *Mol. Catal.* **2023**, *535*, 112896. [CrossRef]
41. Xiu, L.; Wang, Z.; Yu, M.; Wu, X.; Qiu, J. Aggregation-Resistant 3D MXene-Based Architecture as Efficient Bifunctional Electrocatalyst for Overall Water Splitting. *ACS Nano* **2018**, *12*, 8017–8028. [CrossRef] [PubMed]
42. Li, J.; Li, L.; Li, X.; Dong, L.; Wang, Z.; Shen, J.; Van der Bruggen, B. Membranes with ZIF-8 regulated MXene nanosheet stacks for efficient molecular sieving. *Desalination* **2023**, *546*, 116184. [CrossRef]
43. Chi, M.; Sun, X.; Lozano-Blanco, G.; Tatarchuk, B.J. XPS and FTIR investigations of the transient photocatalytic decomposition of surface carbon contaminants from anatase $TiO_2$ in UHV starved water/oxygen environments. *Appl. Surf. Sci.* **2021**, *570*, 151147. [CrossRef]
44. Zhang, Q.; Fan, L.; Yang, Z.; Zhang, R.; Liu, Y.-N.; He, M.; Su, Y.; Jiang, Z. Loose nanofiltration membrane for dye/salt separation through interfacial polymerization with in-situ generated $TiO_2$ nanoparticles. *Appl. Surf. Sci.* **2017**, *410*, 494–504. [CrossRef]
45. Mohamed, A.M.A.; Abdullah, A.M.; Younan, N.A. Corrosion behavior of superhydrophobic surfaces: A review. *Arab. J. Chem.* **2015**, *8*, 749–765. [CrossRef]
46. Roh, I.J.; Greenberg, A.R.; Khare, V.P. Synthesis and characterization of interfacially polymerized polyamide thin films. *Desalination* **2006**, *191*, 279–290. [CrossRef]
47. Ayyaru, S.; Dinh, T.T.L.; Ahn, Y.-H. Enhanced antifouling performance of PVDF ultrafiltration membrane by blending zinc oxide with support of graphene oxide nanoparticle. *Chemosphere* **2020**, *241*, 125068. [CrossRef]
48. Arthanareeswaran, G.; Thanikaivelan, P. Fabrication of cellulose acetate–zirconia hybrid membranes for ultrafiltration applications: Performance, structure and fouling analysis. *Sep. Purif. Technol.* **2010**, *74*, 230–235. [CrossRef]
49. Vetrivel, S.; Rana, D.; Sri Abirami Saraswathi, M.S.; Divya, K.; Kaleekkal, N.J.; Nagendran, A. Cellulose acetate nanocomposite ultrafiltration membranes tailored with hydrous manganese dioxide nanoparticles for water treatment applications. *Polym. Adv. Technol.* **2019**, *30*, 1943–1950. [CrossRef]
50. Jafar Mazumder, M.A.; Raja, P.H.; Isloor, A.M.; Usman, M.; Chowdhury, S.H.; Ali, S.A.; Inamuddin; Al-Ahmed, A. Assessment of sulfonated homo and co-polyimides incorporated polysulfone ultrafiltration blend membranes for effective removal of heavy metals and proteins. *Sci. Rep.* **2020**, *10*, 7049. [CrossRef]
51. Rameesha, L.; Rana, D.; Kaleekkal, N.J.; Nagendran, A. Efficacy of MOF-199 in improvement of permeation, morphological, antifouling and antibacterial characteristics of polyvinylidene fluoride membranes. *New J. Chem.* **2022**, *46*, 7638–7649. [CrossRef]
52. Rong, G.; Zhou, D.; Pang, J. Preparation of high-performance antifouling polyphenylsulfone ultrafiltration membrane by the addition of sulfonated polyaniline. *J. Polym. Res.* **2018**, *25*, 66. [CrossRef]
53. Sri Abirami Saraswathi, M.; Kausalya, R.; Kaleekkal, N.J.; Rana, D.; Nagendran, A. BSA and humic acid separation from aqueous stream using polydopamine coated PVDF ultrafiltration membranes. *J. Environ. Chem. Eng.* **2017**, *5*, 2937–2943. [CrossRef]
54. Lalia, B.S.; Kochkodan, V.; Hashaikeh, R.; Hilal, N. A review on membrane fabrication: Structure, properties and performance relationship. *Desalination* **2013**, *326*, 77–95. [CrossRef]
55. Wang, Y.; Zhang, J.; Bao, C.; Xu, X.; Li, D.; Chen, J.; Hong, M.; Peng, B.; Zhang, Q. Self-cleaning catalytic membrane for water treatment via an integration of Heterogeneous Fenton and membrane process. *J. Membr. Sci.* **2021**, *624*, 119121. [CrossRef]
56. Ademola Bode-Aluko, C.; Pereao, O.; Kyaw, H.H.; Al-Naamani, L.; Al-Abri, M.Z.; Tay Zar Myint, M.; Rossouw, A.; Fatoba, O.; Petrik, L.; Dobretsov, S. Photocatalytic and antifouling properties of electrospun $TiO_2$ polyacrylonitrile composite nanofibers under visible light. *Mater. Sci. Eng. B* **2021**, *264*, 114913. [CrossRef]

57. Khalid, A.; Abdel-Karim, A.; Ali Atieh, M.; Javed, S.; McKay, G. PEG-CNTs nanocomposite PSU membranes for wastewater treatment by membrane bioreactor. *Sep. Purif. Technol.* **2018**, *190*, 165–176. [CrossRef]
58. Ma, T.; Su, Y.; Li, Y.; Zhang, R.; Liu, Y.; He, M.; Li, Y.; Dong, N.; Wu, H.; Jiang, Z. Fabrication of electro-neutral nanofiltration membranes at neutral pH with antifouling surface via interfacial polymerization from a novel zwitterionic amine monomer. *J. Membr. Sci.* **2016**, *503*, 101–109. [CrossRef]

*catalysts*

*Article*

# Facile Preparation of Attapulgite-Supported Ag-AgCl Composite Photocatalysts for Enhanced Degradation of Tetracycline

**Xiaojie Zhang [1,*], Huiqin Wang [1,2,*] and Chenlong Yan [2]**

1 Jiangsu Provincial Key Laboratory of Palygorskite Science and Applied Technology, Huaiyin Institute of Technology, Huai'an 223003, China
2 School of Energy and Power Engineering, Jiangsu University, Zhenjiang 212013, China; ycl1527913077@126.com
* Correspondence: xjzhang@hyit.edu.cn (X.Z.); hqwang@ujs.edu.cn (H.W.)

**Abstract:** In this study, Ag-AgCl/attapulgite (Ag-AgCl/ATP) composites were synthesized via a direct precipitation method using ATP nanorods as a catalyst supporter. ATP nanorods helped to increase the dispersion of Ag-AgCl particles and broaden the light absorption spectrum, which would also help to increase the active site of the catalyst to promote the degradation of tetracycline (TC). The photocatalytic activity of the Ag-AgCl/ATP composites was evaluated through the degradation of TC, identifying the loading amount of Ag-AgCl, the concentration of TC, and the reaction temperature as critical factors influencing activity. Specifically, the optimal conditions were observed when the loading of Ag-AgCl was 75%, resulting in a photocatalytic degradation efficiency of 77.65%. Furthermore, the highest degradation efficiency (85.01%) was achieved with a TC concentration of 20 mg/L at 20 °C. Radical trapping experiments suggested that the superoxide anion radical ($\cdot O_2^-$) was the primary active species in the degradation process, although hydroxyl radicals ($\cdot OH$) and holes ($h^+$) also contributed. Reusability tests confirmed that the Ag-AgCl/ATP composites exhibited excellent stability and could be effectively reused.

**Keywords:** Ag-AgCl; attapulgite; composite photocatalyst; degradation; tetracycline

## 1. Introduction

The country's rapid industrialization has accelerated China's move to the center of the world stage, but at the same time, environmental pollution has become more and more serious [1–4]. In real life, antibiotics are widely used and often treat bacterial infections in humans and animals. Most of the antibiotics are excreted in feces and urine in the form of prototypes or metabolites after use, resulting in generally high concentrations of antibiotics remaining in the feces [5–7]. It will increase the resistance of bacteria and accelerate the spread of antibiotic-resistant genes, causing different degrees of harm to humans and the ecology. The World Health Organization has listed the problem of antibiotic resistance genes caused by the abuse of antibiotics as one of the major challenges facing human health in this century [8–10]. It is urgent to find a new technology to remove antibiotics. Photocatalytic technology is considered to be one of the most effective technologies for degrading and removing antibiotics [11–13]. Photocatalytic technology has been studied by many researchers because of its mild reaction conditions, low price and simple operation [14–16]. Although photocatalysis technology has been fully developed, it has not played a very wide role in practical application because most semiconductor photocatalysts have low surface energy and are prone to agglomeration, resulting which has low photocatalytic activity [17,18]. Therefore, it is very important to select a suitable photocatalyst for TC degradation.

Ag-AgCl nanoparticles could possess photocatalytic activity that changes with the crystal face structure, which could be prepared to catalyze the degradation of much pollution, printing dyes and acetaldehyde, etc. [19–22]. However, Ag-AgCl particles are easy to agglomerate, and finding suitable materials that are too dispersive will help to increase the concentration of the active site to achieve efficient TC degradation. Natural one-dimensional nanomaterial attapulgite ($Mg_5Si_8O_{20}(OH)_2(OH_2)_4{\cdot}4H_2O$) (ATP) is a kind of water-containing magnesium-rich aluminum silicate clay mineral with a chain layered structure, belonging to the 2:1 type clay mineral, which possesses ultra-high internal and external specific surface area, non-toxic, tasteless, chemical properties stability and other advantages, for instance, high adsorption performance, catalytic, chargeability, and stability [23–27]. However, the utilization trend of some ATP is not high, which undoubtedly causes the waste and loss of mineral resources. Accordingly, many researchers have devoted themselves to the development and research of ATP resources as carriers [28–30].

In this paper, ATP was used as the basic material, and a simple experimental method was used to pre-treat it. Ag-AgCl is built into the attapulgite structure to synthesize a series of new composite catalysts. The photocatalytic degradation efficiency of TC was used to evaluate the photodegradation performance of the composite catalyst. At the same time, the optimal special-performance materials were used to conduct recycling experiments to investigate the stability of the materials. ATP possesses a high specific surface area, with attapulgite as the basic material; it can effectively prevent the occurrence of agglomeration and improve the performance of the catalyst. The introduction of Ag-AgCl nanoparticles is deposited on the surface of the photocatalyst, and the photogenerated electrons generated by the light excitation will diffusion from the surface of the catalyst to the surface of the metal. This work provides new ideas for the design and preparation of composite photocatalysts based on ATP. Realizing the effective utilization of attapulgite, especially in water treatment, attapulgite has important theoretical and practical significance.

## 2. Results

### 2.1. XRD Analysis

To determine the crystal structure and composition of the synthesized photocatalytic material, Figure 1 presents the X-ray diffraction (XRD) spectra of ATP, silver chloride (AgCl), AgCl/ATP, and Ag-AgCl/ATP composites. The XRD pattern of attapulgite displays prominent peaks at 2θ values of 13.7°, 16.08°, 20.67°, 21.24°, and 27.5°, corresponding respectively to the (200), (220), (301), (121), and (400) crystallographic planes of ATP [31,32]. Notably, the AgCl peaks appear at 2θ = 27.7°, 32.4°, 46.2°, 54.8°, 57.7°, 67.6°, 74.5°, and 76.9°, aligning with the (111), (200), (220), (311), (222), (400), (331), and (420) planes of AgCl (JCPDS:85-1355) [33,34]. Furthermore, peaks at 38.1° and 44.3° confirm the presence of metallic silver (Ag) corresponding to the (111) and (200) planes (JCPDS: 89-3722), indicative of an in-situ reduction of $Ag^+$ ions on the AgCl surface [35]. This reduction process facilitates the conversion of $Ag^+$ to metallic Ag, confirming the successful preparation of the Ag-AgCl/ATP composite photocatalyst.

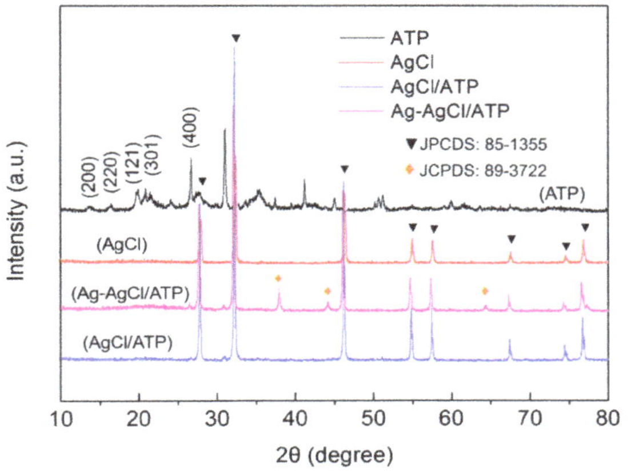


**Figure 1.** XRD patterns of ATP, AgCl, AgCl/ATP and Ag-AgCl/ATP.

## 2.2. Morphology Analysis

To understand and study the morphology of the catalytic materials, we performed SEM and TEM characterizations. Figure 2A is the SEM image of the attapulgite preprocessed with a magnification of 25,000 times. In Figure 2A, it can be seen that the bumpy clay has an acicular or rod-like structure with a random network of rod crystal fibers with a length of about 2 um [36,37]. Figure 2B shows Ag-AgCl, which is granular with a small number of agglomerates and a few fine particles on the surface, which is consistent with the appearance of Ag. Figure 2C is the enlarged SEM images of Ag-AgCl/ATP. From the figure, it can be clearly seen that the attapulgite is loaded with a large number of AgCl particles, and the dispersion is uniform. Attapulgite clay needles or rods can be observed in Figure 2D consistent with scanning. Whereas the spherical shape of AgCl is also shown in Figure 2E, the rod-like structure can be seen interspersed in the spheres in Figure 2F, indicating that the two were successfully composite. Elemental mapping (Figure 3A–F) reveals the distribution of oxygen (O), iron (Fe), chlorine (Cl), silver (Ag), silicon (Si), aluminum (Al), and magnesium (Mg) in the Ag-AgCl/ATP photocatalytic composite, confirming the complex composition of the material.

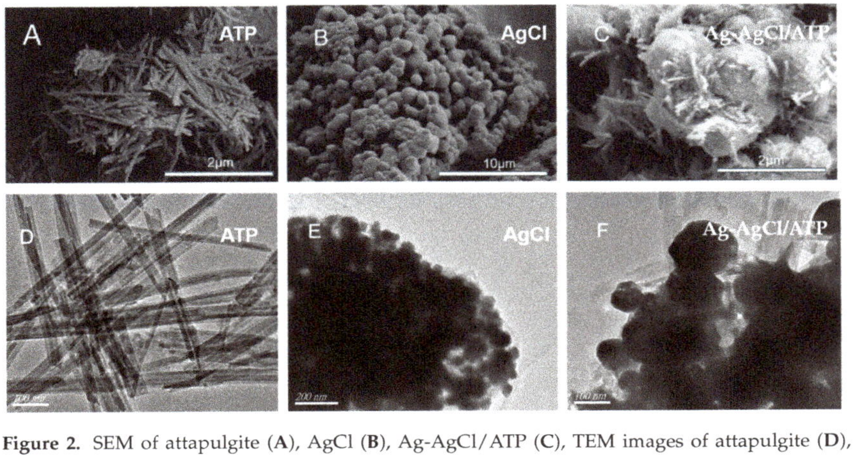


**Figure 2.** SEM of attapulgite (**A**), AgCl (**B**), Ag-AgCl/ATP (**C**), TEM images of attapulgite (**D**), AgCl (**E**), Ag-AgCl/ATP (**F**).

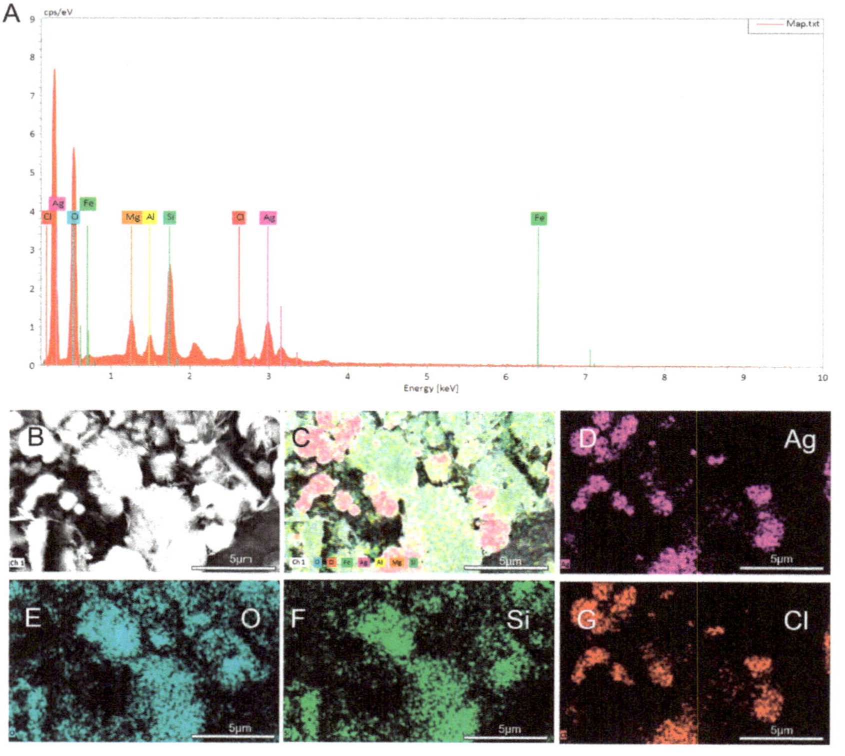


**Figure 3.** Elemental mapping results of 75%-Ag-AgCl/ATP (**A–G**), Ag (**D**), O (**E**), Si (**F**), Cl (**G**).

### 2.3. FT-IR Analysis

Figure 4 shows the FT-IR spectra of ATP, Ag-AgCl, and Ag-AgCl/ATP. ATP primarily consists of palygorskite, characterized by prominent infrared spectral peaks at 510, 980, 1030, 1460, 1640–1660 and 3450–3600 $cm^{-1}$. In the mid-wave number region, the peaks at 980 $cm^{-1}$ and 1030 $cm^{-1}$ correspond to the stretching vibrations of Si–O–Si bonds within the internal structure of ATP [38–40]. The peaks observed at 1460 and 1640 $cm^{-1}$ are attributed to coordinated water (surface-adsorbed water) and zeolite water (pore channel). In the high-wave number region, the peak at 3540 $cm^{-1}$ reflects the stretching vibration of coordinated water, while the peak at 3600 $cm^{-1}$ corresponds to bond stretching in ATP. The FT-IR spectra of Ag-AgCl/ATP composites indicate that the fundamental properties of attapulgite remain unchanged throughout the synthesis process.

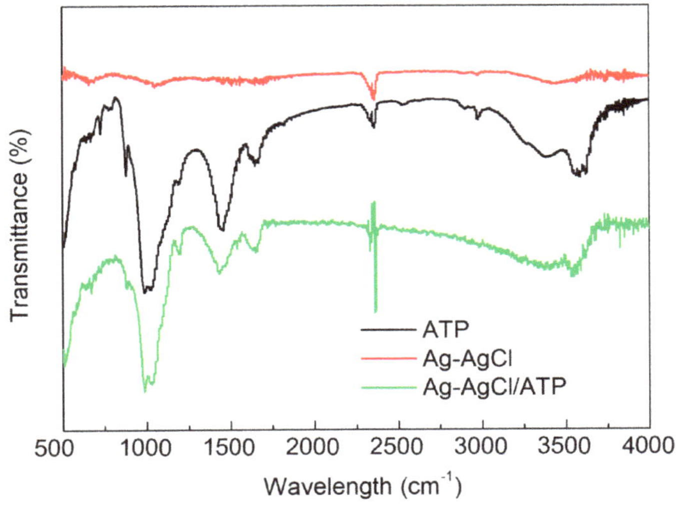


**Figure 4.** FT-IR spectra of ATP, Ag-AgCl and Ag-AgCl/ATP photocatalysts.

### 2.4. Surface Chemical State Analysis

Figure 5 presents the compositional analysis and surface chemical states of the optimized photocatalyst, 75% Ag-AgCl/ATP, as determined by X-ray photoelectron spectroscopy (XPS). The full spectrum shown in Figure 5A revealed more pronounced and intense peaks at 532.69 eV, 367.81 eV, 198.22 eV, and 102.49 eV, which correspond to the presence of oxygen (O), silver (Ag), chlorine (Cl), and silicon (Si), respectively, corroborating the findings from X-ray diffraction (XRD) and elemental mapping analyses. Figure 5B–E detail the XPS spectra for Ag 3d, Cl 2p, Si 2p, and O 1s, respectively. Among them, the composite of Figure 5B shows peaks at 533.64 eV, 532.29 eV, and 531.21 eV, corresponding to the three partial compositions of Si–O–Si, Si–O–H, and Si–O–Mg, respectively. Figure 5D, on the other hand, shows the fitted high-resolution spectrum of AgMN1, which can be divided into two different peaks, Ag and AgCl. Figure 5C illustrates the characteristic peak of Si at 102.84 eV in a two-dimensional orbital profile. After compounding, the binding energy of Si moves in a large direction, showing a blueshift. The binding energies at 199.41 eV and 197.96 eV, shown in Figure 5D, correspond to the Cl $2p_{3/2}$ and Cl $2p_{1/2}$ states, respectively. It can be seen that the binding energy of Cl in the composite is shifted in a smaller direction than that of Cl in the monomer, showing a redshift. Lastly, the binding energies of 373.91 eV and 367.92 eV in Figure 5E are assigned to Ag $3d_{3/2}$ and Ag $3d_{5/2}$, indicating the presence of these elements within the composite [41–43]. These spectral results provide comprehensive insights into the chemical states at the catalyst surface, further substantiating the successful synthesis of the composite catalyst.

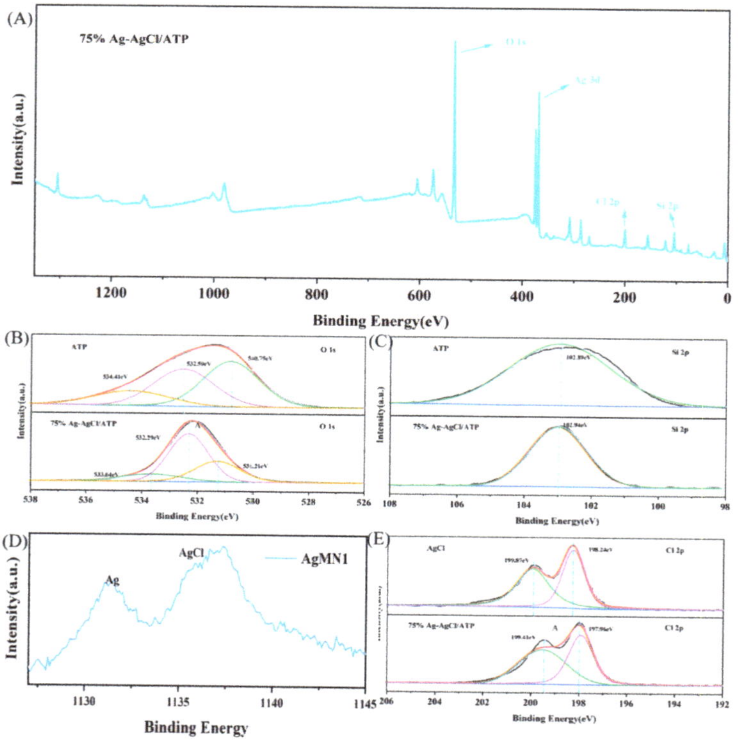


**Figure 5.** (**A**) XPS spectra of survey scan of 75% Ag-AgCl/ATP; High-resolution XPS spectra of (**B**) O 1s, (**C**) Si 2p, (**D**) AgMN1, (**E**) Cl 2p.

### 2.5. UV-Vis DRS

Figure 6 depicts the characterization of the prepared composite photocatalytic materials using an ultraviolet–visible diffuse reflectance spectrometer. The spectral data demonstrate that these materials exhibit pronounced absorption peaks within the wavelength range of 200–400 nm. Notably, ATP and AgCl display distinct absorption edges at 420 nm and 440 nm, respectively. When compared with the ultraviolet-visible diffuse reflectance spectrum of ATP, AgCl and Ag-AgCl/ATP show variations; particularly, the spectrum of Ag-AgCl/ATP exhibits a noticeable blue shift. This shift indicates an increase in the light absorption capacity of the catalyst and an increase in its efficiency in the presence of light.

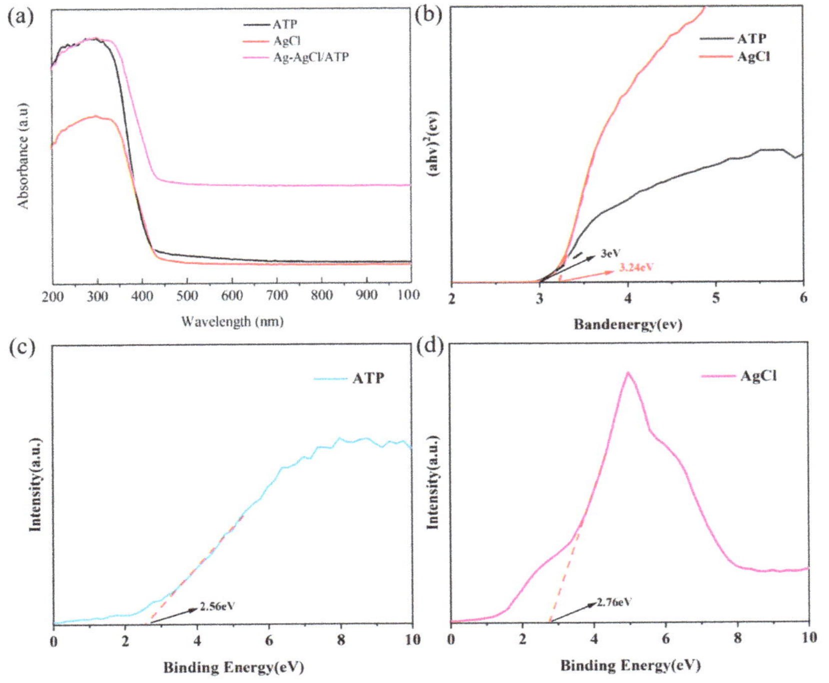


**Figure 6.** DRS pattern of Ag-AgCl/ATP, ATP and AgCl (**a**). Band gap spectra of as-prepared samples (**b**); VB-XPS of ATP (**c**) and AgCl (**d**).

### 2.6. Photo-Electrochemistry Analysis

To further investigate charge transfer and separation effects, electrochemical tests were conducted. Figure 7a illustrates stable photocurrent density profiles across all samples, with composites exhibiting prominent photocurrent signals compared to bare ATP with AgCl; notably, the 75% loaded sample shows optimal photocurrent intensity. This suggests that AgCl addition effectively enhances hole-electron pair separation [44]. Furthermore, impedance spectra (EIS) of composite samples with varying AgCl proportions alongside pure AgCl and ATP are presented in Figure 7b. Results reveal the smallest arc radius for the 75% Ag-AgCl/ATP electrode, indicating lower interfacial resistance conducive to enhanced electron transport [45]. Linear sweep voltammetry (LSV) curves in Figure 7c demonstrate that 75% Ag-AgCl/ATP exhibits the lowest overpotential at identical current densities, underscoring the catalysts' synergistic facilitation of photocatalytic degradation processes. In Figure 6b, the band gaps of ATP and AgCl were tested and found to be 3 eV and 3.24 eV, respectively, and the valence bands (VB) of ATP and AgCl were measured in Figure 6c,d using XPS valence band spectroscopy to be 2.56 eV and 2.76 eV, respectively, and through the empirical formula EVB = ECB + Eg, it can be obtained that the conduction bands (CB) for ATP and AgCl were −0.44 eV and −0.48 eV.

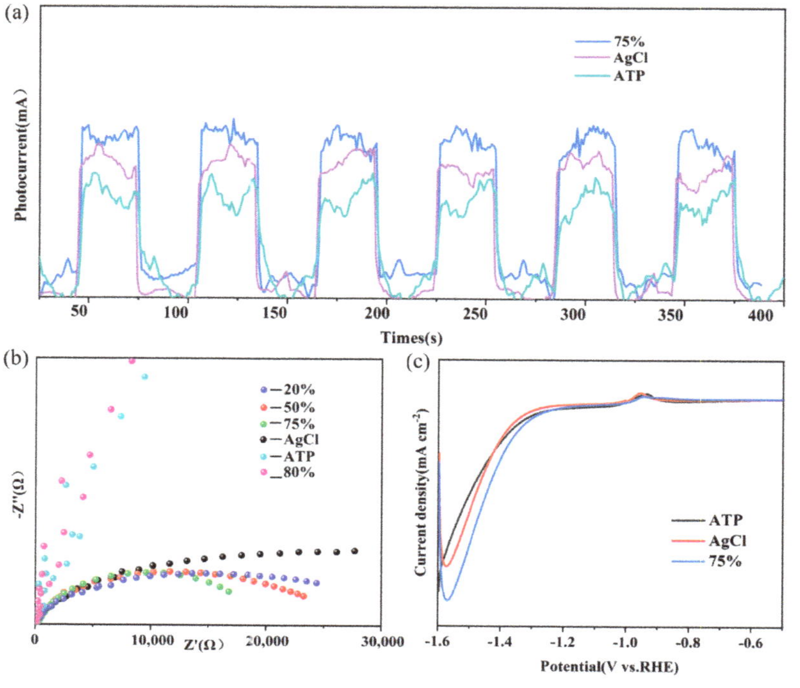


**Figure 7.** (**a**) Photocurrent intensity test for all prepared samples; (**b**) impedance test (EIS); (**c**) linear sweep voltammetry (LSV).

## 3. Photocatalytic Performance

### 3.1. Photocatalytic Activity

This study investigates the photocatalytic properties of prepared materials using TC as a model pollutant. Figure 8A depicts the photocatalytic efficiencies of various loadings of photocatalytic materials (ATP, 2%Ag-AgCl/ATP, 5%Ag-AgCl/ATP, 10%Ag-AgCl/ATP, 20%Ag-AgCl/ATP, 50%Ag-AgCl/ATP, 75%Ag-AgCl/ATP, 80%Ag-AgCl/ATP and AgCl). Upon exposure to a 300 W xenon lamp for 120 min, the photocatalytic efficiencies of Ag-AgCl/ATP photocatalysts were 35.00%, 20.11%, 37.46%, 44.83%, 64.23%, 77.65%, 71.60%, and 76.10%, respectively. Notably, at 2% AgCl loading, Ag-AgCl/ATP exhibits the lowest photocatalytic efficiency, even lower than pure ATP. As the AgCl loading increases within Ag-AgCl/ATP, the photocatalytic efficiency improves significantly. The peak efficiency is achieved at 75% AgCl loading, after which further increases diminish photocatalytic performance, likely due to surface blockage of attapulgite preventing effective pollutant-catalyst interaction and reducing activity. Figure 8B illustrates the photocatalytic degradation rate curve of the optimized Ag-AgCl/ATP photocatalytic material across different concentrations of TC (10, 20, 30, 50, and 100 mg/L). After 120 min of photocatalytic degradation, the degradation rates for each TC concentration are 81.46%, 85.01%, 79.08%, 68.42%, and 35.18%, respectively. Notably, the highest degradation rate is achieved at a TC concentration of 20 mg/L, followed by 10 mg/L, and then 30 mg/L. The lowest degradation rate is observed at 100 mg/L TC concentration. These results indicate that the optimized Ag-AgCl/ATP photocatalytic material exhibits greater efficiency at lower TC concentrations, with degradation rates declining as TC concentration increases. This underscores the importance of considering pollutant concentrations when applying such photocatalytic materials in practical environmental remediation scenarios.

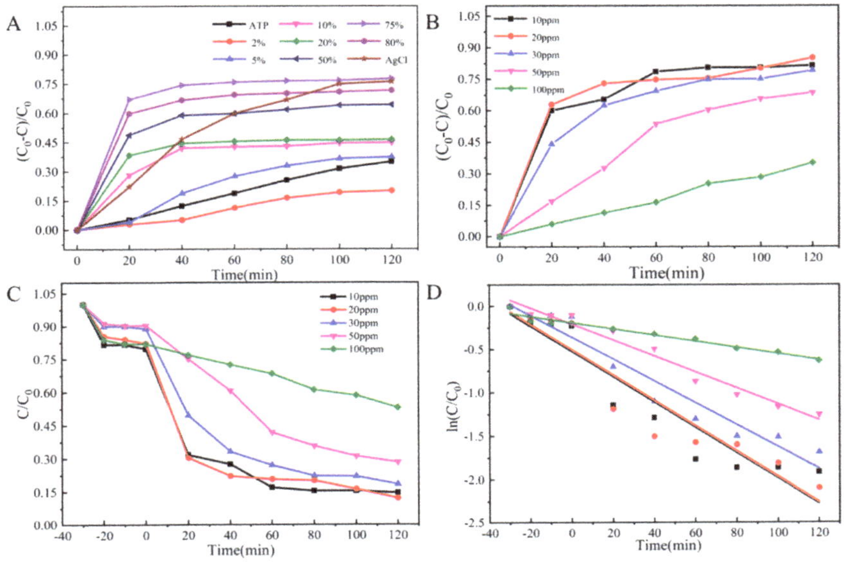

**Figure 8.** (**A**) Different loading amounts of Ag-AgCl/ATP; (**B**) optimal Ag-AgCl/ATP degradation of different concentrations of TC; (**C**) degradation efficiency of different concentrations of TC; (**D**) quasi-first kinetic curve, the trend of $\ln(C_0/C)$ with time under different starting concentrations.

The initial concentration of TC in this experiment is in the range of 10–100 mg $L^{-1}$. The experimental results are shown in Figure 8B,C. The photocatalytic degradation efficiency and rate of TC are negatively correlated with the initial concentration of TC. As the concentration of TC in the solution increases, the total amount of TC also increases. Under the condition of loading with the same AgCl catalyst quality, the degradation rate of high-concentration TC decreases slightly compared with other concentrations. High concentrations of TC cannot absorb photons well and cannot effectively separate photogenerated electrons and holes, which will affect the degradation rate.

The rate equation obtained from the Langmuir–Hinshelwood kinetic equation is expressed as the following Equation (1):

Among them, $C_0$ and $C$ represent the initial and degraded concentration of TC at time $t$, $k_c$ is the pseudo-first rate constant, and $K_{CIP}$ is the equilibrium constant for the adsorption of TC by the catalyst. According to the above equation expression, in the presence of Fe-doped AgCl photocatalyst, the pseudo-first-order kinetic equation of photocatalytic degradation is expressed as the following Equation (2):

Where $k_{obs}$ is the quasi-first order kinetic rate constant, and the above formula can be transformed into (3).

According to Equation (3), the linear relationship between $\ln(C_0/C)$ and time t is shown in Figure 8D. The linear relationship expression between $k_{obs}$ and $C_0$ can be transformed from Equation (1).

It can be seen from Formula (4) that the reciprocal rate ($1/k_{obs}$) and the initial concentration of TC show a linear relationship. At the same time, the photocatalytic degradation rate of TC, $k_{obs}$, changes with the initial concentration, as shown in Figure 9A. According to the slope and intercept of the straight line, the equilibrium adsorption constant of TC KTC = 0.135 L $mg^{-1}$, and the pseudo-second rate constant $k_c$ = 0.933 mg $L^{-1}$ $min^{-1}$. In Table 1, we also show the effects of different concentrations on the degradation rate.

$$Rate = -\frac{dC}{dt} = k_c \frac{K_{CIP}C}{1 + K_{CIP}C_0} \quad (1)$$

$$-\frac{dC}{dt} = k_{obs}C = k_c \frac{K_{CIP}}{1 + K_{CIP}C_0} C \tag{2}$$

$$\ln \frac{C}{C_0} = k_{obs}t \tag{3}$$

$$\frac{1}{k_{obs}} = \frac{1}{k_c K_{CIP}} + \frac{C_0}{k_c} \tag{4}$$

**Table 1.** Effect of TC initial concentration on the degradation rate.

| Initial TC (mg $L^{-1}$) | $k_{obs}$ ($min^{-1}$) | 1/$k_{obs}$ (min) | $R^2$ |
|---|---|---|---|
| 10 | 0.01457 | 68.63 | 0.902 |
| 20 | 0.01461 | 68.45 | 0.906 |
| 30 | 0.01261 | 79.30 | 0.943 |
| 50 | 0.00925 | 108.11 | 0.972 |
| 100 | 0.00358 | 279.33 | 0.960 |

To assess the impact of temperature on the photocatalytic performance of our prepared samples, we conducted experiments on TC solution degradation using the catalysts under different thermal conditions. Figure 9B presents degradation efficiency curves of the optimized Ag-AgCl/ATP photocatalytic material for degrading a 20 mg/L TC solution at temperatures of 20, 30, 40 and 50 °C. Results show that after 120 min of photocatalytic activity, the degradation efficiencies at these temperatures are 85.01%, 70.18%, 67.49%, and 67.97%, respectively. Notably, the highest degradation rate of 85.01% is achieved at 20 °C, indicating superior TC degradation efficiency of the prepared Ag-AgCl/ATP photocatalytic material at lower temperatures. Conversely, efficiency decreases with increasing temperature. Considering that operational environments for these photocatalytic materials typically involve room temperature conditions, the findings highlight the substantial potential for practical applications in TC degradation scenarios.

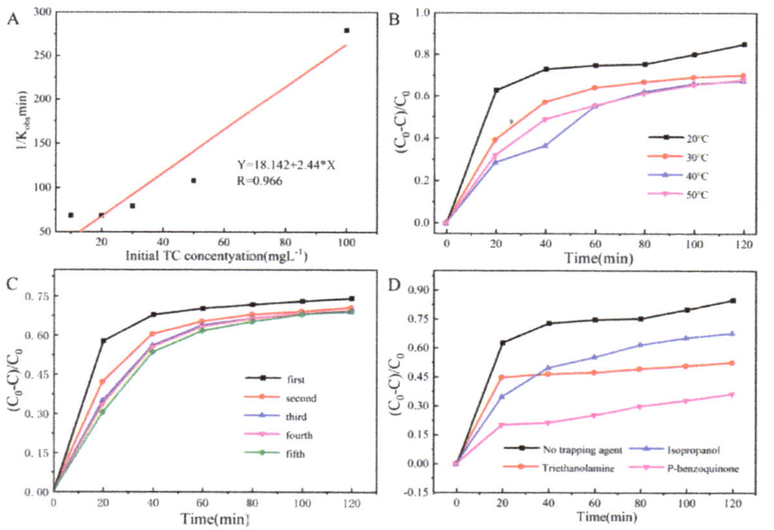


**Figure 9.** (**A**) The linear relationship between the reciprocal of the pseudo-first kinetic rate and the initial concentration of TC; (**B**) optimal Ag-AgCl/ATP degradation of TC at different temperatures; (**C**) Ag-AgCl/ATP cycle degradation of TC; (**D**) Ag-AgCl/ATP photocatalytic degradation of TC capture experiment.

### 3.2. Stability

The stability of a material is a crucial metric for assessing its practical applicability. In particular, the stability of photocatalytic materials is indicative of their quality and potential for broader applications. To evaluate the stability of our composite photocatalytic material, we conducted five successive cycles of photocatalytic experiments. As illustrated in Figure 9C, the photocatalytic degradation efficiency after 120 min of photoreaction was recorded at 74.24%, 70.70%, 69.19%, 69.56%, and 69.32% for each cycle, respectively. These results indicate a slight initial decrease in efficiency, which subsequently stabilizes around 69%. This consistent performance demonstrates that the Ag-AgCl/ATP photocatalytic composite material we developed maintains substantial degradation capability after repeated use, thereby confirming its excellent stability and reinforcing its suitability for practical applications in pollutant degradation.

### 3.3. Capture Experiment

To ascertain the predominant active species involved in the degradation process, this study conducted trapping experiments targeting key reactive species such as holes ($h^+$), hydroxyl radicals ($\cdot OH$), and superoxide anion radicals ($\cdot O_2{}^-$), to elucidate the photocatalytic degradation mechanism of TC (Figure 9D). Triethanolamine, isopropanol (IPA), and p-benzoquinone were utilized as specific scavengers for these active species, respectively. The trapping experiments were integrated into the standard photocatalytic degradation protocol for TC. Specifically, a precise amount of each scavenger was added to the reaction vessel containing the catalyst and 100 mL of TC solution: 0.133 mL of triethanolamine, 0.076 mL of isopropanol, or 0.108 g of p-benzoquinone. Following this, the concentration of TC was monitored by sampling at designated intervals, and the photocatalytic efficiency was quantitatively assessed. This experimental setup was replicated across three separate trials to ensure reproducibility and accuracy in the findings.

### 3.4. Mechanism Analysis

The role of active species in the photodegradation of TC using an Ag-AgCl/ATP composite photocatalyst was elucidated through a series of trapping experiments. Figure 9D illustrates that the addition of isopropanol, a scavenger for $\cdot OH$, has a minimal effect on the degradation rate of TC, indicating a marginal role of $\cdot OH$ radicals in the degradation process. Conversely, the addition of triethanolamine, a scavenger for $h^+$, significantly impacts the TC degradation rate more than isopropanol, suggesting a predominant role of $h^+$ over $\cdot OH$ radicals in the TC degradation mechanism. However, the overall influence of both species is limited, leading to the introduction of p-benzoquinone to further probe the degradation kinetics. In Figure 10, the mechanistic pattern of the reaction can be seen. This addition markedly reduces the photocatalytic degradation efficiency of TC compared to the baseline condition without scavengers, confirming that the $\cdot O_2{}^-$ is the principal active species in the degradation process. Experimental findings reveal that: (1) the inherent high specific surface area of attapulgite provides numerous active adsorption sites, enhancing the photocatalytic activity of the Ag-AgCl/ATP composite; (2) the presence of silver on the AgCl surface induces plasmon resonance effects upon photolysis, facilitating the generation of active species such as $h^+$ and $\cdot O_2{}^-$, which substantially improve the photocatalytic efficiency of the composite material.

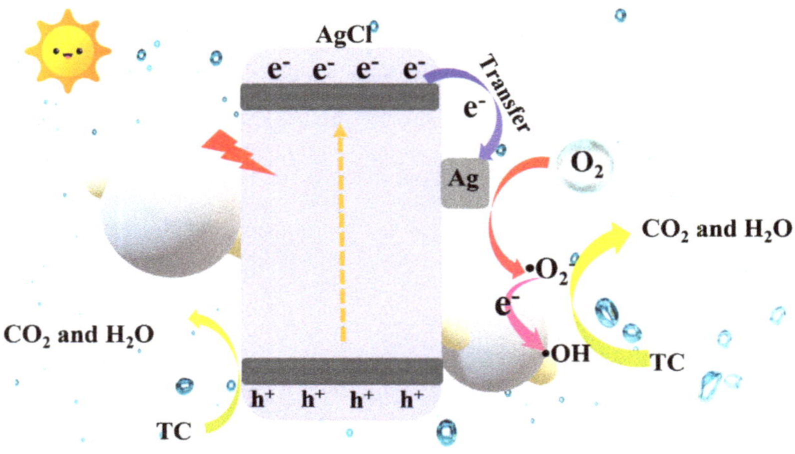


**Figure 10.** Schematic representation for photocatalytic degradation of TC using Ag-AgCl/ATP composites.

## 4. Materials and Methods

### *4.1. Pre-Treatment of the ATP*

Initially, ATP was subjected to pre-treatment by dispersing it into an ample volume of deionized water, followed by stirring at a speed of 500 rpm using a magnetic stirrer for 1 h. Subsequently, the mixture was transferred to a filter for filtration. The filtration was rapidly conducted, and the residue was washed thrice. The reagents were obtained from Sinopharm Chemical Reagent Co., Ltd. (Shanghai, China). The deionized water used throughout the experiments was obtained from locally purified water. All analytical grade reagents are used directly without further purification.

### *4.2. Preparation of Ag-AgCl/ATP Materials*

Initially, 1 g of pre-treated ATP was added to a 100 mL beaker containing 15 mL of deionized water. The mixture was stirred at a speed of 500 r/min for 30 min to ensure complete dispersion of the ATP in the solution. Subsequently, the $AgNO_3$ solution, previously prepared to the required concentration, was added dropwise to the suspension over 5 min. After the addition, stirring was continued for an additional 30 min. Following this, the required amount of NaCl solution was gradually introduced into the beaker over 5 min, using titration for precise control. Upon completion of the titration, the mixture was stirred at 600 r/min for 60 min. The resultant suspension was then filtered, and the retained solid was washed three times with deionized water and three times with ethanol. The washed solid was dried in a blowing drying oven at 60 °C for 10 h. The resulting AgCl/ATP composite with different loading (2%, 5%, 10%, 20%, 50%, 75% and 80%) was ground into a fine powder and stored in a dark place to prevent photodegradation. For subsequent experiments, the AgCl/ATP was exposed to light for 10 min to prepare the Ag-AgCl/ATP composite.

### *4.3. Photocatalytic Degradation of Tetracycline*

To evaluate the photocatalytic activity of the composites, degradation experiments of TC were conducted under various temperatures and concentrations. The experimental procedure is detailed as follows:

Initially, 25 mg of the catalyst was suspended in 100 mL of a 20 mg/L TC solution to conduct the photocatalytic degradation tests. The mixture was stirred magnetically in the dark for 30 min to establish adsorption-desorption equilibrium. During this period, aliquots of 10 mL were sampled every 10 min using a centrifuge tube, collecting three samples in total. Subsequently, a xenon lamp was activated to initiate the photocatalytic reaction, which was continued under illumination for 60 min. Samples were taken every 20 min, resulting in six additional samples. Including the initial sample, ten samples were analyzed in total. These ten samples were then centrifuged at 6000 r/min for 3 min to separate the supernatants for photometric analysis. The absorbance of each sample was recorded to calculate the photocatalytic degradation efficiency. This procedure was repeated using AgCl/ATP composites with varying loadings to identify the composite with the highest photocatalytic activity. The optimal composite was then used to degrade tetracycline at different temperatures (20, 30, 40 and 50 °C) and concentrations (10, 20, 30, 50, and 100 mg/L). For each condition, the procedure mirrored that of the optimal loading tests, and the temperature was strictly controlled during the reaction. The absorbance values were measured, and the corresponding photocatalytic efficiencies were calculated for each set of experimental conditions. In addition, the stirrer was an H01-1A Intelligent Digital Magnetic Stirrer, while the xenon lamp used a power of 300 W (300 mW $cm^{-2}$; full spectrum) (Please refer to Supplementary Information for details).

### 4.4. Stability Test

The recyclability of photocatalysts is critical for the practical application of photocatalytic technologies. To assess whether a photocatalytic material warrants recycling, it is imperative to conduct thorough recycling experiments. These experiments not only evaluate the recyclability of the photocatalyst but also its stability over multiple cycles. The protocol for the recycling experiments mirrors that of the tetracycline photocatalytic degradation tests. Specifically, after each cycle, the nine samples collected are centrifuged to separate the supernatants for photometric analysis to measure their luminosity values. Post-measurement, the supernatants are carefully filtered, and the results are documented. The remaining solution in the centrifuge tubes, along with the sample solution from the degradation container, is recovered. This mixture is then washed and dried to retrieve the catalyst. Subsequently, the recovered catalyst is ground, and its mass is measured. Should there be a shortfall in the catalyst weight, an additional quantity of the original sample is supplemented to match the prescribed mass. This preparation then undergoes another cycle of the tetracycline degradation experiment. This procedure is repeated for a total of five cycles to comprehensively assess the photocatalyst's performance and stability over repeated use.

## 5. Conclusions

The direct precipitation method and the oxidation method were used to successfully prepare the Ag-AgCl/ATP composite photocatalytic material. Through photocatalytic degradation experiments under different conditions and detailed analysis of various characteristics of the composite photocatalytic material, we have drawn a series of conclusions: (1) ATP not only enhances the dispersion of Ag-AgCl nanospheres but also ATP aids in the adsorption of TC molecules, thereby synergistically promoting the degradation of TC with Ag-AgCl nanoparticles. (2) Through the test in the photocatalytic degradation of TC solutions of different concentrations, it was found that when the concentration of TC solution is 20 mg/L, the composite photocatalytic material has the highest photocatalytic degradation efficiency of 85.01%. (3) From the photocatalytic degradation test at different temperatures, we found that the photocatalytic composite material has the highest photocatalytic degradation efficiency at a temperature of 20 °C.

**Supplementary Materials:** The following supporting information can be downloaded at: https://www.mdpi.com/article/10.3390/catal14070464/s1.

**Author Contributions:** Conceptualization, X.Z. and H.W.; methodology, H.W.; investigation, H.W. and C.Y.; writing-original draft preparation, X.Z.; writing-review and editing, X.Z. and H.W.; supervision, X.Z. and H.W.; project administration, X.Z. and H.W. All authors have read and agreed to the published version of the manuscript.

**Funding:** This research was funded by the Key Laboratory for Playgorskite Science and Applied Technology of Jiangsu (HPK202006).

**Data Availability Statement:** Data are contained within the article and Supplementary Materials.

**Conflicts of Interest:** The authors declare no conflicts of interest.

## References


1. Huong, V.T.; Duc, B.V.; An, N.T.; Anh, T.T.P.; Aminabhavi, T.M.; Vasseghian, Y.; Joo, S.W. 3D-Printed $WO_3$-UiO-66@reduced graphene oxide nanocomposites for photocatalytic degradation of sulfamethoxazole. *Chem. Eng. J.* **2024**, *483*, 149277. [CrossRef]
2. Singh, A.; Dhau, J.; Kumar, R.; Badru, R.; Singh, P.; Kumar Mishra, Y.; Kaushik, A. Tailored carbon materials (TCM) for enhancing photocatalytic degradation of polyaromatic hydrocarbons. *Prog. Mater. Sci.* **2024**, *144*, 101289. [CrossRef]
3. Chen, C.; Zhang, X.; Liu, E.; Xu, J.; Sun, J.; Shi, H. Biochar decorated $Bi_4O_5Br_2$/g-$C_3N_4$ S-scheme heterojunction with enhanced photocatalytic activity for Norfloxacin degradation. *J. Mater. Sci. Technol.* **2024**, *198*, 1–11. [CrossRef]
4. Wang, H.; Wan, Y.; Yin, S.; Xu, M.; Zhao, X.; Liu, X.; Song, X.; Wang, H.; Zhu, C.; Huo, P. Boosting mineralized organic pollutants by using a sulfur-vacancy CdS photocatalyst. *Chem. Commun.* **2023**, *59*, 9356–9359. [CrossRef] [PubMed]
5. Yue, Y.; Wu, Q.; Zheng, C.; Sun, Y.; Shah, K.J. Study on the Degradation Effect of Tetracycline Using a Co-Catalyst Loaded on Red Mud. *Catalysts* **2024**, *14*, 133. [CrossRef]
6. Wang, H.; Wan, Y.; Li, B.; Ye, J.; Gan, J.; Liu, J.; Liu, X.; Song, X.; Zhou, W.; Li, X.; et al. Rational design of Ce-doped CdS/N-rGO photocatalyst enhanced interfacial charges transfer for high effective degradation of tetracycline. *J. Mater. Sci. Technol.* **2024**, *173*, 137–148. [CrossRef]
7. Lee, S.; Devarayapalli, K.C.; Kim, B.; Lim, Y.; Lee, D.S. Fabrication of MXene-derived $TiO_2/Ti_3C_2$ integrated with a ZnS heterostructure and their synergistic effect on the enhanced photocatalytic degradation of tetracycline. *J. Mater. Sci. Technol.* **2024**, *198*, 186–199. [CrossRef]
8. Wang, H.; Li, J.; Wan, Y.; Nazir, A.; Song, X.; Huo, P.; Wang, H. Fabrication of Zn vacancies-tunable ultrathin-g-$C_3N_4$@$ZnIn_2S_4$/SWNTs composites for enhancing photocatalytic $CO_2$ reduction. *Appl. Surf. Sci.* **2023**, *613*, 155989. [CrossRef]
9. Veerakumar, P.; Sangili, A.; Chen, S.-M.; Kumar, R.S.; Arivalagan, G.; Firdhouse, M.J.; Hameed, K.S.; Sivakumar, S. Photocatalytic degradation of phenolic pollutants over palladium-tungsten trioxide nanocomposite. *Chem. Eng. J.* **2024**, *489*, 151127. [CrossRef]
10. Cui, Y.; Li, Y.; Liu, Y.; Shang, D.; Liu, Y.; Xie, L.; Zhan, S.; Hu, W. High-efficiency photocatalytic degradation of rhodamine 6G by organic semiconductor tetrathiafulvalene in weak acid-base environment. *Chem. Commun.* **2022**, *58*, 4251–4254. [CrossRef]
11. Mergenbayeva, S.; Abitayev, Z.; Batyrbayeva, M.; Vakros, J.; Mantzavinos, D.; Atabaev, T.S.; Poulopoulos, S.G. $TiO_2$/Zeolite Composites for SMX Degradation under UV Irradiation. *Catalysts* **2024**, *14*, 147. [CrossRef]
12. Wang, H.; Li, J.; Wan, Y.; Nazir, A.; Song, X.; Huo, P.; Wang, H. Synthesis of $AgInS_2$ QDs-$MoS_2$/GO composite with enhanced interfacial charge separation for efficient photocatalytic degradation of tetracycline and $CO_2$ reduction. *J. Alloys Compd.* **2023**, *954*, 170159. [CrossRef]
13. Wang, H.; Liu, Q.; Xu, M.; Yan, C.; Song, X.; Liu, X.; Wang, H.; Zhou, W.; Huo, P. Dual-plasma enhanced 2D/2D/2D g-$C_3N_4$/Pd/$MoO_{3-x}$ S-scheme heterojunction for high-selectivity photocatalytic $CO_2$ reduction. *Appl. Surf. Sci.* **2023**, *640*, 158420. [CrossRef]
14. Arana Juve, J.-M.; Baami González, X.; Bai, L.; Xie, Z.; Shang, Y.; Saad, A.; Gonzalez-Olmos, R.; Wong, M.S.; Ateia, M.; Wei, Z. Size-selective trapping and photocatalytic degradation of PFOA in Fe-modified zeolite frameworks. *Appl. Catal. B Environ. Energy* **2024**, *349*, 123885. [CrossRef]
15. Liu, Y.; Peng, M.; Gao, K.; Fu, R.; Zhang, S.; Xiao, Y.; Guo, J.; Wang, Z.; Wang, H.; Zhao, Y.; et al. Boosting photocatalytic degradation of levofloxacin over plasmonic $TiO_{2-x}$/TiN heterostructure. *Appl. Surf. Sci.* **2024**, *655*, 159516. [CrossRef]
16. Zhang, L.; Chen, Z.; Lu, Y.; Liu, Q.; Kong, L. Oxygen-defected $WO_3$-OVs@$Bi_2MoO_6$ S-scheme micro flowers heterojunctions with promoted photocatalytic degradation of tetracycline. *Appl. Surf. Sci.* **2024**, *657*, 159654. [CrossRef]
17. Li, N.; Gao, X.; Su, J.; Gao, Y.; Ge, L. Metallic $WO_2$-decorated g-$C_3N_4$ nanosheets as noble-metal-free photocatalysts for efficient photocatalysis. *Chin. J. Catal.* **2023**, *47*, 161–170. [CrossRef]
18. Sun, H.; Wang, L.; Wang, X.; Dong, Y.; Pei, T. A magnetically recyclable $Fe_3O_4$/$ZnIn_2S_4$ type-II heterojunction to boost photocatalytic degradation of gemifloxacin. *Appl. Surf. Sci.* **2024**, *656*, 159674. [CrossRef]
19. Wang, Z.; Shi, X.M.; Chen, F.Z.; Fan, G.C.; Zhao, W.W. Ag/AgCl-Like Photogating of a COF-on-MOF Heterojunction in Organic Photoelectrochemical Transistor. *Adv. Funct. Mater.* **2024**, 2404497. [CrossRef]
20. Dai, L.; Liu, R.; Hu, L.; Si, C. Simple and green fabrication of AgCl/Ag-cellulose paper with antibacterial and photocatalytic activity. *Carbohydr. Polym.* **2017**, *174*, 450–455. [CrossRef]
21. Hu, J.; Chen, F.; Mao, J.; Ni, L.; Lu, J. Direction regulation of interface carrier transfer and enhanced photocatalytic oxygen activation over Z-scheme $Bi_4V_2O_{11}$/Ag/AgCl for water purification. *J. Colloid Interface Sci.* **2023**, *641*, 695–706. [CrossRef]
22. Zhu, H.; Chen, D.; Li, N.; Xu, Q.; Li, H.; He, J.; Lu, J. Cyclodextrin-functionalized Ag/AgCl foam with enhanced photocatalytic performance for water purification. *J. Colloid Interface Sci.* **2018**, *531*, 11–17. [CrossRef]
23. Chen, J.; Lu, X.; Wang, D.; Xiu, P.; Gu, X. Effective depolymerization of alkali lignin using an attapulgite-$Ce_{0.75}Zr_{0.25}O_2$(ATP-CZO)-supported cobalt catalyst in ethanol/isopropanol media. *Chin. J. Chem. Eng.* **2023**, *57*, 50–62. [CrossRef]

24. Guan, J.; Wang, H.; Li, J.; Ma, C.; Huo, P. Enhanced photocatalytic reduction of $CO_2$ by fabricating $In_2O_3/CeO_2$/HATP hybrid multi-junction photocatalyst. *J. Taiwan Inst. Chem. Eng.* **2019**, *99*, 93–103. [CrossRef]
25. Fang, Y.; Chen, D. A novel catalyst of Fe-octacarboxylic acid phthalocyanine supported by attapulgite for degradation of Rhodamine B. *Mater. Res. Bull.* **2010**, *45*, 1728–1731. [CrossRef]
26. Wu, F.; Li, X.; Zhang, H.; Zuo, S.; Yao, C. Z-Scheme photocatalyst constructed by natural attapulgite and upconversion rare earth materials for desulfurization. *Front. Chem.* **2018**, *6*, 477. [CrossRef]
27. Deng, L.; Xie, Y.; Zhang, G. Synthesis of C-Cl-codoped titania/attapulgite composites with enhanced visible-light photocatalytic activity. *Chin. J. Catal.* **2017**, *38*, 379–388. [CrossRef]
28. Wang, X.; Mu, B.; An, X.; Wang, A. Insights into the relationship between the color and photocatalytic property of attapulgite/CdS nanocomposites. *Appl. Surf. Sci.* **2018**, *439*, 202–212. [CrossRef]
29. Qi, Y.; Zhao, S.; Jiang, X.; Kang, Z.; Gao, S.; Liu, W.; Shen, Y. Visible-Light-Driven BiOBr-$TiO_2$-Attapulgite Photocatalyst with Excellent Photocatalytic Activity for Multiple Xanthates. *Catalysts* **2023**, *13*, 1504. [CrossRef]
30. Zeng, J.; Han, C.; Wang, B.; Cao, G.; Yao, C.; Li, X. Construction of plasmonic CuS/attapulgite nanocomposites toward photothermal reforming of biomass for hydrogen production. *J. Alloys Compd.* **2024**, *985*, 174038. [CrossRef]
31. Zhu, Z.; Yu, Y.; Dong, H.; Liu, Z.; Li, C.; Huo, P.; Yan, Y. Intercalation effect of attapulgite in g-$C_3N_4$ modified with $Fe_3O_4$ quantum dots to enhance photocatalytic activity for removing 2-mercaptobenzothiazole under visible light. *ACS Sustain. Chem. Eng.* **2017**, *5*, 10614–10623. [CrossRef]
32. Li, X.; He, C.; Zuo, S.; Yan, X.; Dai, D.; Zhang, Y.; Yao, C. Photocatalytic nitrogen fixation over fluoride/attapulgite nanocomposite: Effect of upconversion and fluorine vacancy. *Sol. Energy* **2019**, *191*, 251–262. [CrossRef]
33. Kong, X.; Li, L.; Feng, Q.; Liang, Z.; Huang, J.; Wang, X.; Zhang, J.; Li, J. Soft chemical synthesis and visible light photocatalytic performance of Ag@AgCl/$H_{1.07}Ti_{1.73}O_4$ platelike composite with composition controlling. *J. Alloys Compd.* **2017**, *727*, 311–317. [CrossRef]
34. Feng, Z.; Yu, J.; Sun, D.; Wang, T. Visible-light-driven photocatalysts Ag/AgCl dispersed on mesoporous $Al_2O_3$ with enhanced photocatalytic performance. *J. Colloid Interface Sci.* **2016**, *480*, 184–190. [CrossRef] [PubMed]
35. Sun, L.; Yin, S.; Shen, D.; Zhou, Y.; Li, J.; Li, X.; Wang, H.; Huo, P.; Yan, Y. Fabricating acid-sensitive controlled PAA@Ag/AgCl/CN photocatalyst with reversible photocatalytic activity transformation. *J. Colloid Interface Sci.* **2020**, *580*, 753–767. [CrossRef] [PubMed]
36. Xu, Y.; Zhang, L.; Yin, M.; Xie, D.; Chen, J.; Yin, J.; Fu, Y.; Zhao, P.; Zhong, H.; Zhao, Y.; et al. Ultrathin g-$C_3N_4$ films supported on Attapulgite nanofibers with enhanced photocatalytic performance. *Appl. Surf. Sci.* **2018**, *440*, 170–176. [CrossRef]
37. Zhao, S.; Qi, Y.; Lv, H.; Jiang, X.; Wang, W.; Cui, B.; Liu, W.; Shen, Y. Effect of clay mineral support on photocatalytic performance of BiOBr-$TiO_2$ for efficient photodegradation of xanthate. *Adv. Powder Technol.* **2024**, *35*, 104431. [CrossRef]
38. Liu, C.; Guo, Y.; Li, S.; Xuan, K.; Guo, Y.; Li, J.; Wang, X.; Li, X.; Zhou, Z. Mesoporous sulfur-doped g-$C_3N_4$@attapulgite composite as an advanced photocatalyst for efficiently uranium(VI) recovery from aqueous solutions. *J. Environ. Chem. Eng.* **2024**, *12*, 112886. [CrossRef]
39. Tang, X.; Shen, W.; Li, D.; Li, B.; Wang, Y.; Song, X.; Zhu, Z.; Huo, P. Research on cobalt-doping sites in g-$C_3N_4$ framework and photocatalytic reduction $CO_2$ mechanism insights. *J. Alloys Compd.* **2023**, *954*, 170044. [CrossRef]
40. Cao, G.; Xing, H.; Gui, H.; Yao, C.; Chen, Y.; Chen, Y.; Li, X. Plasmonic quantum dots modulated nano-mineral toward photothermal reduction of $CO_2$ coupled with biomass conversion. *Nano Res.* **2024**, *17*, 5061–5072. [CrossRef]
41. Kaushik, V. XPS core level spectra and Auger parameters for some silver compounds. *J. Electron. Spectrosc.* **1991**, *56*, 273–277. [CrossRef]
42. Mahmoodi, N.M.; Taghizadeh, A.; Taghizadeh, M.; Abdi, J. In situ deposition of Ag/AgCl on the surface of magnetic metal-organic framework nanocomposite and its application for the visible-light photocatalytic degradation of Rhodamine dye. *J. Hazard. Mater.* **2019**, *378*, 120741. [CrossRef]
43. Zeng, Y.; Yin, Q.; Liu, Z.; Dong, H. Attapulgite-interpenetrated g-$C_3N_4$/$Bi_2WO_6$ quantum-dots Z-scheme heterojunction for 2-mercaptobenzothiazole degradation with mechanism insight. *Chem. Eng. J.* **2022**, *435*, 134918. [CrossRef]
44. Bao, Y.; Chen, K. AgCl/Ag/g-$C_3N_4$ Hybrid Composites: Preparation, Visible Light-Driven Photocatalytic Activity and Mechanism. *Nano-Micro Lett.* **2015**, *8*, 182–192. [CrossRef]
45. Zhang, X.; Wang, P.; Meng, W.; Cui, E.; Zhang, Q.; Wang, Z.; Zheng, Z.; Liu, Y.; Cheng, H.; Dai, Y.; et al. Photocatalytic anticancer performance of naked Ag/AgCl nanoparticles. *Chem. Eng. J.* **2022**, *428*, 131265. [CrossRef]

catalysts

MDPI

Article

# Research on Cu-Site Modification of $g$-$C_3N_4$/$CeO_2$-like Z-Scheme Heterojunction for Enhancing $CO_2$ Reduction and Mechanism Insight

**Yiying Zhou [1], Junxi Cai [1], Yuming Sun [1], Shuhan Jia [1], Zhonghuan Liu [1], Xu Tang [1], Bo Hu [1], Yue Zhang [2,*], Yan Yan [1,*] and Zhi Zhu [1,*]**

1 School of Chemistry and Chemical Engineering, Institute for Advanced Materials, Jiangsu University, Zhenjiang 212013, China; cuihua202209@163.com (Y.Z.); 18896660263@163.com (J.C.); s969678125@126.com (Y.S.); jiashuhan999@sina.com (S.J.); lzh0240@sina.com (Z.L.); tangxu@ujs.edu.cn (X.T.); hubo93@bcnu.edu.cn (B.H.)

2 School of Chemistry and Chemical Engineering, Liaoning Normal University, Dalian 116029, China

* Correspondence: zhangyue433@126.com (Y.Z.); dgy5212004@163.com (Y.Y.); zhuzhi@ujs.edu.cn (Z.Z.)

**Abstract:** In this work, the successful synthesis of a Cu@$g$-$C_3N_4$/$CeO_2$-like Z-scheme heterojunction through hydrothermal and photo-deposition methods represents high $CO_2$ reduction activity with remarkable CO selectivity, as evidenced by the impressive CO yield of 33.8 μmol/g for Cu@$g$-$C_3N_4$/$CeO_2$, which is over 10 times higher than that of $g$-$C_3N_4$ and $CeO_2$ individually. The characterization and control experimental results indicate that the formation of heterojunctions and the introduction of Cu sites promote charge separation and the transfer of hot electrons, as well as the photothermal effect, which are the essential reasons for the improved $CO_2$ reduction activity. Remarkably, Cu@$g$-$C_3N_4$/$CeO_2$ still exhibits about 92% performance even after multiple cycles. In situ FTIR was utilized to confirm the production of COOH* at 1472 $cm^{-1}$ and to elucidate the mechanism behind the high selectivity for CO production. The study's investigation into the wide-ranging applicability of the Cu@$g$-$C_3N_4$/$CeO_2$-like Z-scheme heterojunction catalysts is noteworthy, and the exploration of potential reaction mechanisms for CO2 reduction adds valuable insights to the field of catalysis.

**Keywords:** photocatalysis; heterojunction; $CO_2$ reduction; $g$-$C_3N_4$; $CeO_2$

## 1. Introduction

In the face of the escalating energy crisis and environmental degradation spurred by the rapid pace of global industrialization, the quest for sustainable and clean energy solutions has become more critical than ever [1–3]. Solar energy stands out as a beacon of hope, offering a pollution-free, inexhaustible, and renewable source of power that is poised to play a pivotal role in the future landscape of energy production [4]. The harnessing of this abundant energy through semiconductor photocatalytic technology reduction of carbon dioxide ($CO_2$) into useful chemicals or fuel presents a dual-edged sword in tackling not only energy scarcity but also the pressing issue of the environmental greenhouse effect [5].

Graphitic carbon nitride ($g$-$C_3N_4$) has emerged as a promising candidate due to its remarkable stability, non-toxic nature, ready availability from abundant precursors, and economic feasibility [6]. However, the inherent limitations of pure $g$-$C_3N_4$, such as its low solar energy utilization efficiency, modest surface area, and high recombination rates of photogenerated charge carriers, have hindered its full potential as an efficient photocatalyst [7]. To overcome these challenges, researchers have been exploring the construction of $g$-$C_3N_4$-based photocatalysts with enhanced performance [8–12]. One such approach involves the design and fabrication of $g$-$C_3N_4$-based heterojunctions [13], which, through bandgap engineering, can significantly boost the response to visible light [14–16], charge separation efficiency, and overall photocatalytic activity of *g*-$C_3N_4$ [17]. Especially, the Z-scheme heterojunction with high redox potential exhibits excellent photocatalytic performance, which can reduce $CO_2$ to high-value-added hydrocarbon fuels (It refers to hydrocarbons with high market value and application value, which are usually produced through specific chemical or biological processes and have high energy density and/or specific chemical properties, making them widely used in energy, chemical, transportation and other fields).

Cerium dioxide ($CeO_2$) has garnered significant attention due to its exceptional properties, such as low cost, high stability, high oxygen storage capacity (as shown in Figure 1a), rich array of oxygen vacancies, reversible $Ce^{3+}/Ce^{4+}$ redox couples, and robust resistance to photocorrosion [18–20]. The well-matched band structures to *g*-$C_3N_4$ to form $CeO_2$/*g*-$C_3N_4$ Z-scheme heterojunctions has attracted widespread attention [21–23]. As far as we know the $CeO_2$/*g*-$C_3N_4$ Z-scheme has been extensively reported on $CO_2$ reduction, but there are still some issues: 1. The synthesis process require multiple steps and high temperatures, which increases production costs and complicates scalability. 2. The selectivity towards a specific product such as methane, ethylene, or methanol is often not optimal. Unwanted side reactions or competing pathways can lead to a mix of products, reducing the overall efficiency and economic viability. 3. The catalytic activity of the $CeO_2$/*g*-$C_3N_4$ composite under visible light or solar irradiation is not ideal. Therefore, it is necessary to further modify the $CeO_2$/*g*-$C_3N_4$ Z-scheme and systematically study the $CO_2$ reduction performance and mechanism.

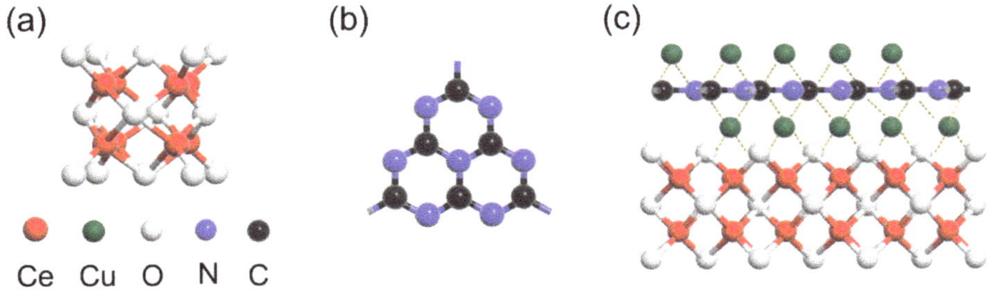


**Figure 1.** Structures of monolayer $CeO_2$ (**a**), *g*-$C_3N_4$ (**b**), Cu@*g*-$C_3N_4$/$CeO_2$ (**c**).

Through literature research, introducing heteroatoms into the *g*-$C_3N_4$ matrix or $CeO_2$ lattice seems to be a very effective method. The structure of *g*-$C_3N_4$ is shown in Figure 1b. The doping of transition metals such as Cu, Co, Ce, etc. [24–26] within *g*-$C_3N_4$ or $CeO_2$ could adjust the electronic structure, band gap and augment the charge separation process of the composite catalyst, as it allows for a sustained supply of electrons and holes that participate in the reduction and oxidation reactions, respectively [27]. Among these transition metals, Cu has attracted much attention because Cu 2p orbitals can hold more electrons, effectively facilitating the charge carrier transfer and further improving the $CO_2$ photoreduction capability of the catalyst. Moreover, the Cu can also be excited and generate the heat to supply electrons, making it easier for these hot electrons to participate in the $CO_2$ reduction reaction [28]. Therefore, the addition of Cu not only has the potential to establish a more effective barrier against electron-hole recombination [29–31], but also offers a novel approach to $CO_2$ reduction by involving thermal electrons in the reaction, thereby further improving the photocatalytic efficiency under visible light irradiation [32].

Thus, a novel Cu@$CeO_2$/*g*-$C_3N_4$-like Z-scheme heterojunction photocatalyst was prepared; by adding Cu sites (as shown in Figure 1c) to increase the thermal reaction pathway, the separation efficiency of electrons and holes can be further improved to optimize $CO_2$ conversion. By combining materials science strategies (TEM, electrochemical) with advanced spectroscopic techniques (such as the in-situ FT-IR and FL), researchers strive to gain a deeper understanding of the complex reaction dynamics at play, with the ultimate goal of advancing sustainable technologies for $CO_2$ utilization and contributing to the global effort against climate change.

## 2. Results and Discussion

### *2.1. Photocatalyst Characterization*

The microstructures of *g*-$C_3N_4$, $CeO_2$, and Cu@*g*-$C_3N_4/CeO_2$ were evaluated using SEM and TEM. The results show that the *g*-$C_3N_4$ exhibits a uniform two-dimensional micro-sheet morphology with a diameter ranging from approximately 1.2–1.5 µm, as depicted in Figure 2a,d. The $CeO_2$ also possess a two-dimensional ultra-thin micro-sheet structure with a smooth surface, as illustrated in Figure 2b,e. Upon the formation of the Cu@*g*-$C_3N_4/CeO_2$, the surface of the $CeO_2$ micro-sheets exhibited a noticeable increase in roughness and thickness. The SEM and TEM images in Figure 2c,f demonstrate the tight binding between the $CeO_2$ and *g*-$C_3N_4$, along with the successful loading copper (Cu) [33,34]. The integration of the two-dimensional *g*-$C_3N_4$ with the two-dimensional $CeO_2$ and the loading of Cu nanoparticles is anticipated to enhance the transfer and separation of photogenerated charges. This synergistic effect is expected to augment the number of active sites available for photocatalysis, thereby improving the overall catalytic activity of the material. Additionally, the elemental distribution and composition within Cu@*g*-$C_3N_4/CeO_2$ were investigated using EDX-mapping. As shown in Figure 2(g1–g5), the mapping reveals a uniform distribution of Cu, C, N, Ce, and O elements across the surface of Cu@*g*-$C_3N_4/CeO_2$. In summary, the comprehensive characterization of the photocatalyst using SEM, TEM, and EDX mapping analysis confirms the successful fabrication of the Cu@*g*-$C_3N_4/CeO_2$. The unique structural features and elemental composition of the composite are expected to contribute to its enhanced photocatalytic performance.

**Figure 2.** TEM images of the *g*-$C_3N_4$ (**a**), $CeO_2$ (**b**), Cu@*g*-$C_3N_4/CeO_2$ (**c**), SEM images of *g*-$C_3N_4$ (**d**), $CeO_2$ (**e**), Cu@*g*-$C_3N_4/CeO_2$ (**f**), EDX mapping of Cu@*g*-$C_3N_4/CeO_2$ (**g1–g5**).

### 2.2. Surface Composition and Photoelectric Analysis

Figure 3a shows the XRD patterns of the as-prepared samples. For $CeO_2$, the diffraction peaks at $2\theta$ = 28.5°, 33°, 47.4°, 56.3°, 59.3°, 69.3°, 76.6° and 79°; these peaks are consistent with the material's cubic fluorite crystal structure, which belongs to the Fm-3m space group and can be indexed to the (111), (200), (220), (311), (222), (400), (331) and (420) planes [35]. $g$-$C_3N_4$ displays a diffraction peak at $2\theta$ = 27.3°, corresponding to the (002) plane and reflecting the interlayer spacing [14]. The Cu@$g$-$C_3N_4/CeO_2$ exhibits diffraction peaks at 28.5°, 33.0°, 47.4°, and 56.3°, which coincide with the known peaks of $CeO_2$ [36]. This alignment suggests the successful formation of a hybrid material through the integration of $CeO_2$ with $g$-$C_3N_4$. The preservation of $CeO_2$'s crystallographic planes in the hybrid is evidenced by the correspondence of these peaks to the (111), (200), (220), and (311) planes of $CeO_2$. The characteristic peak of $g$-$C_3N_4$ at 27.30° may be either influenced or superimposed upon by the hybridization process with $CeO_2$ [37,38]. The minimal presence of Cu peaks should be attributed to the metal's uniform dispersion and low concentration within Cu@$g$-$C_3N_4/CeO_2$. Figure 3b displays the survey spectrum of $CeO_2$, $g$-$C_3N_4$, Cu@$g$-$C_3N_4/CeO_2$, which proved Cu@$g$-$C_3N_4/CeO_2$ was made up of C, N, O, Ce and Cu. And C, N, O, Ce and Cu had atomic percentages of 26.12%, 8.11%, 34.24%, 30.33%, and 1.21%, respectively. The presence of C and N elements were found in $g$-$C_3N_4$, Ce and O elements in $CeO_2$, and C, N, Ce, and O elements in Cu@ $g$-$C_3N_4/CeO_2$; Cu was not effectively detected due to its low concentration. Figure 3c shows the C1s spectra of $g$-$C_3N_4$ and Cu@$g$-$C_3N_4/CeO_2$, featured peaks at 288.4 eV and 284.5 eV, corresponding to $sp^2$ hybridized carbon (C=N–C) and hydroxylated carbon species with C–O and C–OH, or specific out-of-plane sp3 C–N species [13,39,40]. Concurrently, the O1s high-resolution XPS spectrum of $g$-$C_3N_4$ is also displayed in Figure 3d, where the peaks at 529.5 eV and 532.2 eV are assigned to the O=C and O–C groups, respectively. For Cu@$g$-$C_3N_4/CeO_2$, these peaks exhibited a slight shift towards higher binding energy relative to pure $g$-$C_3N_4$ [41], suggesting a significant interaction between $CeO_2$ and $g$-$C_3N_4$. As depicted in Figure 3e, the N1s spectra of $g$-$C_3N_4$ and Cu@$g$-$C_3N_4/CeO_2$ featured peaks at 395.1 eV and 398.7 eV, corresponding to $sp^2$ hybridized carbon (C=N–C). Figure 3f presents the high-resolution XPS spectra of the Ce 3d levels for pure $CeO_2$ and Cu@$g$-$C_3N_4/CeO_2$. The Ce 3d energy level spectrum of the $CeO_2$ sample can be deconvoluted into six contributions at 882.5 eV, 888.9 eV, 898.4 eV, 901.0 eV, 907.9 eV, and 916.7 eV, labeled as $\nu$, $\nu'$, $\nu''$, $\mu$, $\mu'$, and $\mu''$, where $\nu$ and $\mu$ represent the spin-orbit split states of Ce $3d_{5/2}$ and Ce $3d_{3/2}$, respectively. The peaks labeled as $\nu$, $\nu'$, $\nu''$, and $\mu$ are characteristic of $Ce^{4+}$, while the peaks labeled as $\nu''$ and $\mu''$ indicate the presence of $Ce^{3+}$ ions. The coexistence of $Ce^{3+}/Ce^{4+}$ on the surface of Cu@$g$-$C_3N_4/CeO_2$ suggests that under oxygen-deficient conditions, $CeO_2$ can release lattice oxygen, which has an active effect on surface adsorption [42]. Some $Ce^{4+}$ ions are reduced to $Ce^{3+}$, forming oxygen vacancies. Under oxygen-rich conditions, the adsorbed oxygen is stored in the form of lattice oxygen, allowing $Ce^{3+}$ to be re-oxidized to $Ce^{4+}$, which helps improve charge separation efficiency and enhances the photocatalytic activity. Additionally, the characteristic peaks of N 1s for Cu@$g$-$C_3N_4/CeO_2$ also show a slight shift towards higher binding energy compared to pure $g$-$C_3N_4$, further confirming the strong interaction between $CeO_2$ and $g$-$C_3N_4$.

The decrease in photoluminescence (PL) intensity and extended fluorescence (FL) lifetime of Cu@$g$-$C_3N_4/CeO_2$ compared to pure $g$-$C_3N_4$ under excitation light at 375 nm, as illustrated in Figure 4a,b, indicates a reduction in recombination of charge carriers within the heterojunction structure. This phenomenon suggests that the Cu@$g$-$C_3N_4/CeO_2$ Z-scheme catalyst effectively promotes the separation and migration of photoexcited electrons and holes, leading to enhanced photocatalytic performance, which are crucial factors in achieving high catalytic activity and selectivity in $CO_2$ reduction processes [43]. The UV–Vis diffuse reflectance spectra (DRS) analysis depicted in Figure 4c reveals that there is no significant change in the absorption edge and strength of $CeO_2$, $g$-$C_3N_4$, and Cu@$g$-$C_3N_4/CeO_2$ [44]. This observation suggests that the introduction of Cu into the $g$-$C_3N_4/CeO_2$ heterojunction structure does not notably alter the light-absorption properties of the materials. The consistent absorption characteristics across the different samples indicate that the enhanced photocatalytic performance of Cu@$g$-$C_3N_4/CeO_2$ is likely attributed to factors other than changes in light-absorption capacity. Figure 4d presents a detailed depiction of the photocurrent response characteristics of three distinct materials: pristine $CeO_2$, $g$-$C_3N_4$, and the novel Cu@$g$-C3N4/CeO2 composite. Observations indicate that both pristine $CeO_2$ and $g$-$C_3N_4$ exhibit relatively stable photocurrent responses under prolonged illumination, with no significant decay in photocurrent over time. This phenomenon reveals their superior photochemical stability, conducive to maintaining high efficiency under continuous light exposure [45,46]. Moreover, the Cu@$g$-$C_3N_4/CeO_2$ shows a higher photocurrent response compared to individual $CeO_2$ and $g$-$C_3N_4$. This finding not only confirms the significant enhancement in the photochemical activity but also implies Cu@$g$-$C_3N_4/CeO_2$ possesses superior charge carrier separation capabilities. Such improvements in performance are likely attributed to the synergistic effects between $CeO_2$ and $g$-$C_3N_4$ within the composite, as well as the optimization of charge transfer and transport properties brought about by the incorporation of Cu elements [13,14].

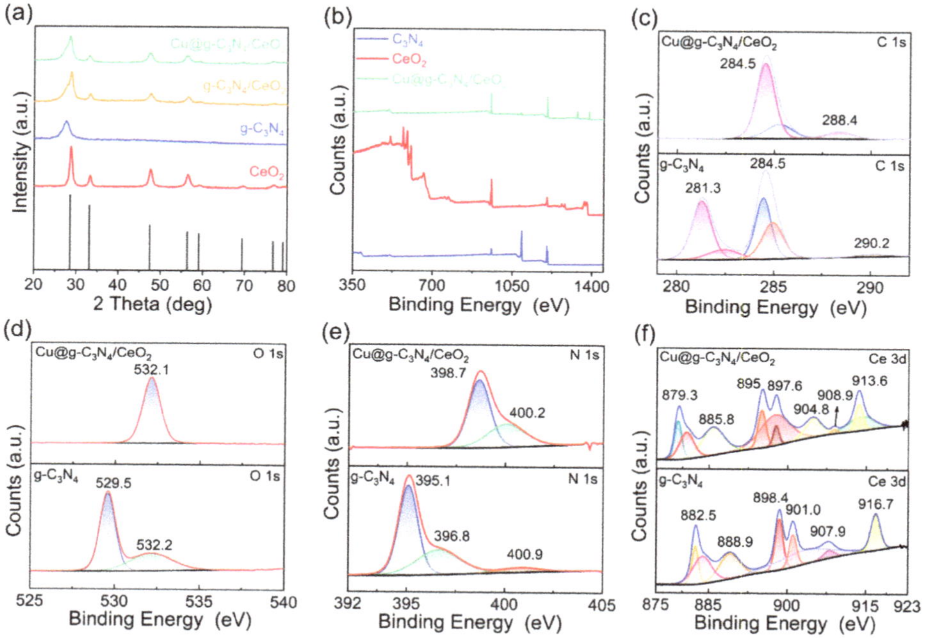


**Figure 3.** XRD patterns (**a**), XPS survey spectra (**b**) of $g$-$C_3N_4$, $CeO_2$ and Cu@$g$-$C_3N_4/CeO_2$, high resolution of C1s of $g$-$C_3N_4$ and Cu@$g$-$C_3N_4/CeO_2$(**c**), O1s XPS spectra of $CeO_2$ and Cu@$g$-$C_3N_4/CeO_2$ (**d**), N1s XPS spectra of $g$-$C_3N_4$ and Cu@$g$-$C_3N_4/CeO_2$ (**e**), Ce 3d XPS spectra of $CeO_2$ and Cu@$g$-$C_3N_4/CeO_2$ (**f**).

As shown in Figure 4e, $CeO_2$ and $g$-$C_3N_4$ exhibit higher impedance values, and the Cu@$g$-$C_3N_4/CeO_2$ exhibits the lowest charge transfer impedance; the low impedance characteristic is likely due to the synergistic effect between $CeO_2$ and $g$-$C_3N_4$, as well as the modification effect of Cu, which collectively promote efficient charge transfer, thereby enhancing the electrochemical activity of the material [47,48]. Additionally, the Mott–Schottky (M-S) plots in Figure 4f, at different frequencies of 2000 Hz, 2500 Hz, and 3000 Hz, of Cu@$g$-$C_3N_4/CeO_2$ consistently show a positive slope, indicating that the material is an n-type semiconductor with electrons as the primary charge carriers. Moreover, the flat-band potential ($E_{fb}$) is determined to be −0.35 V and remains stable across different frequencies, indicating the absence of defect or trap states within Cu@$g$-$C_3N_4/CeO_2$, implying minimal loss of charge carriers during their transport.

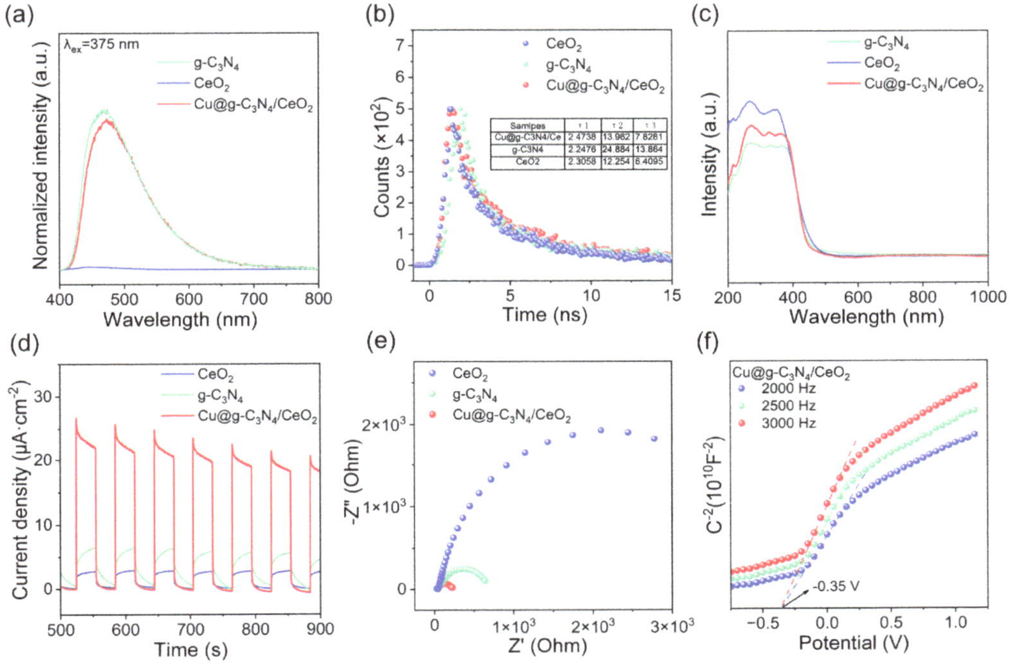


**Figure 4.** PL spectra (**a**), FL spectra (**b**), UV–Vis DRS (**c**), transient photocurrent response (**d**) and EIS Nyquist plots (**e**) of $g$-$C_3N_4$ and $CeO_2$ and Cu@$g$-$C_3N_4/CeO_2$, Mott–Schottky (**f**) of Cu@$g$-$C_3N_4/CeO_2$.

### *2.3. Photocatalytic Activity*

This experiment used gas chromatography to detect the performance of catalysts in $CO_2$ reduction under a xenon lamp (1000 mW/cm$^2$), stirring and adsorption for 30 min to achieve the adsorption and desorption equilibria. CO is our only $CO_2$ reduction product in this system. As shown in Figure 5a, in the presence of condensed water (6 °C), the CO yields of $g$-$C_3N_4$, $CeO_2$ and Cu@$g$-$C_3N_4/CeO_2$ are 0.56, 1.68, 11.44 μmol/g; the CO yield of Cu@$g$-$C_3N_4/CeO_2$ has been significantly improved; the reason should be the formation of the like-Z-scheme and the introducing of Cu reaction sites, allowing charge carriers to be quickly transferred and participate in $CO_2$ reduction reactions. In order to verify the effect of hot electrons on $CO_2$ reduction, the $CO_2$ reduction reaction without temperature control was conducted and it was found that the CO yields of $g$-$C_3N_4$, $CeO_2$ and Cu@$g$-$C_3N_4/CeO_2$ were 1.96, 3.28 and 33.82 μmol/g, respectively [49]; the result is shown in Figure 5b. Through data comparison, it was found that the CO performance without condensate water was more than three times higher than that with condensate water; this is because the addition of Cu makes better use of heat, converting some photo-generated charge carriers into hot electrons. These hot electrons not only participate in the reduction reaction of $CO_2$, but also bombard $g$-$C_3N_4$ at high speed due to the large amount of energy they carry, causing more charge carriers to participate in the chain reaction. This explains why higher CO yields can be achieved without condensation. Cycling experiment without condensate (Figure 5c) was carried out and it was found that after 5 cycles, the performance of Cu@$g$-$C_3N_4/CeO_2$ slightly decreased. XRD (X-ray diffraction) before and after the cycles shows that the crystal of Cu@$g$-$C_3N_4/CeO_2$ has not changed (Figure 6a). In order to determine that CO comes from $CO_2$, the different blank control experiments results are shown in Figure 6b. (a) Under a $N_2$ atmosphere for 4 h of illumination, no significant CO generation was detected. (b) Under a $CO_2$ atmosphere but without light, no significant CO generation was detected after 4 h. (c) Under normal testing conditions. (d) Under a $CO_2$ atmosphere illumination but without a catalyst, no significant CO generation was detected. Therefore, the prepared Cu@*g*-$C_3N_4/CeO_2$ can be identified as a stable photocatalyst.

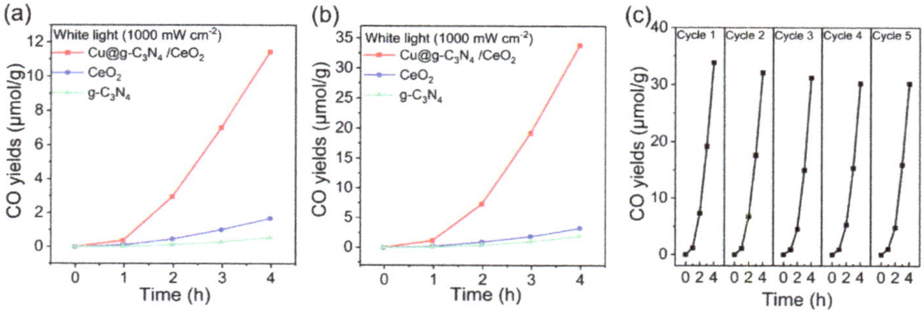


**Figure 5.** CO yields with condensed water ((**a**), Xenon lamp, 1000 mW/cm$^2$, 6 °C); the CO yields over *g*-$C_3N_4$ and $CeO_2$ and Cu@*g*-$C_3N_4/CeO_2$ hybrids without condensed water ((**b**), Xenon lamp, 1000 mW/cm$^2$); the recycling test for $CO_2$ reduction using Cu@*g*-$C_3N_4/CeO_2$ without condensed water ((**c**), Xenon lamp, 1000 mW/cm$^2$).

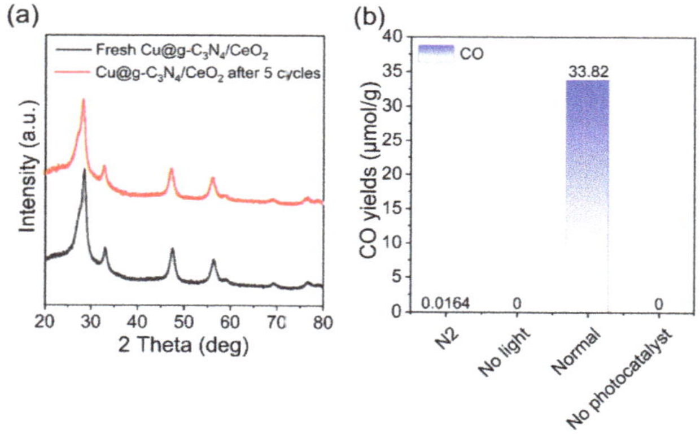


**Figure 6.** (**a**) XRD comparison of photocatalysis before and after photocatalysis; (**b**) Evolution of CO after 4 h of reaction under various reaction conditions. In the case of $N_2$, no light, normal and without photocatalyst, respectively.

### 2.4. Mechanism of $CO_2$ Photoreduction

The in situ FTIR analysis was used to detect the intermediate products, in an attempt to clarify the mechanism of the reaction process of $CO_2$ photoreduction. As shown in Figure 7, a peak at 1398 cm$^{-1}$ is b-$CO_3^{2-}$, and the inverted peak at 1672 cm$^{-1}$ was attributed to the typical bending vibrations of molecular water. These peaks indicate the water molecules participate in the reaction process as the electron donors. With the LED illumination, two distinct positive peaks appeared, which belonged to the intermediates of the $CO_2$ photoreaction process; the peak at 1472 cm$^{-1}$ belonged to the COOH* [13], which is the critical intermediate from $CO_2$ to CO; moreover, the peaks at 1123 and 1063 cm$^{-1}$ belonged to $HCO_3^-$, and the peaks at 1398 cm$^{-1}$ were considered to be the $CO_3^{2-}$ from the hydrolysis of $CO_2$. Based on the in situ FTIR results, the $CO_2$ photoreduction process can be inferred as follows:

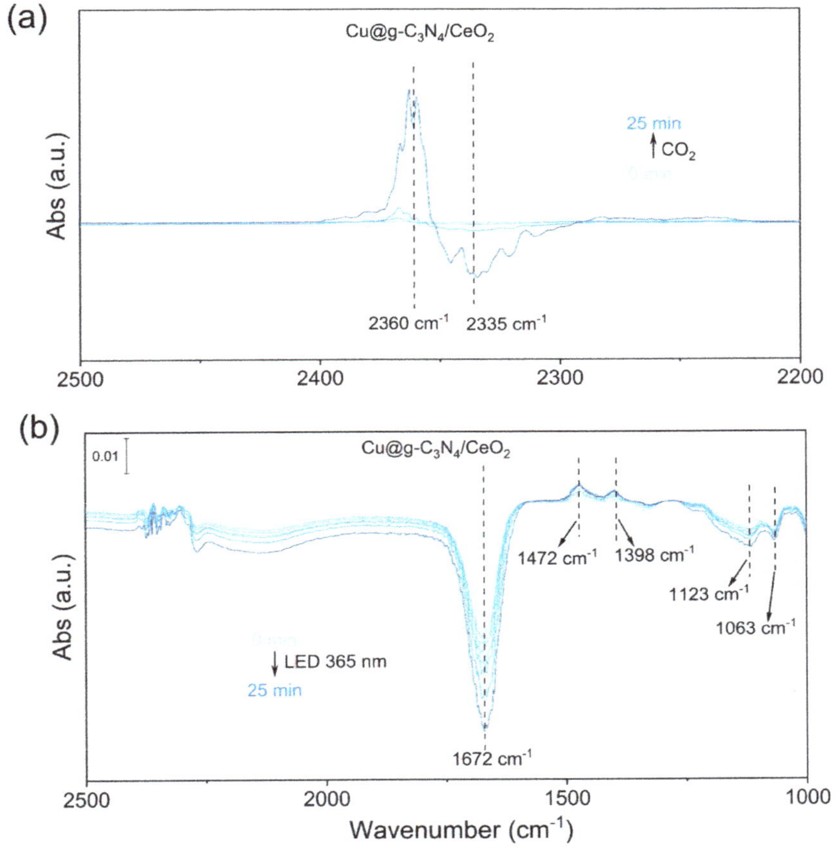


**Figure 7.** In situ FTIR spectra of $CO_2$ adsorption (**a**) and reaction (**b**) of Cu@*g*-$C_3N_4/CeO_2$ collected at different time intervals.

The $CO_2$ reduction mechanism of Cu@*g*-$C_3N_4/CeO_2$ is shown in Figure 8. The light energy generated under xenon lamp irradiation not only stimulates the generation of $e^-/h^+$ pairs but also causes intramolecular motion, resulting in an undeniable thermal effect. The thermal effect of this part further promotes the generation of electron hole pairs. At the same time, this part of electrons is generated by receiving the thermal energy of intramolecular motion, so we call them hot electrons. The addition of Cu reaction sites not only builds a bridge for high-speed electron conduction between the two, but also serves as a high-speed channel for heat conduction. This allows photo-generated electrons to be converted into hot electrons, which are always in a high-speed-motion state, making it difficult to recombine with holes. These electrons are difficult to recombine with holes, and the electron transport channel constructed by Cu@*g*-$C_3N_4/CeO_2$ makes it easier for them to be delivered to the catalyst surface and react with $CO_2$.

$$\text{Cu@}g\text{-}C_3N_4/CeO_2 + \text{Photothermal} \rightarrow \text{Cu@}g\text{-}C_3N_4/CeO_2\ (e^- + h^+) \quad (1)$$

$$CO_2 \rightarrow CO_2{}^* \quad (2)$$

$$H_2O + h^+ \rightarrow \cdot OH + H^+ \quad (3)$$

$$H_2O + \cdot OH + 3h^+ \rightarrow O_2 + 3H^+ \quad (4)$$

$$2H^+ + 2e^- \rightarrow H_2 \quad (5)$$

$$CO_2 + H_2O \rightarrow H_2CO_3 \quad (6)$$

$$H_2CO_3 \rightarrow H^+ + HCO_3^- \quad (7)$$

$$HCO_3^- \rightarrow H^+ + CO_3^{2-} \quad (8)$$

$$CO_2^* + e^- + H^+ \rightarrow COOH^* \quad (9)$$

$$COOH^* + e^- + H^+ \rightarrow CO^* + H_2O \quad (10)$$

$$CO^* \rightarrow CO \quad (11)$$

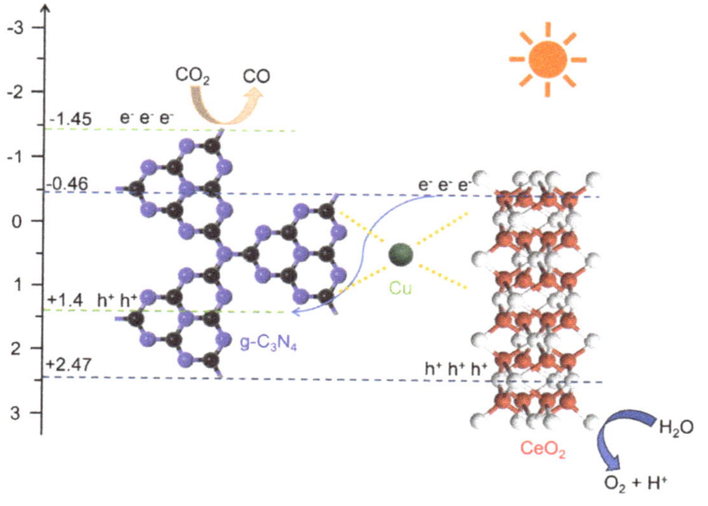


**Figure 8.** Schematic diagram of the mechanism of $CO_2$ reduction by Cu@*g*-$C_3N_4/CeO_2$ under visible light irradiation.

## 3. Experimental Section

### *3.1. Methods*

#### 3.1.1. Materials

The precursors used to prepare Cu@*g*-$C_3N_4/CeO_2$-like Z-scheme heterojunctions are Cupric Nitrate Trihydrate (Cu $(NO_3)_2{\cdot}3H_2O$, ≥99.0%), bought from Macklin (Shanghai, China). Cerium nitrate hexahydrate (Ce $(NO_3)_3{\cdot}6H_2O \geq 99.95\%$) was bought from Aladdin Bio-Chem Technology Co. (Shanghai, China). Ammonium bicarbonate ($NH_4HCO_3 \geq 97.0\%$), Urea ($CO(NH_2)_2 \geq 97.0\%$), Polyvinylpyrrolidone (PVP ≥ 97.0%) and Methyl alcohol (MeOH ≥ 97.0%) were bought from Sinopharm Chemical Reagent Co., Ltd. (Shanghai, China). All experiments were performed with deionized water, and all reagents did not require further purification.

#### 3.1.2. Synthesis of $CeO_2$ Micro-Sheets

$Ce(NO_3)_3{\cdot}6H_2O$ (0.013 mmol) and $NH_4HCO_3$ (0.037 mmol) were dissolved into 200 mL deionized water and stirred thoroughly. Then the mix was statically aged at 30 °C for 24 h and the white precipitate was obtained. After the sediment was separated by filtration, it was washed three times with deionized water and then dried under 60 °C for 8 h. Finally, the dried sample was transferred to a tube furnace and calcined at 500 °C for 4 h to obtain the $CeO_2$ sample.

#### 3.1.3. Synthesis of g-$C_3N_4$ Micro-Sheets

The single-step pyrolysis method was employed to create *g*-$C_3N_4$ by using $CO(NH_2)_2$ as a precursor. In a muffle furnace, 10 g of urea were heated to 550 °C for 240 min to obtain *g*-$C_3N_4$ micro-sheets [13].

#### 3.1.4. Preparation of g-$C_3N_4/CeO_2$ Photocatalysts

$CeO_2$ (0.2 g) and *g*-$C_3N_4$ (0.8 g) were dissolved in a 10 mL methanol solution, stirred thoroughly at 60 °C for 24 h, followed by drying to obtain the solid powder of *g*-$C_3N_4/CeO_2$ [14].

#### 3.1.5. Preparation of Cu@g-$C_3N_4/CeO_2$ Photocatalysts

A total of 0.1 g of the obtained solid powder from the above, along with 0.01 g of $Cu(NO_3)_2$, was dispersed in 20 mL of deionized water and sonicated for 10 min. Subsequently, the dispersion was subjected to photo-deposition under UV light (365 nm) for 30 min. Finally, the resulting Cu@*g*-$C_3N_4/CeO_2$ was obtained after centrifugation and drying.

### 3.2. Characterizations

The crystal structure was examined by X-ray diffraction (XRD model, MAC Science, Yokohama, Japan) using Ni-filtered Cu-K radiation. At 2θ, the scanning range was 10–80° and the scanning speed 10°/min. The measurements of X-ray photoelectron spectroscopy (XPS) were analyzed on a PerkinElmer PHI 5300 (PerkinElmer, Shanghai, China) instrument to check the element's status on the sample surface tested with a monochrome Mg K source. The High0 Resolution Transmission Electron Microscope (HRTEM) images were taken on a field-emission electron microscope (Tecnai G2 F20, Hitachi, Tokyo, Japan) with an acceleration voltage. The XRD patterns were performed on a Shimazu-6100 powder X-ray (Shimadzu, Tokyo, Japan) diffractometer with Cu Kα radiation at a scan rate of 6° $min^{-1}$. The XPS data were obtained using a Shimadzu/Krayos AXIS Ultra DLD (Hitachi, Tokyo, Japan). The scanning electron microscopy (SEM) images were characterized on a field-emission scanning electron microscope (FE-SEM, JSM-7001F, Hitachi, Tokyo, Japan). The ultraviolet–visible diffuse reflectance spectroscopy (UV–Vis DRS) was carried out on a Shimadzu UV-3600 spectrometer. The photoluminescence (PL) spectra for solid power were investigated by an F4500 (Hitachi, Tokyo, Japan) and the xenon (Xe) lamp with an excitation wavelength of 375 nm. In situ Fourier transform infrared spectroscopy (in situ FTIR) was carried out to research the $CO_2$ photoreduction process (Thermo Fisher Nicolet iS-10, Thermo Fisher Scientific, Waltham, MA USA).

### 3.3. $CO_2$ Photoreduction Experiments

$CO_2$ photoreduction was carried out in a sealed self-made 50 mL quartz reactor with a 300 W Xe lamp (1000 mW/$cm^2$, Education Au-light Co., Ltd., Beijing, China) as the white-light source. In a typical procedure, 10 mg of catalyst was dispersed in 10 mL of deionized water. Then, it was subjected to ultrasound for 10 min to make it completely disperse. $CO_2$ was then introduced into the reactor for 20 min to completely remove air. Gas products were detected by a gas chromatography (GC 5890N, Education Au-light Co., Ltd., Beijing, China) equipped with hydrogen flame ionization detector (FID), and the carrier gas was $N_2$.

### 3.4. Photo-Electrochemical Measurements

The corresponding electrode was prepared as follows: 0.05 g of the prepared photocatalyst was dispersed in 3.0 mL of ethanol and 0.03 mL of oleic acid, and then 0.01 g of polyvinylpyrrolidone (PVP) was added. The above mixture was spin-coated on a $1 \times 1$ $cm^2$ fluorine-doped tin oxide glass electrode, and then dried at room temperature. The measurement system includes a standard three-electrode quartz cells with 0.5 M $Na_2SO_4$ electrolyte solution; a Pt wire and a saturated calomel electrode (SCE) were used as the counter electrode and the reference electrodes, respectively. The amplitude of applied sine wave potential in each case was 5 mV, which was carried out using the Chen Hua electrochemical workstation, and all electrochemical signals were recorded by a CHI660 B electrochemical analyzer (Chen Hua Instruments, Shanghai, China).

## 4. Conclusions

The Cu@*g*-$C_3N_4/CeO_2$-like Z-scheme heterojunction was meticulously fabricated through hydrothermal and photo-deposition techniques. It represents charming $CO_2$ reducing activity and product selectivity (CO selectivity 100%), attributed to the internal electric field between *g*-$C_3N_4$ and $CeO_2$, promoting the photogenerated charge transfer, as well as the introduction of Cu, which generates high-energy hot electrons, triggering a cascade of charge carrier generation. The findings expand the photocatalytic mechanism beyond the conventional photoreduction induced by photogenerated carriers to include photothermal reactions. The enhanced charge separation, the introduction of high-energy hot electrons, and the robust stability of the material are key factors that make this heterojunction a promising candidate for $CO_2$ photoreduction under white-light irradiation. This work not only advances the understanding of photocatalytic mechanisms but also opens new avenues for the development of efficient and sustainable photocatalytic systems.

**Author Contributions:** Conceptualization, Y.Y. and Z.Z.; methodology, Y.Z. (Yiying Zhou); software, B.H., J.C. and Y.Z. (Yue Zhang); validation, Y.Z.(Yiying Zhou), Y.S., S.J. and Z.L.; formal analysis, Y.Z. (Yiying Zhou), Y.S., S.J. and Z.L.; investigation, Y.Z. (Yue Zhang); resources, B.H., J.C, X.T. and Y.Z. (Yue Zhang); data curation, Y.Y. and Z.Z.; writing—original draft preparation, Y.Z. (Yiying Zhou) and Z.Z.; writing—review and editing, Y.Z. (Yiying Zhou); supervision, Z.Z.; project administration, Z.Z.; funding acquisition, Y.Y. and Z.Z. All authors have read and agreed to the published version of the manuscript.

**Funding:** This study was funded by the National Natural Science Foundation of China (No. 22208127), the Senior Talent Research Foundation of Jiangsu University (No.22JDG017, No.23JDG030), the RGC Postdoctoral Fellowship Scheme of Hong Kong (RGC-PDFS-2324-2S04), and the Postgraduate Research and Practice Innovation Program of Jiangsu Province (KYCX24_3952, SJCX24_2419, KYCX24_4009). This work was financially supported by the Research project approval of Jiangsu University (23A097, Y23A145).

**Data Availability Statement:** All data are available as download within the database.

**Conflicts of Interest:** The authors declare no conflicts of interest.

## References

1. Yu, J.; Low, J.; Xiao, W.; Zhou, P.; Jaroniec, M. Enhanced Photocatalytic $CO_2$-Reduction Activity of Anatase $TiO_2$ by Coexposed {001} and {101} Facets. *J. Am. Chem. Soc.* **2014**, *136*, 8839–8842. [CrossRef]
2. Li, K.; Peng, B.; Peng, T. Recent Advances in Heterogeneous Photocatalytic $CO_2$ Conversion to Solar Fuels. *ACS Catal.* **2016**, *6*, 7485–7527. [CrossRef]
3. Gong, E.; Ali, S.; Hiragond, C.B.; Kim, H.S.; Powar, N.S.; Kim, D.; Kim, H.; In, S.I. Solar fuels: Research and development strategies to accelerate photocatalytic $CO_2$ conversion into hydrocarbon fuels. *Energy Environ. Sci.* **2022**, *15*, 880. [CrossRef]
4. Wang, X.; Liu, Q.; Bai, Z.; Lei, J.; Jin, H. Thermodynamic investigations of the supercritical $CO_2$ system with solar energy and biomass. *Appl. Energy* **2017**, *227*, 108–118. [CrossRef]
5. Jia, J.; O'Brien, P.G.; He, L.; Qiao, Q.; Fei, T.; Reyes, L.M.; Burrow, T.E.; Dong, Y.; Liao, K.; Varela, M.; et al. Carbon Dioxide Reduction: Visible and Near-Infrared Photothermal Catalyzed Hydrogenation of Gaseous $CO_2$ over Nanostructured Pd@$Nb_2O_5$. *Adv. Sci.* **2016**, *3*, 10. [CrossRef]
6. Hou, S.; Gao, X.; Lv, X.; Zhao, Y.; Yin, X.; Liu, Y.; Fang, J.; Yu, X.; Ma, X.; Ma, T.; et al. Decade Milestone Advancement of Defect-Engineered g-$C_3N_4$ for Solar Catalytic Applications. *Nano-Micro Lett.* **2024**, *16*, 70. [CrossRef] [PubMed]
7. Li, B.; Li, N.; Guo, Z.; Wang, C.; Zhu, Z.; Tang, X.; Andrew Penyikie, G.; Wang, X.; Huo, P. Understanding the effect of multiple configurations in polymeric carbon nitride for efficient solar-driven hydrogen peroxide photosynthesis. *Chem. Eng. J.* **2024**, *491*, 152022. [CrossRef]
8. Ye, Q.; Zhou, Y.; Xu, Y.; Zhang, Q.; Shi, X.; Li, D.; Tian, D.; Jiang, D. Improved charge transfer in polymeric carbon nitride synergistically induced by the aromatic rings modification and Schottky junctions for efficient photocatalytic $CO_2$ reduction. *Chem. Eng. J.* **2023**, *463*, 142395. [CrossRef]

9. Song, X.; Wu, Y.; Zhang, X.; Li, X.; Zhu, Z.; Ma, C.; Yan, Y.; Huo, P.; Yang, G. Boosting charge carriers separation and migration efficiency via fabricating all organic van der Waals heterojunction for efficient photoreduction of $CO_2$. *Chem. Eng. J.* **2020**, *408*, 127292. [CrossRef]
10. Ma, W.; Wang, N.; Guo, Y.; Yang, L.; Lv, M.; Tang, X.; Li, S. Enhanced photoreduction $CO_2$ activity on g-$C_3N_4$: By synergistic effect of nitrogen defective-enriched and porous structure, and mechanism insights. *Chem. Eng. J.* **2020**, *388*, 124288. [CrossRef]
11. Jun, W.; Bi, J.; Chuan, J.; Cheng, J. Hierarchical Porous O-Doped g-$C_3N_4$ with Enhanced Photocatalytic $CO_2$ Reduction Activity. *Small* **2017**, *13*, 1603938.
12. Jiang, T.; Wang, Z.; Wei, G.; Wu, S.; Huang, L.; Li, D.; Ruan, X.; Liu, Y.; Jiang, C.; Ren, F. Defective High-Crystallinity g-$C_3N_4$ Heterostructures by Double-End Modulation for Photocatalysis. *ACS Energy Lett.* **2024**, *9*, 1915–1922. [CrossRef]
13. Liu, Q.; Zhao, X.; Song, X.; Liu, X.; Zhou, W.; Wang, H.; Huo, P. Pd Nanosheet-Decorated 2D/2D g-$C_3N_4$/$WO_3 \cdot H_2O$ S-Scheme Photocatalyst for High Selective Photoreduction of $CO_2$ to CO. *Inorg. Chem.* **2022**, *61*, 4171–4183. [CrossRef]
14. Hu, H.; Hu, J.; Wang, X.; Gan, J.; Su, M.; Ye, W.; Zhang, W.; Ma, X.; Wang, H. Enhanced reduction and oxidation capability over the $CeO_2$/g-C3N4 hybrid through surface carboxylation: Performance and mechanism. *Catal. Sci. Technol.* **2020**, *10*, 4712–4725. [CrossRef]
15. Gao, W.; Li, Y.; Xiao, D.; Ma, D. Advances in photothermal conversion of carbon dioxide to solar fuels. *J. Energy Chem.* **2023**, *83*, 62–78. [CrossRef]
16. Yang, X.; Peng, J.; Zhao, L.; Zhang, H.; Li, J.; Yu, P.; Fan, Y.; Wang, J.; Liu, H.; Dou, S. Insights on advanced g-$C_3N_4$ in energy storage: Applications, challenges, and future. *Carbon Energy* **2024**, *9*, 490. [CrossRef]
17. Yin, S.; Sun, L.; Zhou, Y.; Li, X.; Li, J.; Song, X.; Huo, P.; Wang, H.; Yan, Y. Enhanced electron–hole separation in $SnS_2$/Au/g-$C_3N_4$ embedded structure for efficient $CO_2$ photoreduction. *Chem. Eng. J.* **2020**, *406*, 126776. [CrossRef]
18. Ji, H.; Lin, L.; Chang, K. Plasma-assisted $CO_2$ decomposition catalyzed by $CeO_2$ of various morphologies. *J. CO2 Util.* **2022**, *68*, 102351. [CrossRef]
19. Tran, D.; Pham, M.; Bui, X.; Wang, Y.; You, S. $CeO_2$ as a photocatalytic material for $CO_2$ conversion: A review. *Sol. Energy* **2022**, *240*, 443–466. [CrossRef]
20. Yoko, A.; Wang, H.; Furuya, K.; Takahashi, D.; Seong, G.; Tomai, T.; Frenkel, A.I.; Saito, M.; Inoue, K.; Ikuhara, Y.; et al. Reduction of (100)-Faceted $CeO_2$ for Effective Pt Loading. *Chem. Mater.* **2024**, *36*, 5611–5620. [CrossRef]
21. Zhu, Z.; Chen, C.Y.; Wu, R.J. Hydrocarbon production by addition of Cu-ZnO on g-$C_3N_4$ for $CO_2$ conversion. *J. Chin. Chem. Soc.* **2020**, *67*, 1654–1660. [CrossRef]
22. Tang, X.; Shen, W.; Li, D.; Li, B.; Wang, Y.; Song, X.; Zhu, Z.; Huo, P. Research on cobalt-doping sites in g-$C_3N_4$ framework and photocatalytic reduction $CO_2$ mechanism insights. *J. Alloys Compd.* **2023**, *954*, 170044. [CrossRef]
23. Liang, J.; Zhang, L.; Li, C.; Mo, Z.; Ye, M.; Zhu, Z.; Sun, S.; Wong, J.W.C. Triclocarban transformation and removal in sludge conditioning using chalcopyrite–triggered percarbonate treatment. *J. Hazard. Mater.* **2024**, *463*, 132944. [CrossRef] [PubMed]
24. Wang, J.; Heil, T.; Zhu, B.; Tung, C.-W.; Yu, J.; Chen, H.M.; Antonietti, M.; Cao, S. A Single Cu-Center Containing Enzyme-Mimic Enabling Full Photosynthesis under $CO_2$ Reduction. *ACS Nano* **2020**, *14*, 8584–8593. [CrossRef]
25. Zhu, Z.; Ye, J.; Tang, X.; Chen, Z.; Yang, J.; Huo, P.; Ng, Y.H.; Crittenden, J. Vacancy-Rich CoSx@LDH@Co-NC Catalytic Membrane for Antibiotic Degradation with Mechanistic Insights. *Environ. Sci. Technol.* **2023**, *57*, 16131–16140. [CrossRef]
26. Zhu, Z.; Xing, X.; Qi, Q.; Shen, W.; Wu, H.; Li, D.; Li, B.; Liang, J.; Tang, X.; Zhao, J.; et al. Fabrication of Graphene Modified $CeO_2$/g-$C_3N_4$ Heterostructures for Photocatalytic Degradation of Organic Pollutants. *Chin. J. Struct. Chem.* **2023**, *42*, 100194. [CrossRef]
27. Vikatakavi, A.; Mauri, S.; Rivera-Salazar, M.L.; Dobovičnik, E.; Pelatti, S.; D'Addato, S.; Torelli, P.; Luches, P.; Benedetti, S. Role of Metal Dopants in Hydrogen Dissociation on Cu:$CeO_2$ and Fe:$CeO_2$ Surfaces Studied by Ambient-Pressure X-ray Absorption Spectroscopy. *ACS Appl. Energy Mater.* **2024**, *7*, 2746–2754. [CrossRef]
28. Li, S.; Lu, J.; Wang, G.; Li, X.; Liu, W.; Xin, C.; Zhan, H. Study on the performance of photothermal catalyzed carbon dioxide hydrogenation to $CH_4$ over doped sodium tantalate-based perovskite catalysts. *J. Environ. Chem. Eng.* **2024**, *12*, 113108. [CrossRef]
29. Pan, D.; Wang, Y.; Lin, L.; Zhang, Y.; Zhou, M.; Li, Z.; Xu, S. Cu/Ni-loaded oxygen vacancy-rich $CeO_2$ rod derived from metal organic framework for efficient photothermal $CO_2$ hydrogenation. *J. Mater. Sci. Mater. Electron.* **2024**, *35*, 791. [CrossRef]
30. Wang, X.; Guo, Y.; Li, L.; Chen, L. Carbon nanotubes @Cu/Ni loaded BFO for photothermal catalytic conversion of $CO_2$ by $H_2O$. *Mater. Chem. Phys.* **2023**, *309*, 128417. [CrossRef]
31. Wang, Y.; Ding, J.; Yin, Q.; Zhang, C.; Zeng, Y.; Xu, S.; Liang, Q.; Zhou, M.; Li, Z. Reinforcing the efficiency of photocatalytic $CO_2$ conversion with $H_2O$ vapor through the integration of photothermal Cu/C@Bi/C supported on 3D g-$C_3N_4$ under full-spectrum solar irradiation. *J. Environ. Chem. Eng.* **2024**, *12*, 113008. [CrossRef]
32. Kong, N.; Han, B.; Li, Z.; Fang, Y.; Feng, K.; Wu, Z.; Wang, S.; Xu, A.-B.; Yu, Y.; Li, C.; et al. Ruthenium Nanoparticles Supported on $Mg(OH)_2$ Microflowers as Catalysts for Photothermal Carbon Dioxide Hydrogenation. *ACS Appl. Energy Mater.* **2020**, *3*, 3028–3033. [CrossRef]
33. Zhang, H.; Ding, L.; Long, H.; Li, J.; Tan, W.; Ji, J.; Sun, J.; Tang, C.; Dong, L. Influence of $CeO_2$ loading on structure and catalytic activity for $NH_3$-SCR over $TiO_2$-supported $CeO_2$. *J. Rare Earths* **2020**, *38*, 883–890. [CrossRef]
34. Chenari, H.M.; Riasvand, L.; Khalili, S. The synthesis condition and characterization of thermally treated $CeO_2$ and ZnO $CeO_2$ composite fibers. *Ceram. Int.* **2019**, *45*, 14223–14228. [CrossRef]

35. Zhang, Y.; Feng, W.; Yang, F.; Bao, X. Interface-controlled synthesis of $CeO_2$(111) and $CeO_2$(100) and their structural transition on Pt(111). *Chin. J. Catal.* **2019**, *40*, 204–213. [CrossRef]
36. Smeekens, S.P.; Malireddi, R.K.; Plantinga, T.S.; Buffen, K.; Netea, M.G. Autophagy is redundant for the host defense against systemic Candida albicans infections. *Eur. J. Clin. Microbiol. Infect. Dis.* **2014**, *33*, 711–722. [CrossRef] [PubMed]
37. Wang, R.; Qiu, G.; Xiao, Y.; Tao, X.; Li, B. Optimal construction of $WO_3{\cdot}H_2O$/Pd/CdS ternary Z-scheme photocatalyst with remarkably enhanced performance for oxidative coupling of benzylamines. *J. Catal.* **2019**, *374*, 378–390. [CrossRef]
38. Zhang, N.; Li, X.; Ye, H.; Chen, S.; Ju, H.; Liu, D.; Lin, Y.; Ye, W.; Wang, C.; Xu, Q. Oxide Defect Engineering Enables to Couple Solar Energy into Oxygen Activation. *J. Am. Chem. Soc.* **2016**, *138*, 8928–8935. [CrossRef]
39. Tong, R.; Sun, Z.; Zhong, X.; Wang, X.; Xu, J.; Yang, Y.; Xu, B.; Wang, S.; Pan, H. Enhancement of Visible-Light Photocatalytic Hydrogen Production by $CeCO_3OH$ in g-$C_3N_4$/$CeO_2$ System. *ChemCatChem* **2019**, *11*, 1069–1075. [CrossRef]
40. Saravanan, V.; Lakshmanan, P.; Palanisami, N.; Amalraj, J.; Pyarasani, R.D.; Ramalingan, C. 2D/3D- $C_3N_4$/$CeO_2$ S-scheme heterojunctions with enhanced photocatalytic performance. *Inorg. Chem. Commun.* **2022**, *291*, 898. [CrossRef]
41. Li, M.; Zhang, L.; Fan, X.; Wu, M.; Du, Y.; Wang, M.; Kong, Q.; Zhang, L.; Shi, J. Dual synergetic effects in $MoS_2$/pyridine-modified g-$C_3N_4$ composite for highly active and stable photocatalytic hydrogen evolution under visible light. *Appl. Catal. B Environ.* **2016**, *190*, 36–43. [CrossRef]
42. Michalow-Mauke, K.A.; Lu, Y.; Kowalski, K.; Graule, T.; Nachtegaal, M.; Kröcher, O.; Ferri, D. Flame-Made $WO_3$/$CeO_x$-$TiO_2$ Catalysts for Selective Catalytic Reduction of $NO_x$ by $NH_3$. *ACS Catal.* **2015**, *5*, 5657–5672. [CrossRef]
43. Feng, C.; Luo, J.; Chen, C.; Zuo, S.; Ren, Y.; Wu, Z.-P.; Hu, M.; Ould-Chikh, S.; Ruiz-Martínez, J.; Han, Y.; et al. Cooperative tungsten centers in polymeric carbon nitride for efficient overall photosynthesis of hydrogen peroxide. *Energy Environ. Sci.* **2024**, *17*, 1520–1530. [CrossRef]
44. Feng, C.; Bo, T.; Maity, P.; Zuo, S.; Zhou, W.; Huang, K.W.; Mohammed, O.F.; Zhang, H. Regulating Photocatalytic $CO_2$ Reduction Kinetics through Modification of Surface Coordination Sphere. *Adv. Funct. Mater.* **2023**, *34*, 2309761. [CrossRef]
45. Wu, G.; He, Z.; Wang, Q.; Wang, H.; Wang, Z.; Sun, P.; Mo, Z.; Liu, H.; Xu, H. Non-metal inducing charge rearrangement in carbon nitride to promote photocatalytic hydrogen production. *J. Mater. Sci. Technol.* **2024**, *195*, 1–8. [CrossRef]
46. Mo, Z.; Miao, Z.; Yan, P.; Sun, P.; Wu, G.; Zhu, X.; Ding, C.; Zhu, Q.; Lei, Y.; Xu, H. Electronic and energy level structural engineering of graphitic carbon nitride nanotubes with B and S co-doping for photocatalytic hydrogen evolution. *J. Colloid Interface Sci.* **2023**, *645*, 525–532. [CrossRef]
47. Zhang, X.; Kim, D.; Yan, J.; Lee, L.Y.S. Photocatalytic $CO_2$ Reduction Enabled by Interfacial S-Scheme Heterojunction between Ultrasmall Copper Phosphosulfide and g-$C_3N_4$. *ACS Appl. Mater. Interfaces* **2021**, *13*, 9762–9770. [CrossRef]
48. Cai, S.; Chen, J.; Li, Q.; Jia, H. Enhanced Photocatalytic $CO_2$ Reduction with Photothermal Effect by Cooperative Effect of Oxygen Vacancy and Au Cocatalyst. *ACS Appl. Mater. Interfaces* **2021**, *13*, 14221–14229. [CrossRef]
49. Chen, Y.; Lin, X.; Li, W.; Sun, H.; Jia, S.; Zhou, Y.; Hao, Y.; Liu, Z.; Yin, S.; Guo, C.; et al. Harnessing Photo-to-Thermal Conversion in Sulfur-Vulcanized Mxene for High-Efficiency Solar-to-Carbon-Fuel Synthesis. *Adv. Funct. Mater.* **2024**, *34*, 2400121. [CrossRef]

catalysts

MDPI

Article

# Construction of ZnCdS Quantum-Dot-Modified $CeO_2$ (0D–2D) Heterojunction for Enhancing Photocatalytic $CO_2$ Reduction and Mechanism Insight

Junzhi Yan [1], Yuming Sun [1], Junxi Cai [1], Ming Cai [1], Bo Hu [1], Yan Yan [1,*], Yue Zhang [2,*] and Xu Tang [1,*]

1 School of Chemistry and Chemical Engineering, Institute for Advanced Materials, Jiangsu University, Zhenjiang 212013, China; yanjunzhi0923@163.com (J.Y.); s969678125@126.com (Y.S.); 18896660263@163.com (J.C.); cm1999@yeah.net (M.C.); hubo93@bcnu.edu.cn (B.H.)

2 School of Chemistry and Chemical Engineering, Liaoning Normal University, Dalian 116029, China

* Correspondence: dgy5212004@163.com (Y.Y.); zhangyue433@126.com (Y.Z.); tangxu@ujs.edu.cn (X.T.)

**Abstract:** It is important to improve the separation ability of photogenerated electrons and the adsorption capacity of carbon dioxide ($CO_2$) for efficient photoreduction of $CO_2$. Here, we synthesized ZnCdS quantum dots (ZCS-QDs) and cerium dioxide nanosheets ($CeO_2$) using the solvothermal method and calcination method. We combined $CeO_2$ and ZCS-QDs to effectively enhance the charge separation efficiency, and the lifetime of photogenerated electrons was increased 4.5 times. The CO evolution rate of the optimized composite (ZCS-QDs/$CeO_2$) was up to 495.8 μmol $g^{-1}$ $h^{-1}$, and it had 100% product selectivity. In addition, the stability remained high after five cycles. The $CO_2$ adsorption capacity of the catalyst surface was observed by in situ FTIR. The test results showed that improving $CO_2$ capture ability and promoting photogenic electron separation had positive effects on enhancing photoreduction of $CO_2$. This study provides a reference for constructing a zero-dimensional–two-dimensional (0D–2D) heterojunction and explores potential $CO_2$ reduction reaction mechanisms.

**Keywords:** photocatalysis; $CO_2$ reduction; $CeO_2$; ZnCdS quantum dots

## 1. Introduction

In recent years, the challenge of mitigating $CO_2$ emissions and converting this greenhouse gas into useful and renewable energy sources has been a focal point for contemporary research [1]. Photocatalysis is a promising strategy, particularly due to its potential for sustainability and relatively lower costs compared to other $CO_2$ reduction methods [2,3], and is widely used because of its low cost and sustainable advantages [4–6]. The steps of the photocatalytic $CO_2$ reduction reaction are complicated and affected by many factors, including light absorption, charge carrier generation, separation, and transfer, as well as surface reactions involving the adsorption and reduction of $CO_2$ [7]. In the field of photocatalysis, in order to improve the reaction rate and regulate the reaction pathway, researchers have made a lot of efforts [8], including reducing the particle size [9,10], loading metal onto the surface [11], dispersing to various carriers [12], etc. These methods improve the catalytic activity of the photocatalyst but only to a limited extent. Therefore, selecting a suitable photocatalyst to enhance the light absorption capacity and improve the photoelectron transport efficiency may be an effective means to improve the photocatalytic efficiency. Numerous studies have found that the selection of catalysts and system design plays a pivotal role in enhancing the overall efficiency of this process [13–15].

Two-dimensional (2D) cerium dioxide ($CeO_2$) is a highly versatile material that has garnered significant interest in the realm of photocatalysis due to its excellent chemical and structural stability. In addition, the reversible switch between the Ce(III) and Ce(IV) states confers unique redox properties [16–18]. However, its wide bandgap typically restricts its light absorption in the ultraviolet region, limiting its efficiency under visible light; this limitation hampers the direct conversion of $CO_2$ into fuels. Combining $CeO_2$ with other semiconductors to form $CeO_2$ heterojunctions (such as narrower-bandgap catalysts capable of absorbing visible light) seems to present an opportunity for strategic enhancement of light harvesting and charge separation efficiency. This approach leverages synergistic effects between the components, enhances light absorption, and promotes charge carrier dynamics.

The selection and engineering of catalysts that can not only effectively adsorb $CO_2$ but also facilitate efficient charge transfer are central to improving the photocatalytic conversion of $CO_2$ into renewable energy sources. Zero-dimensional (0D) quantum dot (QD) materials have been extensively investigated as the building blocks of photocatalysts due to their special quantum size effect, electron transfer properties, and their efficient absorption to visible light [19–21]. Ternary ZnCdS creates good conditions for adjusting the optical properties from the ultraviolet to the visible light range [22], and can adjust the appropriate bandgap range [23]. In fact, the use of QDs in combination with semiconductors in the field of photocatalysis has been reported in the past few years. For example, Yang et al. used Ag QDs anchored with $CeO_2$, and the Ag QD/$CeO_2$ composite improved the utilization rate of visible light, and the photocatalytic elimination of acetaldehyde and water decomposition performance improved 7.08 and 6.83 times, respectively [24]. Jiang et al. reported a modification of $CeO_2$ using g-$C_3N_4$ QDs, which improved the photoelectric response and showed significant performance improvement in the test of photoreduction of $CO_2$ [25]. Sre et al. deposited ZnCdS quantum dots on g-$C_3N_4$ for photoinactivation of *E. coli* cells. The prepared composite photocatalyst enhanced the structural stability, as well as increased a large number of reaction sites, and improved the photocatalyst efficiency [26]. So, it can be inferred that if ZnCdS-QDs are introduced on the $CeO_2$ surface and they construct a 0D–2D heterojunction, the photocatalytic $CO_2$ activity and the recycling ability of the photocatalysts will be enhanced.

Inspired by the consideration above, ZnCdS quantum dots were combined with a $CeO_2$ heterojunction (ZCS-QDs/$CeO_2$) and designed to optimize the energy band alignment and $CO_2$ reduction activity. The ZCS-QDs/$CeO_2$ exhibits a substantial increase in the lifetime of photogenerated electrons and achieves a remarkable CO evolution rate of 495.8 $\mu$mol $g^{-1}$ $h^{-1}$ without any sacrificial agents during the $CO_2$ reduction reaction, even showing good stability after five cycles. The in situ DRIFTS result provides direct evidence of the $CO_2$ adsorption dynamics on the $CeO_2$ surface and confirms the vital role of $CeO_2$ in enhancing substrate adsorption. Also, the observation of intermediates and monitoring changes in water and $CO_2$ consumption offer deeper insights into the photocatalytic process, contributing to a better understanding of the mechanism behind the improved efficiency.

## 2. Results and Discussion

### 2.1. Photocatalyst Characterization

Scheme 1 shows the synthesis strategy of ZCS-QDs/$CeO_2$. Figure 1a shows the XRD patterns. For the $CeO_2$, the four major diffraction peaks are located at 28.5°, 33.1°, 47.5°, and 56.3°, corresponding to the (111), (200), (220), and (311) lattice planes of $CeO_2$ (PDF#43-1002), respectively. For the ZCS-QDs, the six major diffraction peaks are located at 26.7°, 28.3°, 45.1°, 47.3°, 53.4°, and 56.2°, corresponding to the (002), (101), (102), (110), (103), and (200) lattice planes of ZnCdS (PDF#40-0834), respectively. The relatively small amount of ZCS-QDs and the partial coincidence of peak positions result in the ZCS-QDs' diffraction peak being covered by the $CeO_2$ diffraction peak; thus, no diffraction peak was observed for the ZCS-QDs in the ZCS-QDs/$CeO_2$ photocatalyst. The TEM image of $CeO_2$ is shown in Figure 1b, exhibiting a smooth two-dimensional sheet structure. The ZnCdS is uniformly dispersed and has a diameter of about 8 nm belonging to the morphology of the quantum dots (Figure 1c,d). From Figure 1e, it can be seen that quantum dots are fully grown on the 2D nanosheets, indicating that our catalyst of 0D–2D ZCS-QDs/$CeO_2$ is well assembled. Moreover, the 12% ZCS-QDs/$CeO_2$ was characterized by HRTEM. As shown in Figure 1f, the measured lattice spacing of 0.33 nm and 0.31 nm corresponds to (002) of ZCS-QDs and (111) of $CeO_2$, respectively. From HRTEM mapping (Figure 1g), the elements of S, Cd, Zn, O, and Ce were detected, providing further proof that the ZCS-QDs/$CeO_2$ has been successfully synthesized.

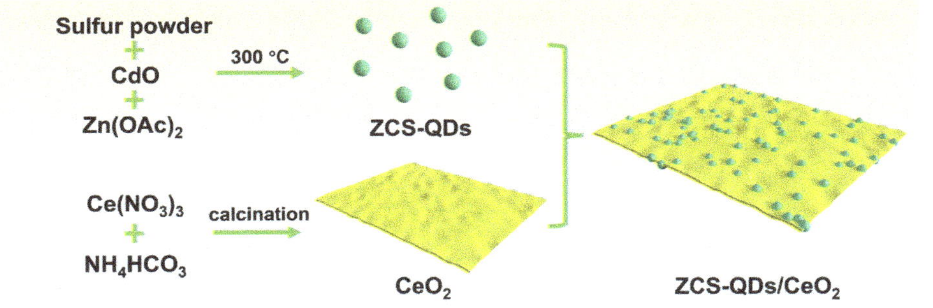


**Scheme 1.** Synthesis diagram of ZCS-QDs/$CeO_2$.

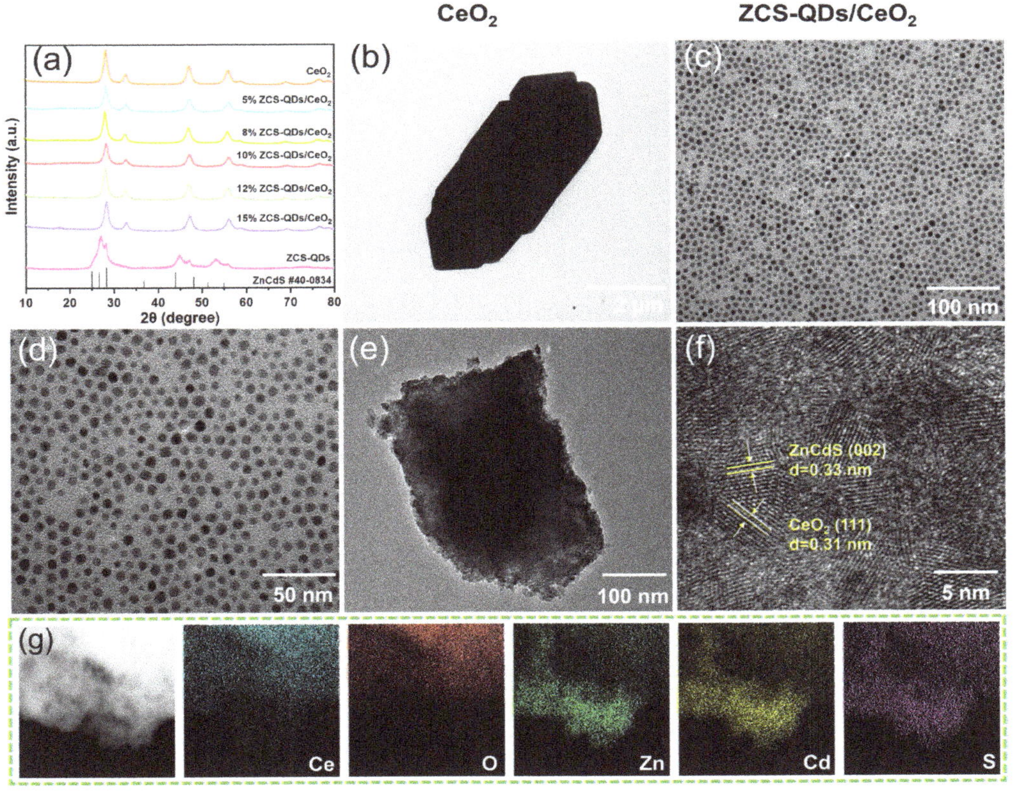


**Figure 1.** (**a**) XRD patterns of $CeO_2$, ZCS-QDs, and x ZCS-QDs/$CeO_2$ (x = 5, 8, 10, 12, and 15%), (**b–d**) TEM image of $CeO_2$ and ZCS-QDs, (**e**) TEM image of 12% ZCS-QDs/$CeO_2$, (**f**) HRTEM image of 12% ZCS-QDs/$CeO_2$, (**g**) EDS elemental mapping images of 12% ZCS-QDs/$CeO_2$.

### 2.2. Surface Composition and Photoelectric Analysis

The chemical state of the catalyst was investigated using X-ray photoelectron spectroscopy (XPS). From the XPS survey spectrum of Figure 2a, it can be seen that the Ce, O, S, Cd, and Zn elements are found in the 12% ZCS-QDs/$CeO_2$ catalyst. Figure 2b shows the Ce 3d levels for $CeO_2$ and 12% ZCS-QDs/$CeO_2$. For $CeO_2$, the peaks at 913.9 and 895.4 eV are from the Ce $3d^9$ $4f^0$ final state, and the peaks at 904.7, 898.2, 886.1, and 879.5 eV are from the Ce $3d^9$ $4f^2$ and Ce $3d^9$ $4f^1$ final states [27]. However, for 12% ZCS-QDs/$CeO_2$, the peak position of 895.4 eV shifted to 896.1 eV. This was attributed to the interaction of ZCS QDs and $CeO_2$, and resulted in the electron transfer in ZCS-QDs/$CeO_2$. For the O 1s of $CeO_2$ (Figure 2c), two characteristic peaks can be observed at 529.1 and 526.8 eV, representing lattice oxygen and adsorbed oxygen, respectively [28], in the spectrum of 12% ZCS-QDs/$CeO_2$. When $CeO_2$ is combined with the nanostructure, the surface oxygen site will greatly affect the reduction reaction of the semiconductor at the interface [29]. In addition, oxygen sites are often involved in constructing the support interface for reactivity, as well as the reaction site of various adsorbents on the carrier [30,31]. Therefore, the new characteristic peak at 530.7 eV is attributed to S-O [32]. For the S 2p spectrum of 12% ZCS-QDs/$CeO_2$ (Figure 2d), 159.5 and 158.6 eV are attributed to S $2p^{1/2}$ and S $2p^{3/2}$ [33], respectively, while the characteristic peak at 162.4 eV was attributed to the newly formed S-O. The formation of S-O fully confirms that a close chemical interaction is formed between ZCS-QDs and $CeO_2$, which provides good conditions for electron transfer during the reaction. Two characteristic peaks at 409.2 and 402.4 eV can be observed in the Cd 3d spectrum (Figure 2e), corresponding to Cd $3d^{3/2}$ and Cd $3d^{5/2}$, respectively [33]. Two characteristic peaks can be observed at 1042.4 and 1019.4 eV in the Zn 2p spectrum (Figure 2f), corresponding to Zn $2p^{1/2}$ and Zn $2p^{3/2}$ [34], respectively.

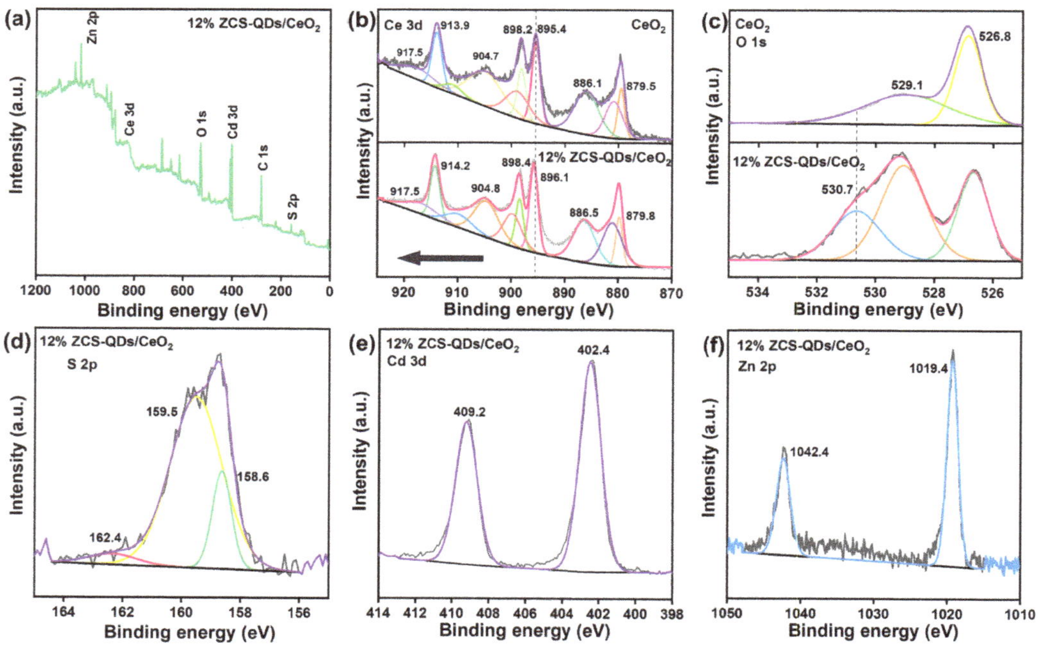


**Figure 2.** XPS spectra of (**a**) survey spectra, (**b**) Ce 3d, (**c**) O 1s, (**d**) S 2p of 12% ZCS-QDs/$CeO_2$, (**e**) Cd 3d of 12% ZCS-QDs/$CeO_2$, and (**f**) Zn 2p of 12% ZCS-QDs/$CeO_2$.

The light absorption capacity of the photocatalyst was measured by UV–Vis DRS [35] (Figure 3a). The $CeO_2$ showed strong absorption properties in the ultraviolet region and an absorption range of ZCS-QDs located at 480 nm. However, 12% ZCS-QDs/$CeO_2$ has a unique enhanced adsorption peak around 450–620 nm. Also, the bandgap value was calculated and showed in Figure 3b using the following formula:

$$(\alpha hv)^{1/n} = A(hv - E_g)$$

where $\alpha$, $hv$, $E_g$, and $A$ represent the absorption coefficient, photon energy, bandgap energy, and proportionality constant, respectively. ZCS-QDs/$CeO_2$ has the narrowest bandgap of 2.61 eV, and the bandgaps of ZCS-QDs and $CeO_2$ were 2.67 eV and 2.98 eV. In addition, the valence band (VB) positions of $CeO_2$ and ZCS-QDs were determined by VB-XPS (Figure 3c,d); the corresponding VBs are 1.45 eV and 2.38 eV. Based on the above results, the band structures of $CeO_2$ and ZCS-QDs were obtained, as shown in Figure 3e.

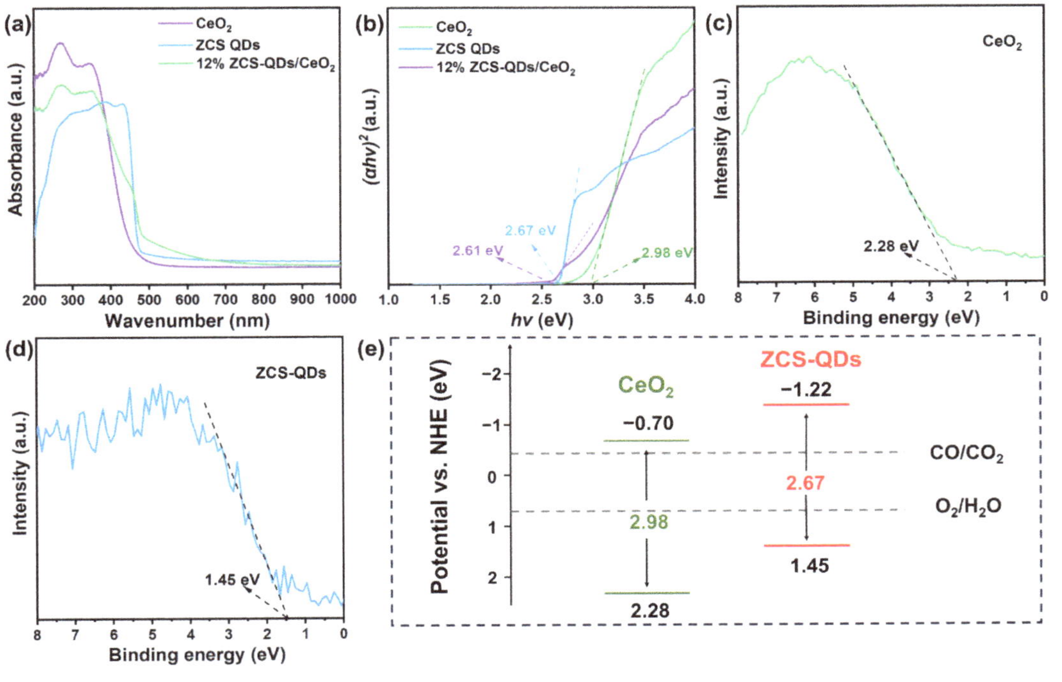


**Figure 3.** (**a**) UV–Vis DRS, (**b**) plots of $(\alpha hv)^{1/2}$ versus $(hv)$, (**c**,**d**) VB-XPS spectra of $CeO_2$ and ZCS-QDs, and (**e**) bandgap structure diagram of $CeO_2$ and ZCS-QDs.

Figure 4a shows the results of TR-PL spectroscopy. The lifetimes of $CeO_2$ and ZCS-QDs are 2.53 ns and 6.67 ns, and 12% ZCS-QDs/$CeO_2$ has the longest average decay life with 30.17 ns, meaning that 12% ZCS-QDs/$CeO_2$ has excellent charge separation ability. As shown in Figure 4b, transient photocurrent responses (TPRs) were recorded, and all the catalysts exhibited stable photocurrent densities. It is worth noting that 12% ZCS-QDs/$CeO_2$ has the highest photocurrent density compared to $CeO_2$ and ZCS-QDs. Also, 12% ZCS-QDs/$CeO_2$ has the smallest EIS radius (Figure 4c). This means that 12% ZCS-QDs/$CeO_2$ has the smallest charge transfer resistance. The above confirmed that 12% ZCS-QDs/$CeO_2$ has the best charge transfer ability and shows the best photocatalytic ability.

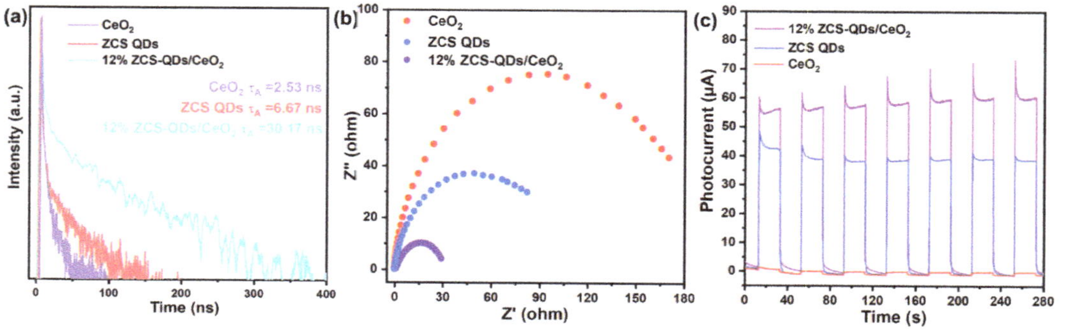


**Figure 4.** (**a**) TR-PL decay spectra of $CeO_2$, ZCS-QDs, and 12% ZCS-QDs/$CeO_2$; (**b**) TPR and (**c**) EIS of $CeO_2$, ZCS-QDs, and 12% ZCS-QDs/$CeO_2$.

### 2.3. Photocatalytic Activity

As shown in Figure 5a, the $CO_2$ photoreduction activity of $CeO_2$, ZCS-QDs, and x-ZCS-QDs/$CeO_2$ (x means the different amount of ZCS-QDs) was determined. The CO yield of ZCS-QDs is 151.8 μmol $g^{-1}$ $h^{-1}$. For the x ZCS-QDs/$CeO_2$, the $CO_2$ reduction performance significantly increased with all values of x, and 12% ZCS-QDs/$CeO_2$ achieved the highest performance (495.8 μmol $g^{-1}$ $h^{-1}$), increasing 3.3 times. However, the $CO_2$ reduction activity of 15% ZCS-QDs/$CeO_2$ decreased because the excessive number of ZCS-QDs cover the $CO_2$ adsorption sites on the $CeO_2$ surface and inhibit $CO_2$ adsorption. In order to exclude the effect of decomposition of organic matter from the catalyst, a controlled experiment under $N_2$ was conducted. No CO was detected (Figure 5b), which confirmed that all of the CO product comes from $CO_2$ conversion. To investigate the stability of 12% ZCS-QDs/$CeO_2$, cyclic experiments were conducted. It can be seen from Figure 5c that, after five cycles, the 12% ZCS-QDs/$CeO_2$ still had high catalytic activity, meaning that the catalyst has good stability. This study was compared with values previously reported for $CeO_2$-based materials, such as $CeO_2/TiO_2$ (CO yield of 61.9 μmol $g^{-1}$ $h^{-1}$) [36], CuO-$CeO_2/TiO_2$ NT (CO yield of 9.2 μmol $g^{-1}$ $h^{-1}$) [37], $CeO_2$/g-$C_3N_4$ (CO yield of 31 μmol $g^{-1}$ $h^{-1}$) [38], and CoAl-LDH/$CeO_2$/RGO (CO yield of 5.5 μmol $g^{-1}$ $h^{-1}$) [39]. Among many $CeO_2$-based photocatalysts, ZCS-QDs have the highest CO evolution rate. It is fully confirmed that ZCS-QDs/$CeO_2$ has great advantages in the field of efficient photocatalytic reduction of $CO_2$.

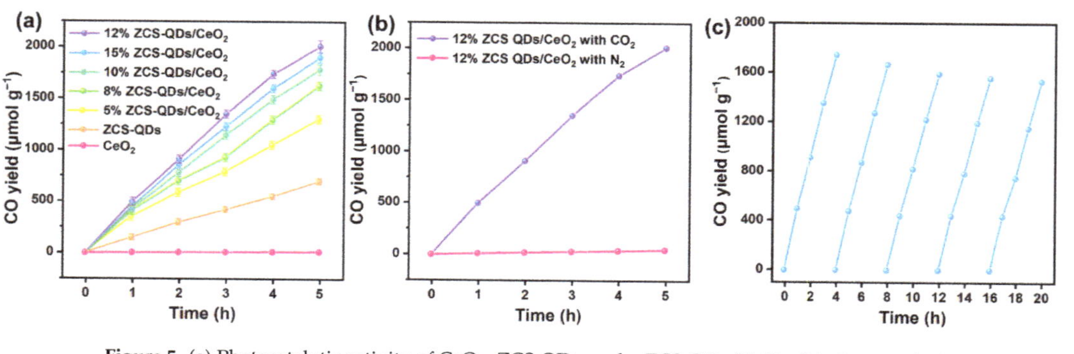


**Figure 5.** (**a**) Photocatalytic activity of $CeO_2$, ZCS-QDs, and x ZCS-QDs/$CeO_2$; (**b**) photocatalytic performance of 12% ZCS-QDs/$CeO_2$ under $CO_2$ and $N_2$; (**c**) cycling test of 12% ZCS-QDs/$CeO_2$.

### 2.4. Mechanism of $CO_2$ Photoreduction

It is known that the $CO_2$ adsorption capacity of catalysts is crucial for the photocatalytic reduction reaction. Thus, $CO_2$ adsorption tests were performed by in situ DRIFTS. As shown in Figure 6a,b, the adsorption capacity of $CeO_2$ for $CO_2$ is significantly higher than that of ZCS-QDs (the peak at 2340 $cm^{-1}$ is a characteristic peak of $CO_2$), indicating that $CeO_2$ has strong $CO_2$ adsorption capacity. However, due to the wide bandgap of $CeO_2$ and the fact that it has fewer active sites, it is difficult to effectively convert the surface-adsorbed $CO_2$ into CO, $CH_4$, and other high value-added products. Although, the $CO_2$ adsorption strength of 12% ZCS-QDs/$CeO_2$ decreased slightly compared with that of $CeO_2$, it was still higher than that of ZCS-QDs (Figure 6c). The weakened $CO_2$ adsorption could be caused by the ZCS-QDs covering some of the adsorption sites of $CeO_2$.

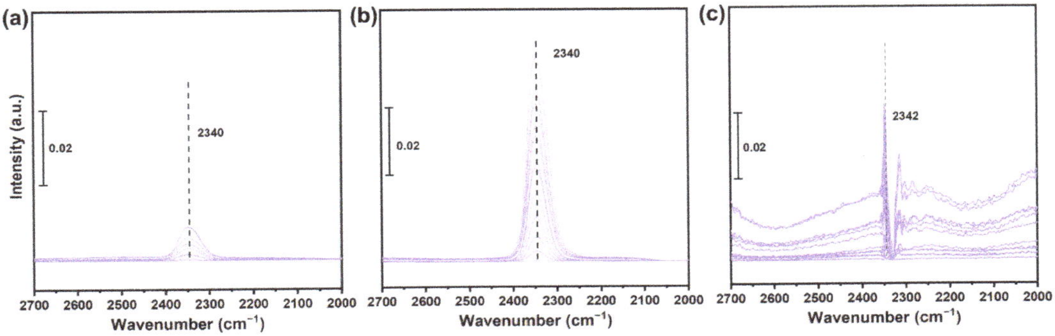


**Figure 6.** DRIFTS spectra of $CO_2$ adsorption of (**a**) ZCS-QDs, (**b**) $CeO_2$, and (**c**) 12% ZCS-QDs/$CeO_2$.

In order to investigate the photoreduction $CO_2$ mechanism, in situ DRIFTS was carried out. Using the pre-reaction equilibrium state as a blank background, positive or negative infrared signals during the reaction were used to represent the increase or disappearance of substances on the interface. Figure 7a,b represent the $CO_2$ photoreduction processes of $CeO_2$ and ZCS-QDs. The peaks at 2339 $cm^{-1}$ and 1658 $cm^{-1}$ correspond to $CO_2$ and $H_2O$ consumption signals, respectively [40]. By comparing the signal strength, the $A_{2339}/A_{1658}$ of $CeO_2$ is 0.27, and the $A_{2339}/A_{1658}$ of ZCS-QDs is 0.31. In 12% ZCS-QDs/$CeO_2$ (Figure 7c), $A_{2344}/A_{1668}$ is 2.08; this result suggests that $CeO_2$ and ZCS-QDs mainly consume $H_2O$ during the photoreaction, while 12% ZCS-QDs/$CeO_2$ consumes $CO_2$ and converts it into CO. The above results show that after $CeO_2$ captures $CO_2$, photogenerated electrons migrate to the catalyst surface and react with $CO_2$, greatly inhibiting competitive reactions of $H_2O$ splitting. Moreover, in Figure 7c, during 12% ZCS-QDs/$CeO_2$ photoreaction, a series of peaks can be observed. The positive signal peak appearing at 2355 $cm^{-1}$ is the protonated $CO_2$ intermediate (O=C=O-$H^+$) [41]. The peaks at 1468 $cm^{-1}$ correspond to COOH* species [42]. In addition, the peak at 1402 $cm^{-1}$ corresponds to carbonate (m-$CO_3^{2-}$), the peak at 1267 $cm^{-1}$ belongs to bidentate carbonate (b-$CO_3^{2-}$,) and the peak at 1145 $cm^{-1}$ represents bicarbonate ($HCO_3^-$) [43–45].

Based on the discussion of the above results, the 0D–2D ZCS-QDs/$CeO_2$ heterojunction is used for selective photoreduction of $CO_2$ to CO. Under visible light (300 W xenon lamp) irradiation, both $CeO_2$ and ZCS-QDs can absorb ultraviolet light and part of the visible light spectrum to produce electrons and holes [46–48]. As $CeO_2$ and ZCS-QDs have a CB position of −0.7 eV and −1.22 eV, they both have the potential for $CO_2$ reduction of CO. When the two materials were combined together, a Z-heterojunction was formed. The electrons on the CB of $CeO_2$ recombine with the holes from the VB of ZCS-QDs; this not only improves the efficiency of photogenerated carrier separation but also the high-reducing-capacity photogenerated electrons on ZCS-QDs react with $CO_2$ and further improve the efficiency of photoreduction of $CO_2$ to CO (Figure 8). The $CO_2$ photoreduction process over ZCS-QDs/$CeO_2$ can be inferred as follows:

$$\text{ZCS-QDs/CeO}_2 + h\nu \rightarrow \text{ZCS-QDs/CeO}_2\ (e^- + h^+) \quad (1)$$

$$CO_2 \rightarrow CO_2{}^* \quad (2)$$

$$CO_2{}^* + e^- + H^+ \rightarrow COOH^* \quad (3)$$

$$COOH^* + e^- + H^+ \rightarrow CO^*\ (g) + H_2O \quad (4)$$

$$CO^* \rightarrow CO \quad (5)$$

$$H_2O + h^+ \rightarrow OH^* + H^+ \quad (6)$$

$$OH^* + h^+ \rightarrow 1/2\ O_2\ (g) + H^+ \quad (7)$$

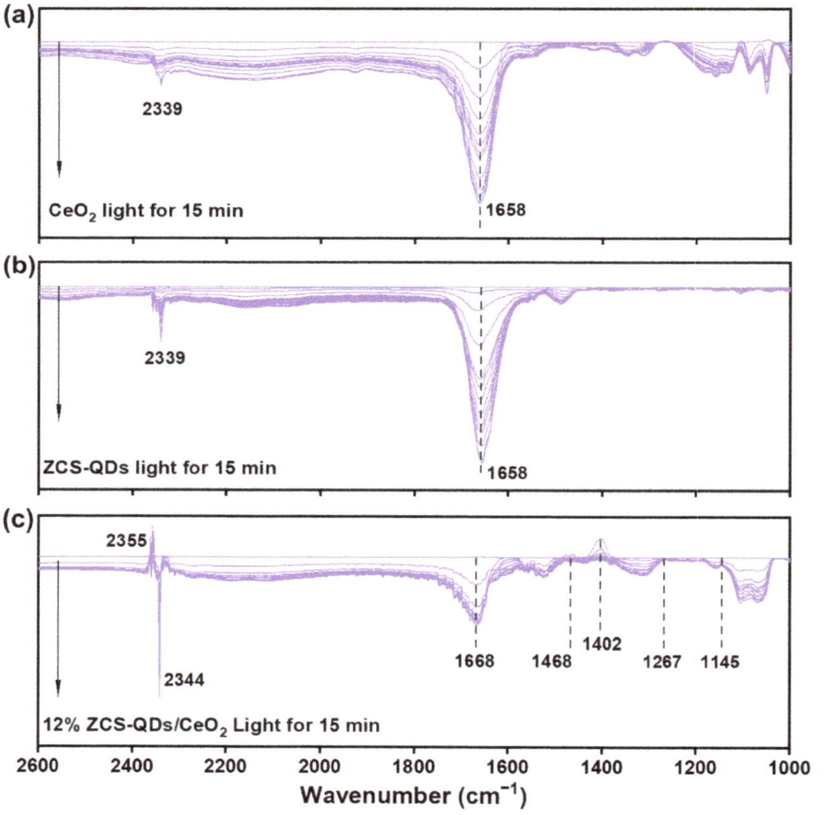


**Figure 7.** DRIFTS spectra of (**a**) $CeO_2$, (**b**) ZCS-QDs, and (**c**) 12% ZCS-QDs/$CeO_2$.

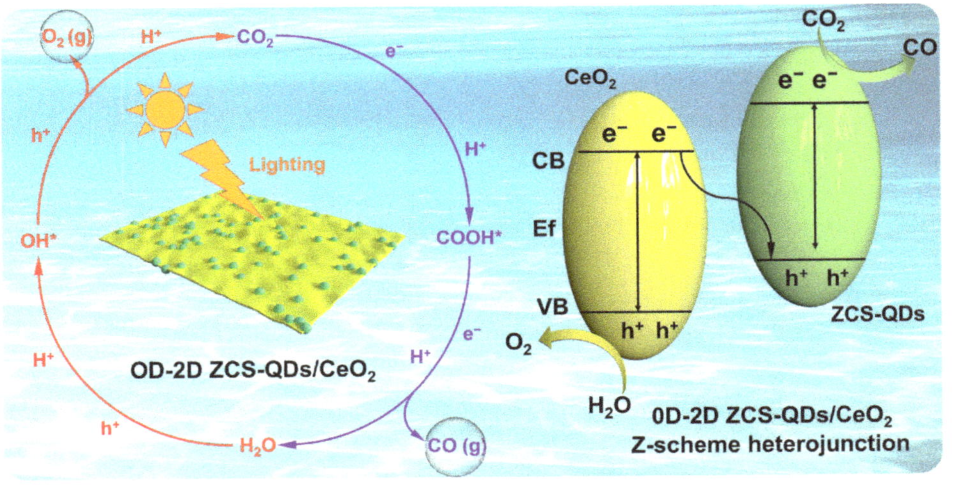


**Figure 8.** Photocatalytic $CO_2$ reduction mechanism of 0D–2D ZCS-QDs/$CeO_2$.

## 3. Experimental Section

### 3.1. Methods

#### 3.1.1. Materials

Sulfur powder (99.999%), 1-octadecence (ODE, 90%), oleic acid (OA, 90%), cadmium oxide (CdO, 99.99%), zinc acetate ($Zn(OAc)_2$, 99.99%), 11-mercaptoundecanoic acid (MUA, 95%), n-hexane (97%) tetramethylammonium hydroxide pentahydrate (97%), cerium nitrate hexahydrate ($Ce(NO_3)_3 \cdot 6H_2O$, 99.5%), ammonium bicarbonate ($NH_4HCO_3$, 99.995%), acetone (99.5%), methyl alcohol (99.5%), and ethyl alcohol (99.7%) were purchased from Sinopharm Chemical Reagent Co., Ltd. (Shanghai, China).

#### 3.1.2. Synthesis of ZCS-QDs

The ZCS-QD synthesis method was improved according to the reported procedure [49]. For the S-precursor, 0.51 g of sulfur powder was dissolved in 16 mL of ODE and heated to 150 °C under $N_2$. For the ZnCd-precursor, 0.256 g of CdO and 0.368 g of $Zn(OAc)_2$ were dissolved in a mixture of 13 mL of ODE and 6 mL of OA and then heated to 300 °C under $N_2$.

A total of 4 mL of S-precursor was injected into the ZnCd-precursor and held at 300 °C for 10 min. When brought to room temperature, the quantum dots were precipitated with acetone and then dispersed in n-hexane for preservation.

#### 3.1.3. Ligand Exchange of ZCS-QDs

For ligand exchange, we used methods that have been previously reported [50]. A total of 20 mg of MUA was dissolved into 15 mL of methanol and the pH of the solution was adjusted to >10 using tetramethylammonium hydroxide pentahydrate. Then, 20 mg worth of ZCS-QDs was added to the above mixed solution, and stirred overnight at 60 °C in a $N_2$ atmosphere. The QDs were precipitated with ethyl acetate, then cleaned with acetone and dispersed in water for preservation.

#### 3.1.4. Synthesis of $CeO_2$

For solution A, 2.78 g of $Ce(NO_3)_3 \cdot 6H_2O$ was added to 100 mL of water and stirred for 30 min. For solution B, 1.50 g of $NH_4HCO_3$ was added to 100 mL of water and stirred for 30 min. The two solutions, A and B, were mixed and stirred for 1h, left stationary at

30 °C for 24 h, and then centrifugally dried. Last calcination was performed at 500 °C for 4 h in an air atmosphere.

#### 3.1.5. Synthesis of x ZCS-QDs/$CeO_2$

A certain amount of $CeO_2$ was added to ZCS-QD solutions of different volumes, stirred away from light for 24 h, and centrifugally dried to obtain x ZCS-QDs/$CeO_2$.

### 3.2. *Characterizations*

Transmission electron microscopy (TEM) analyses were performed on an H-7800 microscope (Hitachi, Tokyc, Japan) with an acceleration voltage of 120 kV. High-resolution transmission electron microscopy (HRTEM) was carried out on a JEM-2010 transmission electron microscope (Tokyo, Japan). The XRD patterns were examined on a Shimazu-6100 powder X-ray diffractometer (Tokyo, Japan) using Cu Kα radiation at a scan rate of 10° $min^{-1}$. Transient fluorescence spectra were produced on an Edinburgh FLS1000 (Edinburgh, UK). The surface chemical states of the samples were studied by X-ray photoelectron spectroscopy (XPS, Thermo Scientific K-Alpha, Waltham, MA, USA). The optical properties of the photocatalyst were studied using UV–Vis diffuse reflectance spectroscopy (DRS, UV-2450 Shimadzu, Kyoto, Japan). Time-resolved photoluminescence (TRPL, Edinburgh FLS1000, UK) attenuation curves were obtained using a photoluminescence detector. In situ diffuse reflection infrared Fourier transform spectroscopy (DRIFTS) experiments were conducted on a Nicolet iS10 (Thermo) machine. In a typical procedure, the catalyst sample was sealed in a reaction chamber with a quartz window. $CO_2$ and $H_2O$ were carried into the reaction chamber by $N_2$ flow until equilibrium. Taking the equilibrium system before reaction as the blank background, IR signals were collected in situ during the incident irradiation of a 365 nm LED lamb (3W, Education Au-light Co., Ltd., Beijing, China) through the quartz glass window. In the in situ $CO_2$ adsorption experiment, $CO_2$ with progressively increasing concentration was introduced into the chamber and infrared signals were collected until the adsorption equilibrium was reached.

### 3.3. *$CO_2$ Photoreduction Experiments*

$CO_2$ photoreduction was carried out in a sealed 150 mL quartz reactor with a 300 W Xenon lamp (1000 mW $cm^{-2}$, Education Au-light Co., Ltd., Beijing, China) as the white light source. In a typical procedure, 10 mg of catalyst was dispersed in 3 mL of deionized water. The resulting mixture was dropped on quartz glass and then dried at 60 °C. The dried catalyst was placed in the quartz reactor with 1 mL of $H_2O$. $CO_2$ was then introduced into the reactor for 30 min to completely remove the air. During the reaction, a Xenon lamp illuminated the catalyst sample through the quartz window (22.1 $cm^2$). Gas products were detected by a gas chromatograph (GC-7920, Beijing, China) equipped with a hydrogen flame ionization detector (FID).

### 3.4. *Photoelectrochemical Measurements*

The photoelectric chemistry experiment was tested on the electrochemistry workstation (CHI 660E, Shanghai Chenhua Instrument Co., Ltd., Shanghai, China). The reaction was carried out in a homemade standard three-electrode cell. The reference electrode and the reverse electrode were saturated Ag/AgCl and Pt electrodes, respectively, and 0.5 M $Na_2SO_4$ solution was used as the electrolyte. The 0.02 g photocatalyst, 0.03 mL oleic acid, 0.01 g PVP were dispersed into 3 mL ethanol and ultrasonic for 1 h. The prepared suspension is then uniformly dropped onto the FTO glass. Using 300 W xenon lamp as light source, the effective area is about 1 $cm^2$.

## 4. Conclusions

In summary, the 0D–2D ZCS-QDs/$CeO_2$ heterojunction was successfully prepared using solvothermal and calcination methods, and it was used in a high-efficiency photocatalytic $CO_2$ reduction reaction. When using no sacrificial agent, the CO evolution rate reached 495.8 μmol $g^{-1}$ $h^{-1}$, which was 3.3 times higher than that of the ZCS-QD monomer, and has 100% selectivity. After five cycles, the catalytic activity was still high. This efficient photoconversion of $CO_2$ can be attributed to the combination of two monomers, ZCS-QDs and $CeO_2$, which promote the effective separation of photogenerated electrons, and the lifetime of photogenerated electrons is increased by 4.5 times. In addition, in situ FTIR was used to observe the changes in $H_2O$ and $CO_2$ consumption during the reaction and monitor the formation of intermediate products. The result shows that the adsorption of $CO_2$ on the surface of $CeO_2$ solves the problem of low porosity on the surface of ZCS-QDs, creating good conditions for efficient photoreduction of $CO_2$. The formation of intermediates such as COOH*, m-$CO_3^{2-}$, b-$CO_3^{2-}$, $HCO_3^-$, and OH* was monitored. Therefore, a well-designed ZCS-QDs/$CeO_2$ heterojunction can be used as an effective composite photocatalyst. In the future design of photocatalysts, synergistically improving both the $CO_2$ capture ability and the photoelectron separation efficiency is an effective means to improve the efficiency of the photocatalyst. This provides a new strategy for efficient $CO_2$ conversion.

**Author Contributions:** Investigation, data curation, and writing of the original draft, J.Y. and X.T.; resources, Y.S. and J.C.; writing—review, M.C., B.H. and Y.Z.; conceptualization, supervision, and writing—review and editing, Y.Y. All authors have read and agreed to the published version of the manuscript.

**Funding:** This research was funded by the National Natural Science Foundation of China (No. 22208127), the Senior Talent Research Foundation of Jiangsu University (No.23JDG030, 22GDG017), the RGC Postdoctoral Fellowship Scheme of Hong Kong (RGC-PDFS-2324-2S04), and the Postgraduate Research & Practice Innovation Program of Jiangsu Province (KYCX24_3952, SJCX24_2419). This work was financially supported by the research project approval of Jiangsu University (23A046, Y23A145).

**Data Availability Statement:** Data are contained within the article.

**Conflicts of Interest:** The authors declare no conflicts of interest.

## References

1. Aresta, M.; Dibenedetto, A.; Angelini, A. Catalysis for the valorization of exhaust carbon: From $CO_2$ to chemicals, materials, and fuels. technological use of $CO_2$. *Chem. Rev.* **2014**, *114*, 1709–1742.
2. Yin, S.; Sun, L.; Zhou, Y.; Li, X.; Li, J.; Song, X.; Huo, P.; Wang, H.; Yan, Y. Enhanced electron–hole separation in $SnS_2$/Au/g-$C_3N_4$ embedded structure for efficient $CO_2$ photoreduction. *Chem. Eng. J.* **2021**, *406*, 126776. [CrossRef]
3. Ma, W.; Wang, N.; Guo, Y.; Yang, L.; Lv, M.; Tang, X.; Li, S. Enhanced photoreduction $CO_2$ activity on g-$C_3N_4$: By synergistic effect of nitrogen defective-enriched and porous structure, and mechanism insights. *Chem. Eng. J.* **2020**, *388*, 124288. [CrossRef]
4. Zhu, Z.; Xing, X.; Qi, Q.; Shen, W.; Wu, H.; Li, D.; Li, B.; Liang, J.; Tang, X.; Zhao, J.; et al. Fabrication of Graphene Modified $CeO_2$/g-$C_3N_4$ Heterostructures for Photocatalytic Degradation of Organic Pollutants. *Chin. J. Struct. Chem.* **2023**, *42*, 100194. [CrossRef]
5. Tang, X.; Shen, W.; Li, D.; Li, B.; Wang, Y.; Song, X.; Zhu, Z.; Huo, P. Research on cobalt-doping sites in g-$C_3N_4$ framework and photocatalytic reduction $CO_2$ mechanism insights. *J. Alloys Compd.* **2023**, *954*, 170044. [CrossRef]
6. Mo, Z.; Miao, Z.; Yan, P.; Sun, P.; Wu, G.; Zhu, X.; Ding, C.; Zhu, Q.; Lei, Y.; Xu, H. Electronic and energy level structural engineering of graphitic carbon nitride nanotubes with B and S co-doping for photocatalytic hydrogen evolution. *J. Colloid Interface Sci.* **2023**, *645*, 525–532. [CrossRef] [PubMed]
7. Li, H.; Cheng, C.; Yang, Z.; Wei, J. Encapsulated CdSe/CdS nanorods in double-shelled porous nanocomposites for efficient photocatalytic $CO_2$ reduction. *Nat. Commun.* **2022**, *13*, 6466. [CrossRef]
8. Chang, X.; Wang, T.; Gong, J. $CO_2$ Photo-reduction: Insights into $CO_2$ Activation and Reaction on Surfaces of Photocatalysts. *Energy Environ. Sci.* **2016**, *9*, 2177–2196.
9. Pathak, P.; Meziani, M.J.; Li, Y.; Cureton, L.T.; Sun, Y.-P. Improving photoreduction of $CO_2$ with homogeneously dispersed nanoscale $TiO_2$ catalysts. *Chem. Commun.* **2004**, *10*, 1234–1235. [CrossRef]
10. Kočí, K.; Obalová, L.; Matějová, L.; Plachá, D.; Lacný, Z.; Jirkovský, J.; Šolcová, O. Effect of $TiO_2$ particle size on the photocatalytic reduction of $CO_2$. *Appl. Catal. B Environ.* **2009**, *89*, 494–502. [CrossRef]
11. Wang, Y.; Li, B.; Zhang, C.; Cui, L.; Kang, S.; Li, X.; Zhou, L. Ordered mesoporous $CeO_2$-$TiO_2$ composites: Highly efficient photocatalysts for the reduction of $CO_2$ with $H_2O$ under simulated solar irradiation. *Appl. Catal. B Environ.* **2013**, *130–131*, 277–284. [CrossRef]
12. Li, Y.; Wang, W.-N.; Zhan, Z.; Woo, M.-H.; Wu, C.-Y.; Biswas, P. Photocatalytic reduction of $CO_2$ with $H_2O$ on mesoporous silica supported Cu/$TiO_2$ catalysts. *Appl. Catal. B Environ.* **2010**, *100*, 386–392. [CrossRef]

13. Zhu, Z.; Ye, J.; Tang, X.; Chen, Z.; Yang, J.; Huo, P.; Ng, Y.; Crittenden, J. Vacancy-Rich CoSx@LDH@Co-NC Catalytic Membrane for Antibiotic Degradation with Mechanistic Insights. *Environ. Sci. Technol.* **2023**, *57*, 16131–16140. [CrossRef]
14. Zhao, X.; Xu, M.; Song, X.; Zhou, W.; Liu, X.; Yan, Y.; Huo, P. Charge separation and transfer activated by covalent bond in UiO-66-$NH_2$/RGO heterostructure for $CO_2$ photoreduction. *Chem. Eng. J.* **2022**, *437*, 135210. [CrossRef]
15. Ye, Q.; Zhou, Y.; Xu, Y.; Zhang, Q.; Shi, X.; Li, D.; Tian, D.; Jiang, D. Improved charge transfer in polymeric carbon nitride synergistically induced by the aromatic rings modification and Schottky junctions for efficient photocatalytic $CO_2$ reduction. *Chem. Eng. J.* **2023**, *463*, 142395. [CrossRef]
16. Hammond, O.S.; Edler, K.J.; Bowron, D.T.; Torrente-Murciano, L. Deep eutectic-solvothermal synthesis of nanostructured ceria. *Nat. Commun.* **2017**, *8*, 14150. [CrossRef]
17. Wang, Z.; Yu, R. Hollow Micro/Nanostructured Ceria-Based Materials: Synthetic Strategies and Versatile Applications. *Adv. Mater.* **2018**, *31*, e1800592. [CrossRef]
18. Huang, X.; Zhang, K.; Peng, B.; Wang, G.; Muhler, M.; Wang, F. Ceria-Based Materials for Thermocatalytic and Photocatalytic Organic Synthesis. *ACS Catal.* **2021**, *11*, 9618–9678. [CrossRef]
19. Selopal, G.S.; Zhao, H.; Wang, Z.M.; Rosei, F. Core/Shell Quantum Dots Solar Cells. *Adv. Funct. Mater.* **2020**, *30*, 1908762. [CrossRef]
20. Ning, Z.; Tian, H.; Yuan, C.; Fu, Y.; Qin, H.; Sun, L.; Ågren, H. Solar cells sensitized with type-II ZnSe–CdS core/shell colloidal quantum dots. *Chem. Commun.* **2011**, *47*, 1536–1538. [CrossRef] [PubMed]
21. Kouhnavard, M.; Ikeda, S.; Ludin, N.; Khairudin, N.A.; Ghaffari, B.; Mat-Teridi, M.; Ibrahim, M.; Sepeai, S.; Sopian, K. A review of semiconductor materials as sensitizers for quantum dot-sensitized solar cells. *Renew. Sustain. Energy Rev.* **2014**, *37*, 397–407. [CrossRef]
22. Yakoubi, A.; Ben Chaabane, T.; Aboulaich, A.; Mahiou, R.; Balan, L.; Medjahdi, G.; Schneider, R. Aqueous synthesis of Cu-doped CdZnS quantum dots with controlled and efficient photoluminescence. *J. Lumin* **2016**, *175*, 193–202. [CrossRef]
23. Saha, S.; Sain, S.; Meikap, A.; Pradhan, S. Microstructure characterization and electrical transport of nanocrystalline CdZnS quantum dots. *Phys. E Low-Dimens. Syst. Nanostruct.* **2015**, *66*, 59–66. [CrossRef]
24. Yang, H.; Xu, B.; Zhang, Q.; Yuan, S.; Zhang, Z.; Liu, Y.; Nan, Z.; Zhang, M.; Ohno, T. Boosting visible-light-driven photocatalytic performance of waxberry-like $CeO_2$ by samarium doping and silver QDs anchoring. *Appl. Catal. B Environ.* **2021**, *286*, 119845. [CrossRef]
25. Jiang, H.; Li, X.; Chen, S.; Wang, H.; Huo, P. g-$C_3N_4$ quantum dots-modified mesoporous $CeO_2$ composite photocatalyst for enhanced $CO_2$ photoreduction. *J. Mater. Sci. Mater. Electron.* **2020**, *31*, 20495–20512. [CrossRef]
26. Sre, V.V.; Okla, M.K.; Janani, B.; Abdel-Maksoud, M.A.; Al-Amri, S.S.; Alaraidh, I.A.; Alatar, A.A.; Khan, S.S. A novel sunlight driven Z scheme ZnCdS QDs deposited over g-$C_3N_4$ photocatalyst for photoinactivation of E. coli cells. *J. Water Process. Eng.* **2024**, *59*, 104957. [CrossRef]
27. Morgan, D.J. Photoelectron spectroscopy of ceria: Reduction, quantification and the myth of the vacancy peak in XPS analysis. *Surf. Interface Anal.* **2023**, *55*, 845–850. [CrossRef]
28. Zhang, Y.; Li, J.; Zhou, W.; Liu, X.; Song, X.; Chen, S.; Wang, H.; Huo, P. Rational design of Ag/CuO@ZnIn2S4 S-scheme plasmonic photocatalyst for highly selective $CO_2$ conversion. *Appl. Catal. B Environ.* **2024**, *342*, 123449.
29. Boaro, M.; Colussi, S.; Trovarelli, A. Ceria-Based Materials in Hydrogenation and Reforming Reactions for $CO_2$ Valorization. *Front. Chem.* **2019**, *7*, 28. [CrossRef]
30. Mullins, D.R. The surface chemistry of cerium oxide. *Surf. Sci. Rep.* **2015**, *70*, 42–85. [CrossRef]
31. Rodriguez, J.A.; Grinter, D.C.; Liu, Z.; Palomino, R.M.; Senanayake, S.D. Ceria-based model catalysts: Fundamental studies on the importance of the metal–ceria interface in CO oxidation, the water–gas shift, $CO_2$ hydrogenation, and methane and alcohol reforming. *Chem. Soc. Rev.* **2017**, *46*, 1824–1841. [CrossRef] [PubMed]
32. Kim, S.S.; Britcher, L.; Kumar, S.; Griesser, H.J. XPS Study of Sulfur and Phosphorus Compounds with Different Oxidation States. *Sains Malays.* **2018**, *47*, 1913–1922. [CrossRef]
33. Wu, L.; Su, F.; Liu, T.; Liu, G.-Q.; Li, Y.; Ma, T.; Wang, Y.; Zhang, C.; Yang, Y.; Yu, S.-H. Phosphorus-Doped Single-Crystalline Quaternary Sulfide Nanobelts Enable Efficient Visible-Light Photocatalytic Hydrogen Evolution. *J. Am. Chem. Soc.* **2022**, *144*, 20620–20629. [CrossRef]
34. Sun, Y.; Hao, Y.; Lin, X.; Liu, Z.; Sun, H.; Jia, S.; Chen, Y.; Yan, Y.; Li, X. Efficient electron transport by 1D CuZnInS modified 2D Ti3C2 MXene for enhanced photocatalytic hydrogen production. *J. Colloid Interface Sci.* **2024**, *653*, 396–404. [CrossRef] [PubMed]
35. Kim, S.-J.; Choi, M.; Hong, G.; Hahn, S.K. Controlled afterglow luminescent particles for photochemical tissue bonding. *Light Sci. Appl.* **2022**, *11*, 314. [CrossRef] [PubMed]
36. Zhao, J.; Wang, Y.; Li, Y.; Yue, X.; Wang, C. Phase-dependent enhancement for $CO_2$ photocatalytic reduction over $CeO_2$/$TiO_2$ catalysts. *Catal. Sci. Technol.* **2016**, *6*, 7967–7975. [CrossRef]
37. Kibar, M.E. Preparation of copper oxide-cerium oxide/nanotube-titanium dioxide photocatalyst for $CO_2$ conversion in solar light. *React. Kinet. Catal. Lett.* **2021**, *134*, 937–950. [CrossRef]
38. Hu, H.; Hu, J.; Wang, X.; Gan, J.; Su, M.; Ye, W.; Zhang, W.; Mac, X.; Huihu, W. Enhanced reduction and oxidation capability over $CeO_2$/g-$C_3N_4$ hybrid through surface carboxylation: Performance and mechanism. *Catal. Sci. Technol.* **2020**, *10*, 4712–4725. [CrossRef]

39. Li, Z.; Liu, Z.; Li, Y.; Wang, Q. Flower-like CoAl layered double hydroxides modified with $CeO_2$ and RGO as efficient photocatalyst towards $CO_2$ reduction. *J. Alloys Compd.* **2021**, *881*, 160650. [CrossRef]
40. Su, B.; Kong, Y.; Wang, S.; Zuo, S.; Lin, W.; Fang, Y.; Hou, Y.; Zhang, G.; Zhang, H.; Wang, X. Hydroxyl-Bonded Ru on Metallic TiN Surface Catalyzing $CO_2$ Reduction with $H_2O$ by Infrared Light. *J. Am. Chem. Soc.* **2023**, *145*, 27415–27423. [CrossRef]
41. Yin, S.; Zhou, Y.; Liu, Z.; Wang, H.; Zhao, X.; Zhu, Z.; Yan, Y.; Huo, P. Elucidating protonation pathways in $CO_2$ photoreduction using the kinetic isotope effect. *Nat. Commun.* **2024**, *15*, 437. [CrossRef] [PubMed]
42. Xu, J.; Ju, Z.; Zhang, W.; Pan, Y.; Zhu, J.; Mao, J.; Zheng, X.; Fu, H.; Yuan, M.; Chen, H.; et al. Efficient Infrared-Light-Driven $CO_2$ Reduction Over Ultrathin Metallic Ni-doped $CoS_2$ Nanosheets. *Angew. Chem. Int. Ed.* **2021**, *60*, 8705–8709. [CrossRef] [PubMed]
43. Liu, L.; Huang, H.; Chen, Z.; Yu, H.; Wang, K.; Huang, J.; Yu, H.; Zhang, Y. Synergistic Polarization Engineering on Bulk and Surface for Boosting $CO_2$ Photoreduction. *Angew. Chem. Int. Ed.* **2021**, *60*, 18303–18308. [CrossRef] [PubMed]
44. Song, X.; Wu, Y.; Zhang, X.; Li, X.; Zhu, Z.; Ma, C.; Yan, Y.; Huo, P.; Yang, G. Boosting charge carriers separation and migration efficiency via fabricating all organic van der Waals heterojunction for efficient photoreduction of $CO_2$. *Chem. Eng. J.* **2021**, *408*, 127292. [CrossRef]
45. Zhu, K.; Zhu, Q.; Jiang, M.; Zhang, Y.; Shao, Z.; Geng, Z.; Wang, X.; Zeng, H.; Wu, X.; Zhang, W.; et al. Modulating $Tit_{2g}$ Orbital Occupancy in a Cu/$TiO_2$ Composite for Selective Photocatalytic $CO_2$ Reduction to CO. *Angew. Chem. Int. Ed.* **2022**, *61*, e202207600. [CrossRef]
46. Chen, Y.; Yu, S.; Fan, X.-B.; Wu, L.-Z.; Zhou, Y. Mechanistic insights into the influence of surface ligands on quantum dots for photocatalysis. *J. Mater. Chem. A* **2023**, *11*, 8497–8514. [CrossRef]
47. Xiao, Y.; Jiang, Y.; Zhou, E.; Zhang, W.; Liu, Y.; Zhang, J.; Wu, X.; Qi, Q.; Liu, Z. In-suit fabricating an efficient electronic transport channels via S-scheme polyaniline/Cd0.5Zn0.5S heterojunction for rapid removal of tetracycline hydrochloride and hydrogen production. *J. Mater. Sci. Technol.* **2023**, *153*, 205–218. [CrossRef]
48. Xu, S.; Jiang, G.; Zhang, H.; Gao, C.; Chen, Z.; Liu, Z.; Wang, J.; Du, J.; Cai, B.; Li, Z. Boosting Photocatalytic $CO_2$ Methanation through Interface Fusion over CdS Quantum Dot Aerogels. *Small* **2024**, e2400769. [CrossRef]
49. Zhang, Z.; Rogers, C.R.; Weiss, E.A. Energy Transfer from CdS QDs to a Photogenerated Pd Complex Enhances the Rate and Selectivity of a Pd-Photocatalyzed Heck Reaction. *J. Am. Chem. Soc.* **2019**, *142*, 495–501. [CrossRef]
50. Aldana, J.; Wang, Y.A.; Peng, X. Photochemical Instability of CdSe Nanocrystals Coated by Hydrophilic Thiols. *J. Am. Chem. Soc.* **2001**, *123*, 8844–8850. [CrossRef]

catalysts

Article

# Efficient Catalytic Oxidation of 5-Hydroxymethylfurfural to 2,5-Furandicarboxylic Acid Using a Natural Mineral Vermiculite-Loaded Gold–Palladium Bimetallic Catalyst

**Hongke Li [1,2,†], Yiwang Li [1,†], Qinghua Xia [1], Yiran Liu [1], Wen Guan [1] and Yao Chen [3,*]**

1 Institute of Green Chemistry and Chemical Technology, School of Chemistry and Chemical Engineering, Jiangsu University, Zhenjiang 212013, China; 2222212084@stmail.ujs.edu.cn (H.L.); 3211306088@stmail.ujs.edu.cn (Y.L.); 3221306014@stmail.ujs.edu.cn (Q.X.); 2112412237@stmail.ujs.edu.cn (Y.L.); 2112212008@stmail.ujs.edu.cn (W.G.)
2 College of Physics, Jilin Normal University, Siping 130023, China
3 School of the Environment and Safety, Jiangsu University, Zhenjiang 212013, China
* Correspondence: chenyao1390@ujs.edu.cn; Tel./Fax.: +86-511-88830099
† These authors contributed equally to this work and should be considered co-first authors.

**Abstract:** Upgrading the bio-derived platform chemical 5-hydroxymethylfurfural (HMF) into the high value-added bioplastic monomer 2,5-furandicarboxylic acid (FDCA) is a promising pathway for biomass conversion. In this work, the natural and abundant available mineral vermiculite was employed as a carrier for loading a Au-Pd bimetal catalyst. Due to the high dispersion of bimetallic nanoparticles, this synthesized vermiculite-supported Au-Pd bimetal catalyst revealed excellent catalytic performance for the aerobic oxidation of HMF to FDCA. By adjusting the ratio of Au and Pd metals, the catalytic performance of the catalyst can be optimized. Finally, 100% HMF conversion and 99.9% FDCA yield could be obtained under the conditions of Au/Pd = 2/1, 2 h, 2 MPa $O_2$, and 100 °C. The catalyst revealed good stability, and the FDCA yield can be maintained at 90.1% after five recycle usages. The physicochemical properties of the synthesized catalysts were characterized by various characterization methods. It could be concluded that the high dispersion and alloying effect of bimetallic nanoparticles promoted the activation of reactants and intermediates, resulting in the effective production of FDCA. This study could provide ideas and references for the development and utilization of natural minerals and also offer a new way to realize the efficient conversion of HMF to FDCA under green conditions.

**Keywords:** 5-hydroxymethylfurfural; 2,5-furandicarboxylic acid; natural mineral vermiculite; gold–palladium bimetal

## 1. Introduction

The extensive use of non-renewable fossil resources has led to an energy crisis and contributed to several environmental problems, including global warming and the depletion of the ozone layer [1–5]. The development of sustainable resources is imminent. Biomass has been widely considered the fourth largest resource in the world because of its rich source and green and renewable characteristics [6–8]. Among these, 5-hydroxymethylfurfural (HMF) has been prepared from biomass carbohydrates and has a particularly high potential for application in many platform compounds [9–11]. Through catalytic oxidation, it can be converted into a variety of high value-added chemicals and biofuels, including 5-hydroxymethyl-2-furandicarboxylic acid (HMFCA), 5-formyl-2-furanoic acid (FFCA), 2,5-furandicarboxaldehyde (DFF), and 2,5-furandicarboxylic acid (FDCA) [12]. Among them, FDCA can replace terephthalic acid to make new polymer materials such as polyester and polyamide because of its similar structure and properties to terephthalic acid monomers derived from Petroleum [13–15]. FDCA can be polymerized with ethylene glycol to synthesize polyethylene 2,5-furandicarboxylate (PEF). PEF can be used to make eco-friendly, recyclable plastic packaging, and its biodegradability can effectively alleviate environmental pollution problems [16]. Therefore, the efficient conversion of HMF to FDCA is of great significance for the fine utilization of biomass resources.

In the existing research, compared with the homogeneous catalyst, the heterogeneous modifier has become the focus and main research direction in the catalytic oxidation reaction of HMF because of its strong stability, easy separation, and recyclability [17–20]. Further studies displayed that under different catalysts and reaction conditions, the conversion path of HMF to FDCA was different, and the rate-controlling steps of the reaction were also different [21]. Therefore, selecting a suitable catalyst and constructing a corresponding catalytic system to minimize the activation energy of the whole reaction process is the key to improving the efficiency of catalytic oxidation of HMF to FDCA [22,23]. Precious metal catalysts in multiphase catalysts have received much attention due to their early research time and good catalytic performance [24,25]. For example, Au [26,27], Pt [28–30], Pd [31–33], and Ru [34–36] have high catalytic activity in catalytic systems using $O_2$ as oxidants. A further study found that the single-component metal catalysts had their own unique catalytic characteristics in the process of preparing FDCA from HMF. Au-based catalysts have relatively high catalytic activity for the conversion of the aldehyde group to a carboxyl group, while Pd and Pt-based catalysts have higher catalytic ability for the conversion of alcohol hydroxyl group [37,38]. Therefore, a bi-precious metal catalytic system can be constructed to effectively improve the catalytic activity of selective oxidation of HMF to FDCA by making full use of the advantages of different metal active centers [39]. However, for bimetallic catalytic systems, researchers have focused their attention on the study of materials to improve the catalytic performance. Neglecting the problem of aggregation of bimetallic nanoparticles severely limits their catalytic activity [40]. The main carriers that have been explored are metal oxides, carbon materials, and more [41–43]. However, these support materials are inherently synthetic, leading to higher cost and complexity, and may even be harmful to the environment during preparation. Therefore, the development of a new type of catalyst with green, inexpensive, and highly dispersed bimetallic carriers is imperative.

Vermiculite is a common mineral in nature, with huge reserves, low cost, and simple processing. Vermiculite is composed of two layers of silica tetrahedron and one layer of aluminum oxide octahedron, forming a mica stack structure with a specific structure [44,45]. Due to the formation of interlayer hydrogen bonds and van der Waals bonds, 2-D vermiculite has excellent thermal stability and proton conduction function, and nanochannels in its structure can regulate molecular transport [46,47]. Heavy metal cations can be easily adsorbed on the surface of vermiculite through interlayer ion exchange, surface electrostatic interaction, and surface complexation of the vermiculite. And because vermiculite can expand more than 10–30 times when heated, its adsorption efficiency is also ideal [48]. The double precious metal catalyst adopts vermiculite as the carrier because of its traits such as both green, renewable and excellent properties to highlight the green and efficient characteristics of the material.

In order to achieve efficient FDCA production, a natural mineral-supported Au-Pd bimetallic particle was constructed in this study and employed for the highly selective catalytic oxidation of HMF. The effect of metals on the catalytic oxidation of HMF to FDCA reaction was further investigated by kinetics. Combined with the ESR technique, it was shown that an increase in the concentration of superoxide radicals is essential to increase the catalytic oxidation rate. This study provides ideas and references for the development and utilization of natural minerals and also offers a new way for the synthesis of FDCA under green conditions.

## 2. Results and Discussion

### 2.1. Characterization of the Physicochemical Properties of Catalysts

The synthesis process of the $Au_xPd_y$/VMT catalyst is exhibited in Figure 1a. First, VMT was ground and dried in an N-methylpyrrolidone environment to activate it, and then the VMT active carrier was prepared and loaded with AuPd nanoparticles. Figure S1a reveals the SEM image of VMT, and the samples are irregular and thin. Figure S1b illustrates the SEM diagram of the prepared $Au_2Pd_1$/VMT catalyst. Compared with the SEM diagram of the VMT, it can be seen that the overall morphology of the material will not be affected when metal particles are loaded on the surface of the VMT. At the same time, the distribution of elements on the surface of the $Au_2Pd_1$/VMT catalyst was surveyed using TEM mapping, and the results are revealed in Figure S2. It was known that VMT was composed of Si, Mg, Al, and Fe elements, and loaded Au and Pd metal particles were evenly distributed on the surface of VMT. In order to observe the morphology of the catalyst, $Au_2Pd_1$/VMT was characterized by the HR-TEM technique. This is displayed in Figure 1b. The $Au_2Pd_1$/VMT catalyst showed a two-dimensional layered structure of VMT and revealed agglomeration of Au and Pd nanoparticles in some locations. The lamellar structure of VMT in $Au_2Pd_1$/VMT catalysts was still preserved. VMT plays a good supporting role for the catalyst, which can effectively improve the dispersion degree of the catalyst and make the bimetallic particles evenly distributed on the surface of the catalyst. When taking 100 particles, the average size of $Au_2Pd_1$/VMT is 3.74 nm. The smaller the size of the metal nanoparticles, the more active sites they were exposed to, thus improving catalytic performance. The feature crystal plane spacing of VMT, Pd (111), and Au (111) is, respectively, 0.1551, 0.221, and 0.235 nm. The overlapping gold and palladium lattices can be observed using HRTEM image measurements of $Au_2Pd_1$/VMT (Figure 1c). The EDS pattern of $Au_2Pd_1$/VMT displayed that the AuPd nanoparticles were evenly distributed on the VMT carrier (Figure 1d). The highly dispersed nature of VMT as a carrier was further confirmed.

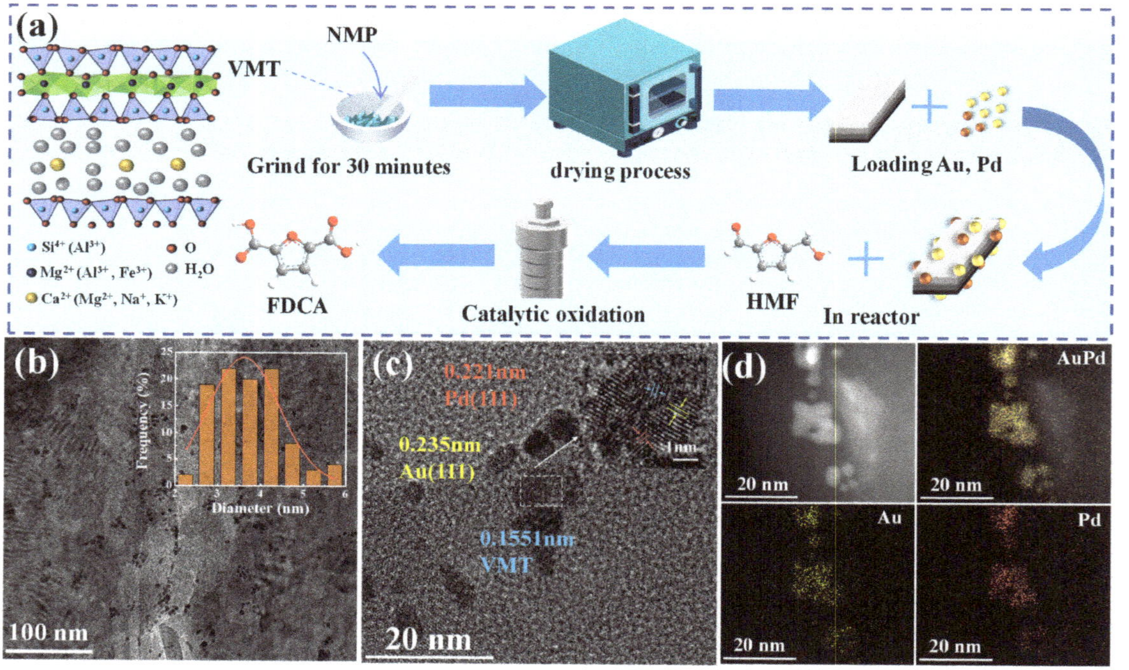


**Figure 1.** The flow chart for the preparation of $Au_2Pd_1$/VMT catalysts (**a**), TEM images of $Au_2Pd_1$/VMT (**b**), HR-TEM images of $Au_2Pd_1$/VMT (**c**), and EDS elemental (Au,Pd) mapping of $Au_2Pd_1$/VMT (**d**).

The XRD of different AuPd ratios are given in Figure 2a,b. As a natural mineral material, the characteristic peaks of VMT's main element composition are as follows [49]. The (110) crystal face of Fe is at 44.67°, the (111) crystal face of Al is at 44.73°, the (110) crystal face of $Fe_2Si$ is at 44.69°, the (100) face of Si is at 27.07°, and the (100) face of Mg is at 32.19°. However, the characteristic diffraction peaks of metal particles Au (JCPDS, No. 04-0784) and Pd (JCPDS, No. 46-1043) cannot be observed, which was due to the low metal content of the catalyst loading and the small size of the metal particles and also indicates that the nanoparticles were highly dispersed [50]. In addition, the (100) plane of Mg is located at 27.24°, and the intensity of characteristic peaks increases with the increase in metal proportion. The specific surface area and pore structure of VMT catalysts with different metal ratios were surveyed by nitrogen adsorption–desorption experiments. From Figure 2c, it can be seen that the adsorption–desorption isotherms of several samples are close to typical type IV adsorption–desorption isotherms, and the adsorption curves of the isotherms do not coincide with the desorption curves, and there is a hysteresis back to the line, and the observed hysteresis corresponds to interspecies pores, which may be macropores. According to the pore size distribution curves in Figure 2d, the pores of the surface samples were mostly microporous with a few meso- and macropores. Meanwhile, the specific surface area, average pore size, and total pore volume data of the samples are shown in Table 1. The addition of different ratios of Au and Pd particles to the VMT surface increases the specific surface area and total pore volume of the samples, and the average pore size of some samples decreases.

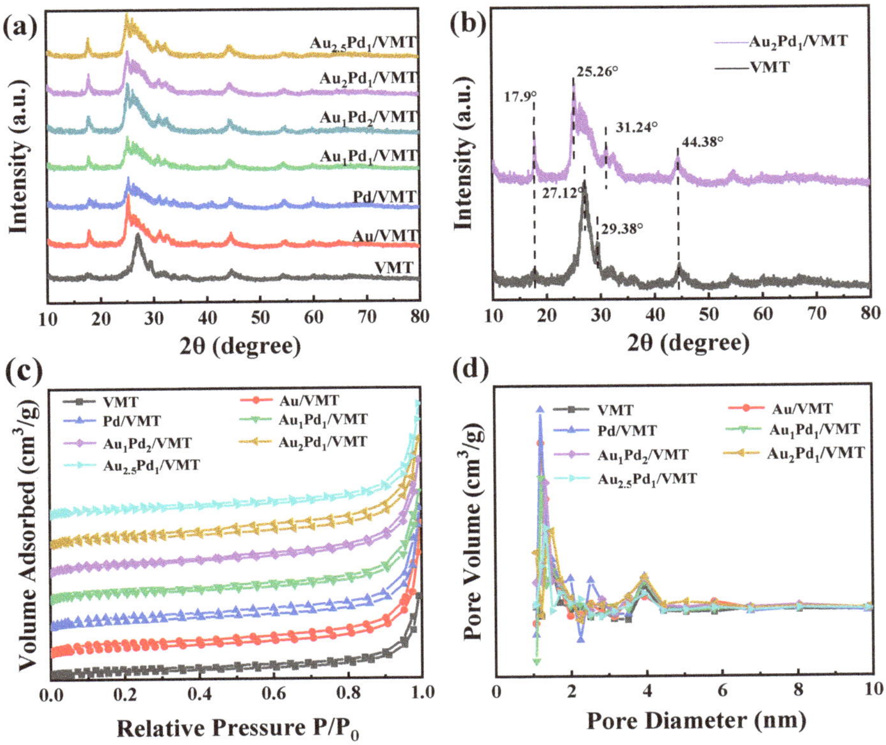


**Figure 2.** XRD patterns of VMT and catalysts with different metal ratios (**a**,**b**), isothermal adsorption–desorption curves (**c**), and average pore size distribution (**d**) of VMT and catalysts with different metal ratios.

**Table 1.** VMT and $Au_xPd_y$/VMT specific surface area and aperture size list.

| Sample | $S_{BET}$ [a] ($m^2/g_{cat}$) | Mean Aperture Size (nm) | $D_{BJH}$ [b] (nm) | $V_p$ [c] |
|---|---|---|---|---|
| VMT | 6.3033 | 11.0170 | 1.3425 | 0.0372 |
| Au/VMT | 10.1610 | 10.2590 | 1.2175 | 0.0628 |
| Pd/VMT | 10.8701 | 9.2395 | 1.2175 | 0.0596 |
| $Au_1Pd_1$/VMT | 11.3160 | 12.8840 | 1.2175 | 0.0511 |
| $Au_1Pd_2$/VMT | 12.4440 | 8.0373 | 1.3425 | 0.0562 |
| $Au_2Pd_1$/VMT | 11.8170 | 7.6418 | 1.5001 | 0.0543 |
| $Au_{2.5}Pd_1$/VMT | 10.4020 | 11.8440 | 1.3425 | 0.0542 |

$S_{BET}$ [a] is the specific surface area of BET; $D_{BJH}$ [b] is the average pore diameter of BJH calculated according to the desorption curve of the nitrogen adsorption isotherm; $V_p$ [c] is the total pore volume measured at a relative pressure of 0.99.

In order to understand the chemical states of Au and Pd on $Au_2Pd_1$, XPS was used to study Au/VMT, Pd/VMT, and $Au_2Pd_1$/VMT, and the results are displayed in Figure 3a as a full XPS spectrum, which visually expresses the position distribution of Mg, O, Si, Al, and loaded metal. As reflected in Figure 3b, there are two Au 4f peaks in the sample, which can be decomposed into Au $4f_{7/2}$ (84.01 eV in $Au_2Pd_1$/VMT, 83.76 eV in Au/VMT) and Au $4f_{5/2}$ spin–orbit splitting peaks (88.44 eV in $Au_2Pd_1$/VMT and 87.58 eV in Au/VMT). As indicated In Figure 3c, the combinations of Pd 3d are 337.86 eV, 336.32 eV ($Au_2Pd_1$/VMT) and 337.77 eV, 336.78 eV (Pd/VMT), respectively. Among them, the peaks at 337.86 eV and 337.77 eV corresponded to the $Pd^0$, and the peaks at 336.32 eV and 336.78 eV corresponded to the $Pd^{2+}$. The peak of Au 4f of $Au_2Pd_1$/VMT was shifted to the direction of high binding energy, and the peak of Pd 3d was shifted to the direction of low binding energy compared to that of Au/VMT and Pd/VMT, which suggests that there is an electron transfer between AuPd and VMT with a strong interplay [51]. Figure 3d,e show the 2p orbital spectrum of Si, the 2p orbital spectrum of Al, and the 1s orbital spectrum of Mg, respectively.

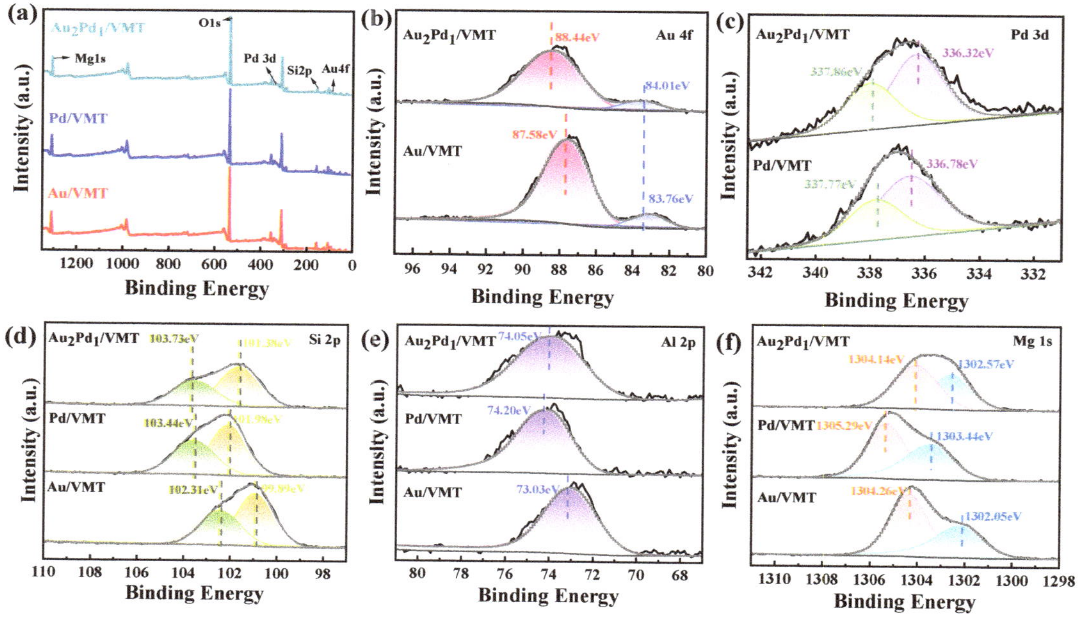


**Figure 3.** Full-range (**a**), Au 4f (**b**), Pd 3d (**c**), Si 2p (**d**), Al 2p (**e**), and Mg 1s (**f**) XPS spectra of Au/VMT, Pd/VMT, and $Au_2Pd_1$/VMT catalysts.

### 2.2. Influence of Different Factors on Oxidation Performance of Catalyst

In order to obtain the highest yield, the effect of different factors on the performance of the catalyst was investigated initially. The reaction conditions were 50 mg HMF (0.3965 mmol), 40 mL deionized water, base/HMF = 4/1 molar ratio, 50 mg catalysts. The theoretical value of metal content was 2 wt%, while the value determined by ICP is 1.77 wt% as referenced in Table S5. $Au_2Pd_1$/VMT catalyst was used to optimize the reaction temperature (Figure 4a). After the reaction at 90 °C, 100 °C and 110 °C for 120 min, the yields of FDCA are 75.4%, 99.9%, and 99.9%, respectively, indicating that temperature has a certain influence on the catalytic performance of $Au_2Pd_1$/VMT catalyst below 100 °C. However, when the reaction temperature was further increased to 110 °C, the final yield of FDCA did not change significantly. Therefore, 100 °C is the best reaction temperature of the catalyst. As displayed in Figure 4b, when 0.5 MPa oxygen was used, the yield of FDCA was 85.6%, and with the increase of oxygen pressure, the yield of FDCA also increased (92% at 1 MPa and 99.9% at 2 MPa), indicating that $O_2$ plays a key role in the catalytic oxidation of HMF to FDCA. The fixed oxygen pressure is 2 MPa, and the catalyst dosage was changed to continue the catalytic experiment. As reflected in Figure 4c, when the amount of catalyst was increased from 30.0 mg to 50.0 mg, the yield of FDCA increased from 64.7% to 99.9%, and the catalyst dosage continued to increase to 70.0 mg, the yield of FDCA was not significantly improved. Therefore, the optimal dosage of the catalyst is 50.0 mg, and the optimal reaction oxygen pressure is 2.0 MPa. Finally, the effects of different base types on the yield of FDCA were studied to explore the catalytic activity of $Au_2Pd_1$/VMT catalyst for converting HMF to FDCA under different base types. As shown in Figure 4d, the molar ratios of different types of bases to HMF remained unchanged at 4:1. After 120 min of reaction, the highest yield of FDCA was 99.9% with NaOH as the base. When $Na_2CO_3$ and $NaHCO_3$ were used as the bases, the FDCA yields were only 94.1% and 6.0%. The results show that different types of bases can directly affect the yield of FDCA formation during HMF oxidation.

In addition, we studied the effect of different AuPd molar ratios on the yield of FDCA (Figure 5a). Through the experimental results, it was exhibited that the yield of HMFCA increased with the increase in the Au ratio, while HMF reflected the lowest oxidation of the loaded mono-Pd catalyst. Davis et al. [38]. found that Au-based catalysts are more capable of oxidizing aldehydes when under alkaline conditions. And Lei et al. [37]. found that Pd-based catalysts are more favorable for the oxidation of alcohols. Combining the advantages of AuPd nanoparticles, the bimetallic-loaded catalysts have excellent catalytic performance. When the ratio of Au in the bimetallic catalyst is 2:1, the catalytic effect of the catalyst is the best, and the ratio of Au to Pd was further increased to 2.5:1, and the yield of FDCA was slightly decreased, which may be related to the metal clusters observed in TEM. Therefore, in this catalytic system, the optimal ratio of gold-palladium bimetal is 2:1, and the bimetal exhibits the best synergistic effect, greatly improving the catalytic activity of HMF oxidation to FDCA. Then, the $Au_2Pd_1$/VMT catalyst was used as the representative catalyst to explore the best reaction time of catalytic oxidation of HMF to FDCA. As exhibited in Figure 5b, when the reaction was carried out for 5 min, the conversion rate of HMF reached 100%. With the extension of reaction time, the yield of intermediate products HMFCA and FFCA showed a trend of first increasing and then decreasing, while the yield of target product FDCA gradually increased, and the yield of FDCA reached 99.9% at 120 min. Compared with the previously reported studies, $Au_2Pd_1$/VMT showed relatively favorable FDCA yield after two hours of reaction, and VMT as the carrier has the advantages of low cost and easy availability (Table S1). The HMF oxidation process on $Au_xPd_y$/VMT catalysts was better understood by the kinetic studies conducted in Figure 5c–e. First, the catalytic reaction with $Au_2Pd_1$/VMT as the catalyst was carried out. The data are indicated in Figure 5c. It can be seen that when HMF was oxidized to HMFCA in the presence of AuPd nanoparticles, the Ea was 42.53 KJ·mol$^{-1}$. The Ea of the oxidation process of HMFCA and FFCA is 69.08 KJ·mol$^{-1}$ and 61.39 KJ·mol$^{-1}$, respectively, which indicates that the oxidation of HMFCA is the RDS of the reaction. To further investigate the effect of metals on the oxidative preparation of FDCA from HMF, the kinetics of the reaction were studied with Au/VMT and Pd/VMT as catalysts under the best conditions. As revealed in Figure 5d, the activation energy of Au catalyzed the oxidation of HMF is the lowest, Ea is 60.99 KJ·mol$^{-1}$, and the activation energy of HMFCA is the highest, Ea is 84.16 KJ·mol$^{-1}$, indicating that Au had a high oxidation activity on aldehyde groups in alkaline environment. As displayed in Figure 5e, Pd catalyzed the oxidation of HMFCA with the lowest activation energy, Ea is 72.76 KJ·mol$^{-1}$, and the catalytic activation energy of HMF is the highest, Ea is 104.62 KJ·mol$^{-1}$, indicating that Pd had a high oxidation activity for hydroxyl in alkaline environment.

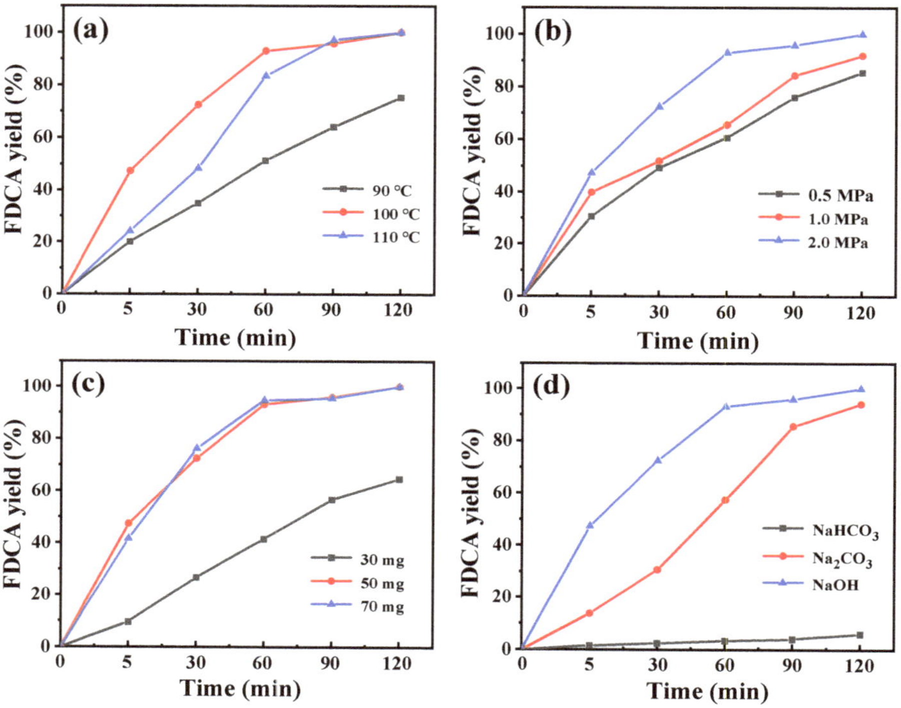


**Figure 4.** Properties of $Au_2Pd_1$/VMT at different temperatures (**a**), oxygen pressures (**b**), catalyst dosages (**c**) and bases (**d**). Reaction conditions (reaction time 120 min, base/HMF = 4/1 molar ratio, amount of HMF 50 mg, 40 mL deionized water). When working on one parameter, the rest were fixed.

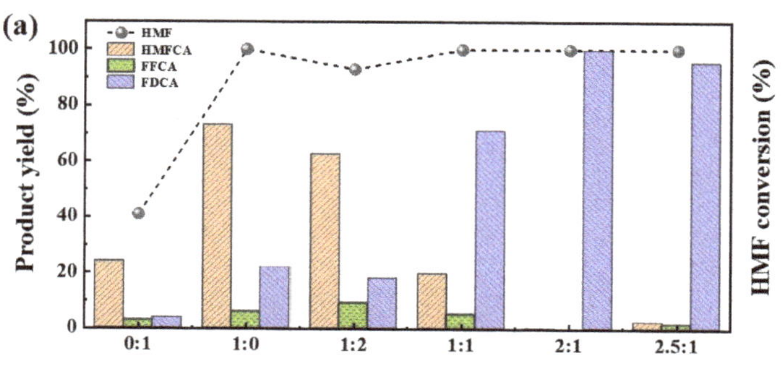


**Figure 5.** *Cont.*

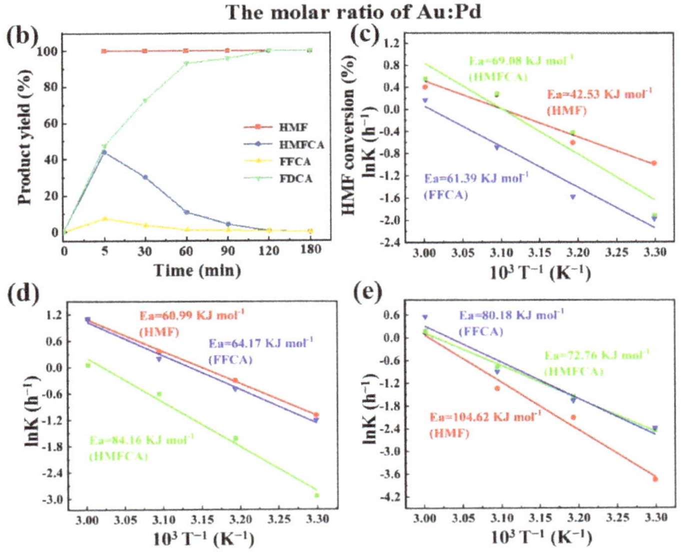


**Figure 5.** The catalytic oxidation performance of $Au_xPd_y$/VMT (**a**), effect of reaction time on the catalytic performance of $Au_2Pd_1$/VMT catalyst. Reaction conditions: 50.0 mg HMF, 50.0 mg catalyst, nNaOH: nHMF= 4:1, 2.0 MPa $O_2$, T = 100 °C (**b**), substrate conversions plotted by Arrhenius plots when on $Au_2Pd_1$/VMT catalysts (**c**), substrate conversions plotted by Arrhenius plots when on Au/VMT catalysts (**d**), substrate conversions plotted by Arrhenius plots when on Pd/VMT catalyst (**e**).

The role of $O_2$ molecules in the catalytic oxidation reaction of HMF is significant. To investigate the capacity of oxygen molecules adsorbed on metal nanoparticle surfaces to generate superoxide free radicals ($\cdot O_2^-$) through electron removal activation, three catalysts were characterized using ESR to determine the types and concentrations of reactive oxygen species. As depicted in Figure 6a, the characteristic peak of DMPO-$O_2^-$ was observed. The results demonstrate that the $Au_2Pd_1$/VMT catalyst exhibits the highest signal intensity compared to the other two catalysts, indicating that an increased concentration of superoxide radicals is crucial for enhancing the rate of catalytic oxidation.

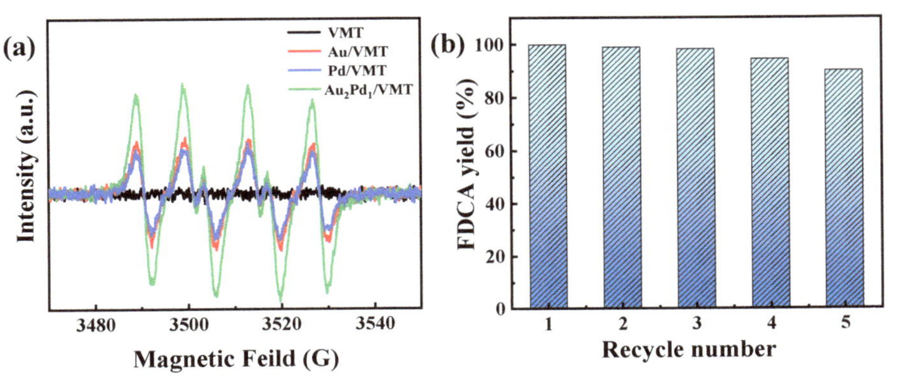


**Figure 6.** ESR images of $Au_2Pd_1$/VMT, Au/VMT, and Pd/VMT (**a**); recycle number (**b**).

The stability of the $Au_2Pd_1$/VMT catalyst was surveyed by regeneration experiments. As shown in Figure 6b, after repeated use of the catalyst three times, the FDCA yield did not decrease significantly until the fifth cycle test; the FDCA yield decreased from the initial 99.9% to 90.1%, which may be related to the metal leaching supported on the catalyst surface. The metal loading of the fresh catalyst before the reaction and the catalyst after five cycles was tested by ICP, and the AuPd content after the reaction was slightly reduced by 0.08 wt% compared to the metal content before the cycle (Table S5). Therefore, after five catalytic cycles, the performance of the $Au_2Pd_1$/VMT catalyst was only slightly decreased, indicating that it has good stability.

The HMF was transformed into the target product FDCA through the "HMF-HMFCA-FFCA-FDCA path" (Figures S5 and S6). Based on the investigation of rate-determining steps and analysis of active substances, a plausible oxidation mechanism for HMF catalyzed by $Au_2Pd_1$/VMT was postulated. As shown in Scheme 1, base was introduced into the reaction system to create an alkaline environment, and the reaction process can be divided into three stages. Firstly, HMF molecules were adsorbed onto the catalyst surface, where the aldehyde group reacts with $OH^-$ ions on the AuPd nanoparticles adsorbed on the catalyst to form an oxy-alcohol ion intermediate. Subsequently, under the influence of superoxide free radicals, this bisdiol intermediate undergoes dehydration to yield HMFCA. Then, both $OH^-$ ions and $O_2{}^-$ species aggregated at the active center of the catalyst act upon the hydroxyl group of HMFCA, leading to the formation of aldehyde groups and subsequent production of FFCA. Finally, FFCA undergoes a similar reaction pathway as that observed for HMFCA, resulting in the formation of FDCA.

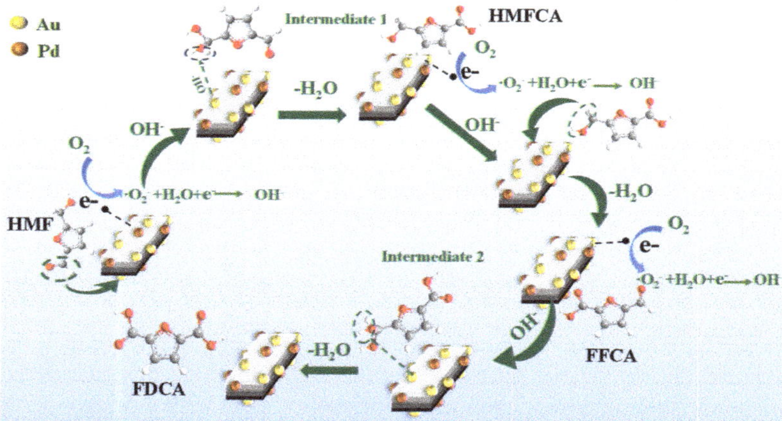


**Scheme 1.** Possible mechanism for the selective oxidation of HMF to FDCA over $Au_2Pd_1$/VMT.

## 3. Experimental Section

### 3.1. Reagents and Materials

Polyvinyl alcohol (AR), N-methylpyrrolidone (AR), chloroauric acid (AR), palladium chloride (AR), and sodium borohydride (AR) were purchased from Aladdin Chemical Reagents (Shanghai, China). Hydrochloric acid (AR), sodium hydroxide (AR), sodium carbonate (AR), sodium bicarbonate (AR), ethanol (AR), acetic acid (HPLC), acetonitrile (HPLC), methanol (HPLC grade) were obtained from Sinopharm Chemical Reagent (Shanghai, China). 5-hydroxymethylfurfural (AR), 5-hydroxymethyl-2-furanoic acid (AR), 5-formyl-2-furanoic acid (AR), and 2,5-furanodicarboxylic acid (AR) were supplied by Macklin (Shanghai, China). All materials and reagents were used in the condition in which they were received and no further purification was required.

Instrumentation and detailed procedures for catalyst characterization are detailed in the Supplementary Materials.

### 3.2. Preparation of AuxPdy/VMT Catalyst

First, vermiculite (VMT) was dispersed into N-methylpyrrolidone (NMP) and milled for 30 min, washed several times with deionized water and ethanol, respectively, and dried and milled for stockpile use. Solutions of 1.57 mL $HAuCl_4 \cdot 3H_2O$ (4 mg/mL) and 0.355 mL $PdCl_2$ (4 mg/mL) were added to a 100 mL round-bottled flask, followed by 0.13 mL PVA solution (0.1 g/mL) and 25 mL deionized water. The solution was stirred well for 15 min at room temperature (25 °C) to make a homogeneous mixture, and then a freshly prepared sodium borohydride solution (NaBH4/metal = 5/1, mol/mol) was added slowly to produce a colloidal solution of metal. After the solution had fully reacted for 20 min, the pH of the solution was adjusted to 3–4. Subsequently, 0.2 g of vermiculite carrier was added to the solution, and rapid stirring was continued for 2 h at room temperature (25 °C). The resulting product was centrifuged at 10,000 rpm for 3 min and washed with deionized water at least three times and then with ethanol once to ensure the complete removal of residual $Na^+$, $Cl^-$, and so forth. The centrifugally washed catalyst was dried overnight in a blast drying oven at 70 °C and labeled as $Au_2Pd_1$/VMT.

To explore the optimal metal ratio, a series of AuPd/VMT catalysts were synthesized by changing the metal ratios in the above method. The corresponding samples were labeled $Au_xPd_y$/VMT, where "x/y" indicates the mole ratio of Au to Pd (i.e., 1/0, 0/1, 1/1, 1/2, 2/1, and 2.5/1).

### 3.3. Catalytic Reaction

Oxidation of HMF was performed in a Teflon-lined high-pressure reactor (100 mL capacity) with a magnetic stirrer, internal thermostat, and pressure display gauge. The high-pressure reactor core was added with 50.0 mg HMF, 50.0 mg catalyst, NaOH (NaOH/HMF = 4/1, mol/mol), and 40.0 mL deionized water. The high-pressure reactor was first blown three times with $O_2$ to remove as much air as possible and then pressurized to 2.0 MPa. The temperature was raised to a set point (typically 100 °C), and the reaction mixture was reacted at 600 rpm for 2 h. During the reaction, samples were taken once at 5 min, 30 min, 60 min, 90 min, and 120 min, respectively. After reaction, the catalyst in the reactor was first centrifuged and then washed with ethanol and deionized water to remove the residual organic matter and base on the catalyst surface, followed by drying in an oven at 60 degrees Celsius for 12 h when the catalyst was ready for the next cycle. Because a small amount of catalyst is lost during centrifugation, a little bit of fresh catalyst was added before the next cycle to keep the catalyst at 50 mg.The samples were analyzed for HMF conversion and FDCA yield using an Agilent (Santa Clara, CA, USA) 1260 HPLC and a TC-C18 column (4.6 × 250.0 mm, 5.0 µm). The mobile phase consisted of acetonitrile in vial B and 0.1% acetic acid solution in vial D in a ratio of 8:92. The mobile phase flow rate was 0.6 mL/min, the temperature of the column was kept at 30 °C, and the UV wavelength was set at 280 nm. In order to identify and quantify the compounds, standard curves were drawn using standard solutions prepared from purchased pure HMF, HMFCA, FFCA, and FDCA, and the instrument was calibrated. The actual test sample was the sample solution diluted 100 times and filtered through a 0.22 µm × 13 mm microporous filter membrane. The sampler had a sampling volume of 10.0 µL, and the sample was tested in 15 min. After the test was completed, the following formula was used to calculate the content changes in HMF and FDCA during the reaction process:

$$\text{HMF conversion } (\%) = \frac{\text{mole of converted HMF}}{\text{mole of initial HMF}} \times 100\% \quad (1)$$

$$\text{HMFCA, FFCA or FDCA yield } (\%) = \frac{\text{mole of obtained HMFCA, FFCA or FDCA}}{\text{mole of initial HMF}} \times 100\% \quad (2)$$

## 4. Conclusions

In summary, we successfully prepared a series of $Au_xPd_y$/VMT bimetallic catalysts with different proportions using the sol–gel method with natural mineral vermiculite used as the carrier, among which $Au_2Pd_1$/VMT showed the best catalytic performance. Using $Au_2Pd_1$/VMT as a catalyst, the reaction time, temperature, $O_2$ pressure, catalyst dosage, and base type were optimized. A 100% HMF conversion and 99.9% FDCA yield were obtained, and the stability of the catalyst was good, with the FDCA yield remaining at 90.1% after five catalytic cycles. In this catalytic system, HMF was transformed into the target product FDCA through the "HMF-HMFCA-FFCA-FDCA path". The high dispersion and alloy effect of the bimetallic nanoparticles promoted the activation of the reactants and intermediates, and the reaction substrates and intermediates were quickly converted into FDCA. The synthesis method of $Au_2Pd_1$/VMT catalyst prepared in this study is simple, and the introduction of VMT carrier reduces the cost of precious metal catalyst, which is expected to be applied to actual industrial production, providing ideas and references for the development and utilization of natural minerals, and also providing a new way to realize the efficient conversion of HMF to FDCA under green conditions.

**Supplementary Materials:** The following supporting information can be downloaded at https://www.mdpi.com/article/10.3390/catal14120949/s1. Figure S1. SEM images of VMT (a); SEM diagram of $Au_2Pd_1$/VMT catalyst (b). Figure S2. TEM mapping of $Au_2Pd_1$/VMT catalyst. Figure S3. TEM mapping of $Au_2Pd_1$/VMT catalyst after reaction. Figure S4. TEM images of catalysts with different metal ratios. Au/VMT (a); Pd/VMT (b); $Au_1Pd_2$/VMT (c); $Au_1Pd_1$/VMT (d); $Au_2Pd_1$/VMT (e); $Au_{2.5}Pd_1$/VMT (f). Figure S5. The reaction path of HMF to FDCA over $Au_2Pd_1$/VMT catalyst. Figure S6. Reaction path diagram of the chemical equation for the conversion of HMF to FDCA over $Au_2Pd_1$/VMT catalyst. Table S1. Compare the catalytic performance with other references. Table S2. Kinetic parameters of HMF ($lnk_1$), HMFCA ($lnk_2$) and FFCA ($lnk_3$) conversion in water over $Au_2Pd_1$/VMT catalyst. Table S3. Kinetic parameters of HMF ($lnk_4$), HMFCA ($lnk_5$) and FFCA ($lnk_6$) conversion in water over Au/VMT catalyst. Table S4. Kinetic parameters of HMF ($lnk_7$), HMFCA ($lnk_8$) and FFCA ($lnk_3$) conversion in water over Pd/VMT catalyst. Table S5. Metal loading was determined by ICP-AES method. References [40,52–56] are cited in the Supplementary Materials.

**Author Contributions:** Conceptualization: Y.C.; Methodology, Y.L. (Yiran Liu) and W.G.; Validation, Q.X.; Investigation, H.L. and Y.L. (Yiwang Li); Data curation, H.L. and Y.L. (Yiwang Li); Writing—original draft, H.L. and Y.L. (Yiwang Li); Funding acquisition, Y.C. All authors have read and agreed to the published version of the manuscript.

**Funding:** This research was funded by the Collaborative Innovation Center for Water Treatment Technology and Materials, the Special Fund of Henan Key Laboratory of Water Pollution Control and Rehabilitation Technology (CJSZ2024010), and the College Students Innovative Practice Plan of Jiangsu University (202410299160Y).

**Data Availability Statement:** All data are available to download within the database.

**Conflicts of Interest:** The authors declare no conflicts of interest.

## References

1. Biniek-Poskart, A.; Sajdak, M.; Skrzyniarz, M.; Rzacki, J.; Skibinski, A.; Zajemska, M. The Application of Lignocellulosic Biomass Waste in the Iron and Steel Industry in the Context of Challenges Related to the Energy Crisis. *Energies* **2023**, *16*, 6662. [CrossRef]
2. Vávrová, K.; Knápek, J.; Weger, J. Short-term boosting of biomass energy sources—Determination of biomass potential for prevention of regional crisis situations. *Renew. Sustain. Energy Rev.* **2017**, *67*, 426–436. [CrossRef]
3. Zandalinas, S.I.; Fritschi, F.B.; Mittler, R. Global Warming, Climate Change, and Environmental Pollution: Recipe for a Multifactorial Stress Combination Disaster. *Trends Plant Sci.* **2021**, *26*, 588–599. [CrossRef]
4. Wei, L.; Yang, H.; Niu, Y.; Zhang, Y.; Xu, L.; Chai, X. Wheat biomass, yield, and straw-grain ratio estimation from multi-temporal UAV-based RGB and multispectral images. *Biosyst. Eng.* **2023**, *234*, 187–205. [CrossRef]
5. Zhou, M.; Fakayode, O.A.; Ren, M.; Li, H.; Liang, J.; Zhou, C. Green and sustainable extraction of lignin by deep eutectic solvent, its antioxidant activity, and applications in the food industry. *Crit. Rev. Food Sci. Nutr.* **2024**, *64*, 7201–7219. [CrossRef]
6. Ji, Q.; Yu, X.; Yagoub, A.G.A.; Chen, L.; Zhou, C. Efficient removal of lignin from vegetable wastes by ultrasonic and microwave-assisted treatment with ternary deep eutectic solvent. *Ind. Crops Prod.* **2020**, *149*, 112357. [CrossRef]

7. Voigt, C.; Zwikowics, C. Glucanocellulosic biomass: Learning from marine biomass to optimize terrestrial biomass conversion. *N. Biotechnol.* **2014**, *31*, S159–S160. [CrossRef]
8. Dutta, S.; De, S.; Saha, B. Advances in biomass transformation to 5-hydroxymethylfurfural and mechanistic aspects. *Biomass Bioenergy* **2013**, *55*, 355–369. [CrossRef]
9. Li, M.; Jiang, H.; Zhang, L.; Yu, X.; Liu, H.; Yagoub, A.A.; Zhou, C. Synthesis of 5-HMF from an ultrasound-ionic liquid pretreated sugarcane bagasse by using a microwave-solid acid/ionic liquid system. *Ind. Crops Prod.* **2020**, *149*, 112361. [CrossRef]
10. Gong, C.; Meng, X.; Jin, C.; Yang, M.; Liu, J.; Sheng, K.; Pu, Y.; Ragauskas, A.; Ji, G.; Zhang, X. Green synthesis of cellulose formate and its efficient conversion into 5-hydroxymethylfurfural. *Ind. Crops Prod.* **2023**, *192*, 115985. [CrossRef]
11. He, O.; Zhang, Y.; Wang, P.; Liu, L.; Wang, Q.; Yang, N.; Li, W.; Champagne, P.; Yu, H. Experimental and Kinetic Study on the Production of Furfural and HMF from Glucose. *Catalysts* **2021**, *11*, 11. [CrossRef]
12. Bello, S.; Salim, I.; Méndez-Trelles, P.; Rodil, E.; Feijoo, G.; Moreira, M.T. Environmental sustainability assessment of HMF and FDCA production from lignocellutosic biomass through life cycle assessment (LCA). *Holzforschung* **2019**, *73*, 105–115. [CrossRef]
13. Wang, J.; Liu, X.; Jia, Z.; Sun, L.; Zhu, J. Highly crystalline polyesters synthesized from furandicarboxylic acid (FDCA): Potential bio-based engineering plastic. *Eur. Polym. J.* **2018**, *109*, 379–390. [CrossRef]
14. Heo, J.B.; Lee, Y.S.; Chung, C.H. Conversion of inulin-rich raw plant biomass to 2,5-furandicarboxylic acid (FDCA): Progress and challenge towards biorenewable plastics. *Biotechnol. Adv.* **2021**, *53*, 107838. [CrossRef]
15. Kim, H.; Choi, J.; Won, W. Process synthesis and analysis of green plastic monomer production from cellulose. *J. Clean. Prod.* **2020**, *277*, 124072. [CrossRef]
16. Hwang, K.R.; Jeon, W.; Lee, S.Y.; Kim, M.S.; Park, Y.K. Sustainable bioplastics: Recent progress in the production of bio-building blocks for the bio-based next-generation polymer PEF. *Chem. Eng. J.* **2020**, *390*, 124636. [CrossRef]
17. Hou, Y.; He, Z.; Ren, S.; Wu, W. Catalytic Oxidation of Biomass to Formic Acid under $O_2$ with Homogeneous Catalysts. *Acta Phys.-Chim. Sin.* **2023**, *39*, 2212065. [CrossRef]
18. Ren, M.; Kong, F.; Zhou, C.; Fakayode, O.A.; Liang, J.; Li, H.; Zhou, M.; Fan, X. Green, one-pot biomass hierarchical utilization strategy for lignin-containing cellulose nanofibrils and fractionated lignin preparation. *Ind. Crops Prod.* **2023**, *203*, 117193. [CrossRef]
19. Chen, C.; Wang, L.; Zhu, B.; Zhou, Z.; El-Hout, S.I.; Yang, J.; Zhang, J. 2,5-Furandicarboxylic acid production via catalytic oxidation of 5-hydroxymethylfurfural: Catalysts, processes and reaction mechanism. *J. Energy Chem.* **2021**, *54*, 528–554. [CrossRef]
20. Han, X.; Li, C.; Liu, X.; Xia, Q.; Wang, Y. Selective oxidation of 5-hydroxymethylfurfural to 2,5-furandicarboxylic acid over MnOx-$CeO_2$ composite catalysts. *Green Chem.* **2017**, *19*, 996–1004. [CrossRef]
21. Rass, H.A.; Essayem, N.; Besson, M. Selective aqueous phase oxidation of 5-hydroxymethylfurfural to 2,5-furandicarboxylic acid over Pt/C catalysts: Influence of the base and effect of bismuth promotion. *Green Chem.* **2013**, *15*, 2240–2251. [CrossRef]
22. Zhang, Y.; Cao, Y.; Yan, C.; Liu, W.; Chen, Y.; Guan, W.; Wang, F.; Liu, Y.; Huo, P. Rationally designed Au-ZrOx interaction for boosting 5-hydroxymethylfurfural oxidation. *Chem. Eng. J.* **2023**, *459*, 141644. [CrossRef]
23. Wang, T.; Song, Y.; Zhao, W.; Zhou, C.; Jin, Y.; Wan, X.; Dai, Y.; Yang, Y. Electro-catalytic oxidation of HMF to FDCA over $RuO_2/MnO_2$/CNT catalysts in base-free solution. *New J. Chem.* **2021**, *45*, 21285–21292. [CrossRef]
24. Zhang, Y.; Liu, Y.; Guan, W.; Cao, M.; Chen, Y.; Huo, P. Strong metal-support interaction between AuPd nanoparticles and oxygen-rich defect $ZrO_2$ for enhanced catalytic 5-hydroxymethylfurfural oxidation. *Chin. Chem. Lett.* **2024**, *35*, 108932. [CrossRef]
25. Jing, T.; Yang, S.; Feng, Y.; Li, T.; Zuo, Y.; Rao, D. Selective and effective oxidation of 5-hydroxymethylfurfural by tuning the intermediates adsorption on Co-Cu-$CN_x$. *Nano Res.* **2023**, *16*, 6670–6678. [CrossRef]
26. Rapado, P.; Lois, C.; Cano, M.; Faba, L.; Ordóñez, S. Biomass to plastics: From HMF to fructose for the synthesis of 2,5-furandicarboxylic acid over Au/$TiO_2$. *Catal. Today* **2025**, *445*, 115036. [CrossRef]
27. Liu, Y.; Chen, Y.; Guan, W.; Cao, Y.; Wang, F.; Zhang, Y. Mechanistic Studies into the Selective Production of 2,5-furandicarboxylic Acid from 2,5-bis(hydroxymethyl)furan Using Au-Pd Bimetallic Catalyst Supported on Nitrated Carbon Material. *Catalysts* **2023**, *13*, 435. [CrossRef]
28. Cheng, X.; Li, S.; Liu, S.; Xin, Y.; Yang, J.; Chen, B.; Liu, H. Highly efficient catalytic oxidation of 5-hydroxymethylfurfural to 2,5-furandicarboxylic acid using bimetallic Pt-Cu alloy nanoparticles as catalysts. *Chem. Commun.* **2022**, *58*, 1183–1186. [CrossRef]
29. Zhang, Y.; Xue, Z.; Wang, J.; Zhao, X.; Deng, Y.; Zhao, W.; Mu, T. Controlled deposition of Pt nanoparticles on $Fe_3O_4$@ carbon microspheres for efficient oxidation of 5-hydroxymethylfurfural. *Rsc Adv.* **2016**, *6*, 51229–51237. [CrossRef]
30. Seehamongkol, T.; Rungtaweevoranit, B.; Chakthranont, P.; Butburee, T.; Nimsaila, W.; Faungnawakij, K. Roles of Metal-Organic Framework Supports in Base-Free Oxidation of 5-Hydroxymethylfurfural to Furan-2,5-Dicarboxylic Acid over Pt-Based Catalysts. *Chemnanomat* **2024**, *10*, e202400037. [CrossRef]
31. Liao, W.; Zhu, Z.; Chen, N.; Su, T.; Deng, C.; Zhao, Y.; Ren, W.; Lü, H. Highly active bifunctional Pd-$Co_9S_8$/S-CNT catalysts for selective hydrogenolysis of 5-hydroxymethylfurfural to 2,5-dimethylfuran. *Mol. Catal.* **2020**, *482*, 110756. [CrossRef]
32. Gupta, K.; Rai, R.K.; Singh, S.K. Catalytic aerial oxidation of 5-hydroxymethyl-2-furfural to furan-2,5-dicarboxylic acid over Ni-Pd nanoparticles supported on $Mg(OH)_2$ nanoflakes for the synthesis of furan diesters. *Inorg. Chem. Front.* **2017**, *4*, 871–880. [CrossRef]
33. Zou, B.; Chen, X.; Xia, J.; Zhou, C. Alkaline Ionic Liquid Modified Pd/C Catalyst as an Efficient Catalyst for Oxidation of 5-Hydroxymethylfurfural. *J. Chem.* **2018**, *2018*, 1–9. [CrossRef]

34. Wang, T.; Zhang, J.; Xie, W.; Tang, Y.; Guo, D.; Ni, Y. Catalytic Transfer Hydrogenation of Biobased HMF to 2,5-Bis-(Hydroxymethyl) Furan over Ru/$Co_3O_4$. *Catalysts* **2017**, *7*, 92. [CrossRef]
35. Feng, L.; Li, X.; Lin, Y.; Liang, Y.; Chen, Y.; Zhou, W. Catalytic hydrogenation of 5-hydroxymethylfurfural to 2,5-dimethylfuran over Ru based catalyst: Effects of process parameters on conversion and products selectivity. *Renew. Energy* **2020**, *160*, 261–268. [CrossRef]
36. Zhang, J.; Chen, P.; Liu, H.; Zhang, R.; Jia, W.; Zhang, J.; Peng, L. Electron-rich ruthenium nanoparticles on nitrogen-doped carbon for the efficient catalytic oxidation of 5-hydroxymethylfurfural into 2,5-furandicarboxylic acid. *Appl. Catal. A-Gen.* **2024**, *676*, 119663. [CrossRef]
37. Lei, D.; Yu, K.; Li, M.; Wang, Y.; Wang, Q.; Liu, T.; Liu, P.; Lou, L.; Wang, G.; Liu, S. Facet Effect of Single-Crystalline Pd Nanocrystals for Aerobic Oxidation of 5-Hydroxymethyl-2-furfural. *ACS Catal.* **2017**, *7*, 421–432. [CrossRef]
38. Davis, S.E.; Houk, L.R.; Tamargo, E.C.; Datye, A.K.; Davis, R.J. Oxidation of 5-hydroxymethylfurfural over supported Pt, Pd and Au catalysts. *Catal. Today* **2011**, *160*, 55–60. [CrossRef]
39. Cho, A.; Byun, S.; Cho, J.H.; Kim, B.M. AuPd-$Fe_3O_4$ Nanoparticle-Catalyzed Synthesis of Furan-2,5-dimethylcarboxylate from 5-Hydroxymethylfurfural under Mild Conditions. *Chemsuschem* **2019**, *12*, 2310–2317. [CrossRef]
40. Zhong, X.; Yuan, P.; Wei, Y.; Liu, D.; Losic, D.; Li, M. Coupling Natural Halloysite Nanotubes and Bimetallic Pt-Au Alloy Nanoparticles for Highly Efficient and Selective Oxidation of 5-Hydroxymethylfurfural to 2,5-Furandicarboxylic Acid. *Acs Appl. Mater. Interfaces* **2022**, *14*, 3949–3960. [CrossRef] [PubMed]
41. German, D.; Pakrieva, E.; Kolobova, E.; Carabineiro, S.A.C.; Stucchi, M.; Villa, A.; Prati, L.; Bogdanchikova, N.; Corberán, V.C.; Pestryakov, A. Oxidation of 5-Hydroxymethylfurfural on Supported Ag, Au, Pd and Bimetallic Pd-Au Catalysts: Effect of the Support. *Catalysts* **2021**, *11*, 115. [CrossRef]
42. Wan, X.; Zhou, C.; Chen, J.; Deng, W.; Zhang, Q.; Yang, Y.; Wang, Y. Base-Free Aerobic Oxidation of 5-Hydroxymethyl-furfural to 2,5-Furandicarboxylic Acid in Water Catalyzed by Functionalized Carbon Nanotube-Supported Au-Pd Alloy Nanoparticles. *Acs Catal.* **2014**, *4*, 2175–2185. [CrossRef]
43. Zhang, B.; Zhu, H.; Zou, M.; Liu, X.; Yang, H.; Zhang, M.; Wu, W.; Yao, J.; Du, M. Design and fabrication of size-controlled Pt-Au bimetallic alloy nanostructure in carbon nanofibers: A bifunctional material for biosensors and the hydrogen evolution reaction. *J. Mater. Sci.* **2017**, *52*, 8207–8218. [CrossRef]
44. Shcherbina, N.F.; Kochetkova, T.V.; Bastrygina, S.V.; Grishin, N.N. Synthesis of High-Temperature Phases on Vermiculite Filler. *Glass Ceram.* **2014**, *71*, 201–204. [CrossRef]
45. Petersen, R.R.; Christensen, J.F.S.; Jorgensen, N.T.; Gustafson, S.; Lindbjerg, L.A.; Yue, Y.Z. Preparation and thermal properties of commercial vermiculite bonded with potassium silicate. *Thermochim. Acta* **2021**, *699*, 178926. [CrossRef]
46. De Haro, M.C.J.; Perez-Maqueda, L.A.; Stepkowska, E.T.; Martínez, J.M.; Pérez-Rodríguez, J.L. The influence of exchangeable cation on thermal behaviour of ground vermiculite. *J. Therm. Anal. Calorim.* **2003**, *71*, 761–771. [CrossRef]
47. Perez-Maqueda, L.A.; Maqueda, C.; Perez-Rodriguez, J.L.; Subrt, J.; Cerny, Z.; Balek, V. Thermal behaviour of ground and unground acid leached vermiculite. *J. Therm. Anal. Calorim.* **2012**, *107*, 431–438. [CrossRef]
48. Yamaguchi, A.; Tanaka, M.; Kurihara, Y.; Takahashi, Y. Local structure of strontium adsorbed on 2:1 clay minerals and its comparison with cesium by XAFS in terms of migration of their radioisotopes in the environment. *J. Radioanal. Nucl. Chem.* **2018**, *317*, 545–551. [CrossRef]
49. Balek, V.; Pérez-Rodríguez, J.L.; Pérez-Maqueda, L.A.; Subrt, J.; Poyato, J. Thermal behaviour of ground vermiculite. *J. Therm. Anal. Calorim.* **2007**, *88*, 819–823. [CrossRef]
50. Ulrich, V.; Moroz, B.; Pyrjaev, P.; Sinev, I.; Bukhtiyarov, A.; Gerasimov, E.; Bukhtiyarov, V.; Roldan Cuenya, B.; Grünert, W. Three-way catalysis with bimetallic supported Pd-Au catalysts: Gold as a poison and as a promotor. *Appl. Catal. B-Environ.* **2021**, *282*, 119614. [CrossRef]
51. Liu, Y.; Chen, Y.; Li, Y.; Guan, W.; Xia, Q.; Cao, M.; Huo, P.; Zhang, Y. Oxygen vacancy-driven strong metal-support interactions on AuPd/$TiO_2$ catalysts for high-efficient air-oxidation of 5-hydroxymethylfurfural. *Chem. Eng. J.* **2023**, *476*, 146874. [CrossRef]
52. Xia, H.; An, J.H.; Hong, M.; Xu, S.; Zhang, L.; Zuo, S. Aerobic oxidation of 5-hydroxymethylfurfural to 2,5-difurancarboxylic acid over Pd-Au nanoparticles supported on Mg-Al hydrotalcite. *Catal. Today* **2019**, *319*, 113–120. [CrossRef]
53. Wang, F.; Yan, C.; Jiang, R.; Chen, Y.; Wei, Y.; Cao, Y.; Guan, W.; Huo, P.; Zhang, Y. Highly efficient aerobic oxidation of 5-hydroxymethylfurfural to 2,5-fur- andicarboxylic acid over halloysite nanotubes-templated nitrogen-doped carbon supported bimetallic AuPd catalyst. *Appl. Clay Sci.* **2023**, *235*, 106872. [CrossRef]
54. Gao, Z.; Xie, R.; Fan, G.; Yang, L.; Li, F. Highly Efficient and Stable Bimetallic AuPd over La-Doped Ca-Mg-Al Layered Double Hydroxide for Base-Free Aerobic Oxidation of 5-Hydroxymethylfurfural in Water. *ACS Sustain. Chem. Eng.* **2017**, *5*, 5852–5861. [CrossRef]
55. Gao, T.; Yin, Y.; Fang, W.; Cao, Q. Highly dispersed ruthenium nanoparticles on hydroxyapatite as selective and reusable catalyst for aerobic oxidation of 5-hydroxymethylfurfural to 2, 5-furandicarboxylic acid under base-free conditions. *Mol. Catal.* **2018**, *450*, 55–64. [CrossRef]
56. Zhong, X.; Wei, Y.; Sadjadi, S.; Liu, D.; Li, M.; Yu, T.; Zhuang, G.; Yuan, P. Base-free oxidation of 5-hydroxymethylfurfural to 2, 5-furandicarboxylic acid over palygorskite-supported bimetallic Pt-Pd catalyst. *Appl. Clay Sci.* **2022**, *226*, 106574. [CrossRef]

catalysts

MDPI

Review

# Advances in the Preparation of Carrier-Based Composite Photocatalysts and Their Applications

Huiqin Wang [1], Chenlong Yan [1], Mengyang Xu [2] and Xianghai Song [2,*]

[1] School of Energy and Power Engineering, Jiangsu University, Zhenjiang 212013, China; hqwang@ujs.edu.cn (H.W.); ycl1527913077@126.com (C.Y.)
[2] Institute of Green Chemistry and Chemical Technology, School of Chemistry & Chemical Engineering, Jiangsu University, Zhenjiang 212013, China; xumy13243188936@163.com
* Correspondence: songxianghai@ujs.edu.cn

**Abstract:** Photocatalytic technology offers significant advantages in addressing water pollution and energy regeneration challenges. Notably, photocatalytic $CO_2$ reduction technology can convert $CO_2$ into stable, efficient, and clean carbon compounds such as carbon monoxide, methane, ethylene, and other high-value compounds, providing a novel approach to mitigating the global energy crisis and maintaining the carbon balance. However, traditional semiconductor photocatalytic materials face limitations in photocatalytic degradation and reduction due to their low light energy utilization, severe photocorrosion, rapid photogenerated carrier recombination, and slow electron transport rates. Recent studies have shown that introducing various carrier materials can effectively address these issues. Carrier materials, with their unique properties, enhance semiconductor composite photocatalyst systems, promoting photogenerated carrier separation and improving light energy utilization. This review introduces different carrier materials used in photocatalyst fabrication, systematically explains the preparation strategies for carrier-based composite photocatalysts, and summarizes their applications. Finally, future developments in this field are discussed. This review aims to provide diverse strategies for designing carrier-based photocatalysts, leveraging the special effects of carrier materials to control semiconductor composite modes, interface behaviors, and energy band structures.

**Keywords:** photocatalytic technology; carrier effect; carrier based photocatalyst

## 1. Introduction

Among the most pressing challenges for the sustainable development of global human society, environment and resource issues remain a central focus [1–4]. As human society becomes increasingly complex, large quantities of pesticides, persistent pollutants, and toxic wastes from industrial pollution are discharged, severely endangering water and soil environments. The global pandemics has further exacerbated this issue, leading to the widespread use of antibiotics, which pose a threat to public health and the environment [5–8]. The growing global population and increasing carbon consumption have resulted in a per capita carbon footprint exceeding safe limits [9]; the uncontrolled combustion of fossil energy fuels, primarily petroleum and coal, has plunged humanity into a dual crisis of resource depletion and global warming, leading to extreme weather events and ecosystem damage caused by the greenhouse effect [10–13]. Solar energy-driven green photocatalytic technology not only efficiently degrades organic pollutants in water but also provides a novel approach for reducing and converting $CO_2$ into hydrocarbon

 https://doi.org/10.3390/catal15030286

products [14–16]. The $CO_2$ molecule is highly stable (804.4 kJ $mol^{-1}$), and the $CO_2$ reduction reaction ($CO_2$ RR) is non-spontaneous ($\Delta G > 0$), making the process challenging. Traditional $CO_2$ conversion technologies, such as electrocatalysis, are energy-intensive, unstable, complex, and highly polluting [17–19]. Photocatalytic $CO_2$ reduction, however, can effectively utilize solar energy to convert $CO_2$ into stable, clean carbon-containing compounds such as carbon monoxide (CO), methane ($CH_4$), methanol ($CH_3OH$), ethylene ($C_2H_4$), and other carbonaceous energy sources used in daily life. Thus, photocatalytic technology not only purifies water by degrading organic pollutants but also helps reduce atmospheric $CO_2$ levels, alleviate the greenhouse effect, and produce clean energy, and it also enables the storage of renewable energy in reduction products for secondary use, helping to mitigate the global energy crisis and maintaining the natural carbon balance [20–22]. Therefore, there is an urgent need to develop high-efficiency photocatalyst materials for clean energy storage and conversion to accelerate the industrialization of environmental purification and clean energy substitution processes [23,24].

Constructing a composite photocatalytic material system is an effective way to address the limitations of single semiconductor photocatalysts, such as narrow light absorption ranges, mismatched energy band structures, severe photocorrosion, and rapid photogenerated carrier recombination [25–27]. Whether for the photocatalytic degradation of water pollutants or light-driven reduction of atmospheric $CO_2$, the effective contact area with the target pollutants or carbon source, as well as the adsorption capacity of the composite material, are critical factors influencing photocatalysts' performance [28–30]. The design of loaded composite catalysts has become a hot research topic in photocatalysis, as most carrier materials offer advantages such as large specific surface areas, strong adsorption capacities, and excellent physicochemical properties [31–33]. Traditionally, catalyst carriers have been used as skeletons to support and disperse active components, enhancing the material strength. However, common carrier materials such as alumina and silicon oxide and some natural products such as diatomaceous earth, pumice, and activated carbon or molecular sieve carriers are typically coupled with semiconductors through physicochemical methods, which only fix the semiconductor catalysts without strong interfacial interactions; these traditional carriers are also difficult to degrade, and often cause secondary pollution, limiting their ability to enhance a photocatalyst's activity [34–36]. Therefore, developing new semiconductor photocatalyst carrier materials with high specific surface area, adsorption property, and unique material properties is crucial for enhancing the interfacial carrier effect with the active semiconductor components [37–40].

Herein, this review begins by discussing the classification of photocatalyst carrier materials, followed by an exploration of different preparation strategies for carrier material-based composite photocatalysts and their application in environmental photocatalysis (Scheme 1). Finally, we present remarks and perspectives on the further development of carrier material-based composite photocatalysts. Through this review, we hope to provide useful insights for designing carrier-based composite photocatalysts.

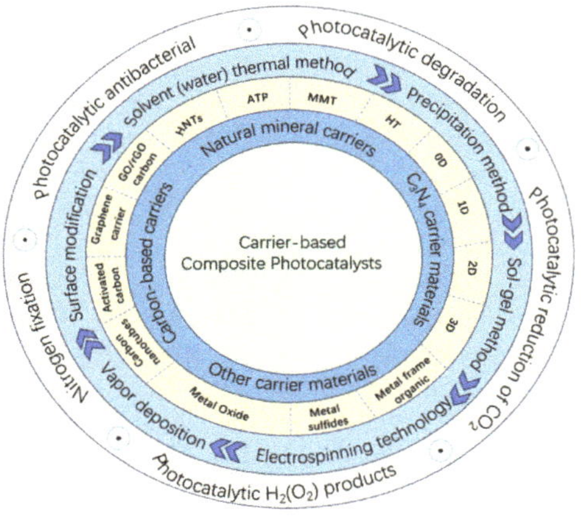


**Scheme 1.** A brief overview of the main applications over carrier material-based composite photocatalysts.

## 2. Classification of Carrier Materials for Photocatalysts

In recent years, an increasing number of carrier materials suitable for synthesizing compound semiconductor photocatalysts have been used in photocatalytic system design. Different preparation methods achieve various coupling effects, aiming to enhance the catalytic activity and overall performance of compound semiconductor photocatalysts. Generally, factors such as the catalytic efficiency, activity, carrier effect, loading stability, cycle life, cost, and price must be considered when selecting carrier materials for loaded photocatalysts. Based on published studies, carrier materials can be broadly classified into natural mineral carriers [41–43], carbon-based carriers [44–46], $C_3N_4$ carrier materials [47–49], and other types, including metal oxides [50–52], metal sulfides [53–55], metal frame organic compounds [56–58], and organic polymers [59–61]. Numerous studies have demonstrated the crucial role of carrier materials in eco-environmental and energy conversion applications such as degrading organic pollutants in water, photocatalytic hydrogen production, photocatalytic reduction, and the conversion of $CO_2$ [62–64], as well as nitrogen fixation and antimicrobial activity [65–68].

### 2.1. Natural Mineral Carriers

Natural minerals, widely distributed in nature, often possess excellent physicochemical properties and are commonly used as catalyst carrier materials. The natural mineral carriers used for synthesizing loaded photocatalytic materials are primarily layered inorganic silicate clay minerals, classified into kaolinite, monazite, vermiculite, hydromica, and fibrous barite based on the layer type, family, and subfamily [69]. These layered clay minerals, composed of aluminum–oxygen octahedral sheets and silica–oxygen tetrahedral sheets connected by shared oxygen, exhibit unique physical and chemical properties such as diverse pore structures, large specific surface areas, hydroxyl-rich inner and outer surfaces, high cation exchange capacities, and negative charges. These properties enable strong adsorption of heavy metals, organic pollutants, and gaseous molecules in water [70,71]. Selecting suitable natural mineral carriers and modifying them through washing, acidification, column support, and calcification can enhance the performance of loaded composite semiconductor photocatalysts, which are widely studied and applied in the photocatalytic degradation of pollutants and wastewater, as well as energy conversion [72,73]. The high-value natural mineral carriers include halloysite nanotubes (HNTs) [74], attapulgite (ATP) [75], montmorillonite (MMT) [76], and hydrotalcite (HT/LDHs) [77].

### 2.1.1. Halloysite Nanotubes

Halloysite nanotubes (HNTs) are naturally occurring aluminosilicate minerals with a layered structure, primarily found in the form of nanotubes. Due to differences in geological environments and crystallization conditions, HNTs can form flakes, tubes, and spherical structures. The tubular structure of HNTs consists of aluminum–oxygen octahedra and silica–oxygen tetrahedra, forming the inner and outer layers of the nanotubes [78]. HNTs are hydrated, with a single water molecule present between the layers, resulting in a layer spacing of approximately 10 Å (1 nm). The dimensions and morphologies of HNTs vary, with their inner diameters ranging from 10 to 100 nm, outer diameters from 30 to 190 nm, and lengths from 0.02 to 30 μm. The inner and outer walls of the nanotubes are exposed to Si-O-Si, Al-OH, and Si-OH groups [79].

In addition to their abundance, low cost, and environmental compatibility, HNTs possess a unique mesoporous nanotubular structure, a large specific surface area, and rich surface groups [80,81]. The opposite charges on the inner and outer surfaces enhance their adsorption efficiency, making HNTs suitable for various applications. HNTs have been extensively utilized in the treatment of water pollution, exhibiting remarkable adsorption capabilities for multiple contaminants, including antibiotic wastewater (e.g., hygromycin, tetracycline hydrochloride, and ciprofloxacin), organic dyes (e.g., methylene blue, rhodamine B, and gentian violet), heavy metal ions (e.g., chromium and lead), ammonium ions, and radioactive elements (e.g., uranium). HNTs can reach adsorption equilibrium quickly, making them a cost-effective adsorbent and carrier material with significant potential [82,83]. Modifying HNTs through physical or chemical methods can further enhance their adsorption performance and facilitate the construction of composite photocatalyst systems. Additionally, the natural hollow nanostructure of HNTs is beneficial for gas molecule adsorption and storage, making them suitable for photocatalytic hydrogen production, hydrogen storage, and $CO_2$ capture and reduction [84,85].

Lin et al. [86] successfully prepared $Cd_{0.5}Zn_{0.5}S$@HNT composites using HNTs as a matrix. They first grafted the hole-trapping agent EDTA onto the surfaces of HNTs and then loaded $Cd^{2+}/Zn^{2+}$ and $S^{2-}$ to form $Cd_{0.5}Zn_{0.5}S$ nanoparticles through electrostatic assembly. The hydrogen production rate of the EDTA-grafted nanotube composites was significantly higher than those of composites modified with NaCl and CTAB. Under visible light ($\lambda > 420$ nm) irradiation, the optimal hydrogen production rate of $Cd_{0.5}Zn_{0.5}S$@HNTs reached 25.67 mmol $g^{-1}$ $h^{-1}$, 7.03 times higher than that of pure $Cd_{0.5}Zn_{0.5}S$. The quantum efficiency (AQY) at $\lambda = 420 \pm 10$ nm was 32.29%. A characterization analysis showed that the good dispersibility of $Cd_{0.5}Zn_{0.5}S$@HNTs and efficient charge separation due to charge-directed migration contributed to the enhanced photocatalytic water splitting performance. Furthermore, the natural hollow nanostructure of HNTs serves as an excellent hydrogen storage container, with the composites capable of storing up to 0.042% hydrogen at ambient temperature (25 °C) and atmospheric pressure (2.65 MPa), achieving integrated hydrogen production and storage.

### 2.1.2. Attapulgite

Attapulgite (ATP), also known as palygorskite, is a hydrated magnesium–aluminum silicate with a unique fibrous crystal structure. ATP exhibits excellent adsorption, catalysis, and physicochemical stabilization properties [87–89]. Its internal and external surfaces contain Si-OH groups, and the substitution of $Si^{4+}$ by $Al^{3+}$ results in residual negative charges, enabling the effective adsorption of heavy metal ions, radionuclides, and gaseous substances such as $CO_2$ [90,91]. As a carrier material for composite photocatalysts, ATP can be functionally activated or modified to impart unique properties to composite materials, expanding their applications in photocatalysis [92,93]. Ma's team [94] utilized ATP's characteristics as a high-quality mineral carrier to uniformly deposit $Ag_3PO_4$ nanoparticles on the ATP's surface. The ATP-loaded $Ag_3PO_4$ nanoparticles effectively catalyzed the removal of Orange II from water under sunlight, reducing the Ag content in the composite photocatalyst from 77.3 wt% in pure $Ag_3PO_4$ to 18.8 wt%, significantly lowering the industrial costs while enhancing the photocatalytic degradation activity and stability. ATP's excellent $CO_2$ adsorption and trapping abilities have also been exploited in $CO_2$ catalytic conversion. Guo et al. [95] introduced ATP as a carrier in $CO_2$ immobilization reactions, designing and preparing Ag/ATP nanocomposite photocatalysts. These catalysts demonstrated high catalytic activity in the synthesis of asymmetric carbonates, with TOF values of up to 132 $h^{-1}$. The ATP-loaded composite photocatalysts maintained their activity over 10 consecutive reactions, showcasing their potential for industrial applications. This study not only provides an efficient and stable catalyst for carbon dioxide fixation but it highlights the role of ATP as an excellent mineral carrier and provides new ideas for the development of related multiphase catalysts.

#### 2.1.3. Montmorillonite

Montmorillonite is an excellent photocatalyst carrier due to its regular pore structure, large specific surface area, and superior adsorption properties [96–98]. Montmorillonite-supported composite photocatalysts, such as $TiO_2$/montmorillonite and g-$C_3N_4$/montmorillonite, combine the adsorption performance of the carrier with the photocatalytic performance of the semiconductor, resulting in higher photodegradation rates and broader application prospects. For example, $TiO_2$/montmorillonite composite photocatalysts achieve high degradation rates for organic pollutants such as methyl orange under visible light [99]. Similarly, g-$C_3N_4$/montmorillonite composites exhibit excellent visible light absorption and high antibacterial performance. Montmorillonite-supported composite photocatalysts represent a new type of photocatalytic material with significant potential for environmental treatment and energy conversion. For example, Liu et al. [100] achieved 90% degradation of fluoroquinolone antibiotics by stripping montmorillonite-supported S-scheme BiOBr/$Bi_2MO_6$ heterojunctions.

#### 2.1.4. Hydrotalcite

Hydrotalcite (LDHs), with its layered structure, provides a large specific surface area, facilitating the uniform dispersion and efficient utilization of photocatalysts. The interlayer ions in hydrotalcite can be exchanged, allowing photocatalytic active substances to be inserted into the interlayer, forming an intercalation structure that enhances the photocatalytic performance. Hydrotalcite carrier composite photocatalysts combine the layered structure and adsorption properties of hydrotalcite with the efficient photocatalytic performance of the photocatalyst, resulting in excellent photocatalytic activity. The common preparation methods for hydrotalcite composite photocatalysts include coprecipitation and hydrothermal methods, which regulate the structure and properties of the composite materials to achieve optimal photocatalytic effects. Hydrotalcite carrier composite photocatalysts have broad application prospects in the photocatalytic degradation of organic pollutants, water splitting for hydrogen production, and photocatalytic $CO_2$ reduction. Their high photocatalytic efficiency and stable chemical properties make them promising materials for environmental governance and energy conversion [101]. Zheng et al. [102] thermally designed ZnCdS/NiAl hydrotalcite S-type heterojunctions with highly efficient photocatalytic hydrogen precipitation activity levels, resulting in a several-fold improvement in hydrogen precipitation performance, about 7 times that of ZnCdS and 130 times that of NiAl LDH.

### 2.2. Carbon-Based Carriers

Carbonaceous materials have received wide attention in recent years due to their diversified structure, strong tunability, moderate price, good stability, and excellent conductivity characteristics, which have made carbon-based carriers a hot spot of nanocarrier research both at home and abroad, and they have played a significant role in various fields such as catalysis, energy storage and conversion, and biomedicine [103–105]. The use of carbon-based heteroatom coordination has stabilized transition metal single-atom catalysts, which have garnered significant attention in recent years. These catalysts present promising applications in the field of electrocatalysis related to energy conversion, owing to the excellent electrical conductivity of carbon-based carriers, the optimization of single-atom utilization rates, and the unique interactions between metal atoms and carrier materials. Carbon-based carriers have adjustable types and forms of heteroatom doping, which makes it possible to achieve the ideal design of highly active single-atom electrocatalysts by modulating the local atom distribution and coordination environment on the carbon-based carriers. Wu and Mei's group [106] jointly published a review to classify and summarize the research work on the modulation of the electronic structures of atomically dispersed transition metal sites on carbon-based carriers, introduce the modulation mechanisms involved, and electrocatalyze the oxygen reduction reaction (ORR), hydrogen precipitation reaction (HER), oxygen precipitation reaction (OER), and carbon dioxide reduction reaction ($CO_2$RR) based on carbon-based monatomic catalysts and the nitrogen reduction reaction (NRR), further describing the related research work on catalytic performance enhancement through local coordination environment modulation. The authors concluded that the development of a more intuitive structure–activity descriptor, as well as the full use of synergistic effects between atomically dispersed polymetallic sites, would help to rationalize the design and further improve the activity of such single-atom catalysts. Meanwhile, the stability of transition metal monatomic systems based on carbon-based carriers is another major challenge for the practical application of such materials, and the search for a balance between the high degree of graphitization of the carbonaceous carriers, the dense distribution of the metal sites, and the optimization of the coordinated atomic environment of the metal monatomic requires further exploration and the refinement of previous research efforts.

Based on the outstanding physicochemical properties of carbon-based materials, numerous carbons have been developed and designed as excellent carrier materials for various applications in photocatalysis research. The preferred carbon-based carrier materials are mainly oxidized and reduced graphene oxide (GO and rGO, respectively), single- and multi-walled carbon nanotubes (SWNTs and MWNTs, respectively), and graphyne developed from fullerene derivation, as well as carbon nanofibers, carbon nanospheres, and other geometrically shaped carbon-based nanocages ranging from 0D randomly filled nanocages (or hollow spheres) to 3D assemblies, which have opened up several new branches of research on carbon-based nanocarrier materials [107]. In addition, biomass carbon, chitosan, activated carbon, carbon black, dicyandiamide, and melamine, among others, can also be prepared as a class of highly promising carbon-based carriers for efficient CRR and HER catalysts [108].

#### 2.2.1. Graphene Carrier Materials

As a single-layer two-dimensional crystalline material with a unique honeycomb structure formed by Sp2 hybridized C atoms, graphene is fully utilized to enhance and improve the weak visible-light responsiveness and high photogenerated electron–hole complexation rates of semiconductor photocatalytic materials due to a series of properties, such as its large specific surface area, excellent electrical conductivity, the half-integer Hall effect, the unique quantum-limited-domain effect, and the bipolar electric field effect [109].

By compounding it with graphene, the forbidden bandwidth can be effectively regulated, and graphene-based carrier materials can not only be used as the carrier and behavioral receptor of semiconductors, precious metals, and other photoexcitation elements but also as highly efficient carriers of the electron transport network, which further creates favorable conditions for the transfer of electrons. Graphene can also be used as a photocatalyst or a co-catalyst to participate in the catalytic reaction, which can help to solve the problems of photocatalytic degradation, cracking, water hydrogen production, and reductive conversion. This provides a feasible way to solve the bottlenecks in photocatalytic degradation, hydrogen production from water, and reductive conversion. Therefore, the development and study of loaded graphene and graphene oxide-modified nanocomposites are of great value in the field of photocatalysis. Zhang's team at Tianjin Polytechnic University [110] reported a concise and effective synthesis strategy, using GO as a 2D template, firstly doping $Co^{2+}$ ions in GO, then grafting and stabilizing the metal coordination layer of CoMOF on the 2D GO template and preparing Co-MOL with $H_2L$ ligands (Co-MOL@GOs) using the in situ growth method, successfully constructing Co-MOL@GO with three metal coordination layers, while ultrathin 2D MOLs were also successfully constructed. As shown in Figure 1 for the related work and characterization results, it can be seen from the TEM images that small- and medium-sized nanosheets (15–20 nm) of Co-MOL@GO are uniformly distributed on the GO template, and the synergistic interaction between the ultrathin MOLs and the GO electron conductor greatly enhances the intrinsic photocatalytic $CO_2$ reduction activity, whereas reduced GO can be used as a 2D template to load and stabilize MOLs of uniform thickness (~1.5 nm). In addition, GO can act as an effective electron mediator to bridge the gap between multiphase catalysts and homogeneous antenna molecules, which greatly contributes to the high activity. Thanks to these advantages, Co-MOL@GO exhibits excellent photocatalytic activity and high selectivity in visible-light-driven $CO_2$ to CO conversion, with an all-time high 12 h total CO yield of 3133 $mmolg^{-1}mol^{-1}$ among all state-of-the-art MOF and MOL catalysts. This work opens a new path for the low-cost preparation of ultrathin MOLs with excellent performance and demonstrates the key role of carbon-based carrier graphene as an electron mediator in significantly promoting various aspects of photocatalytic performance.

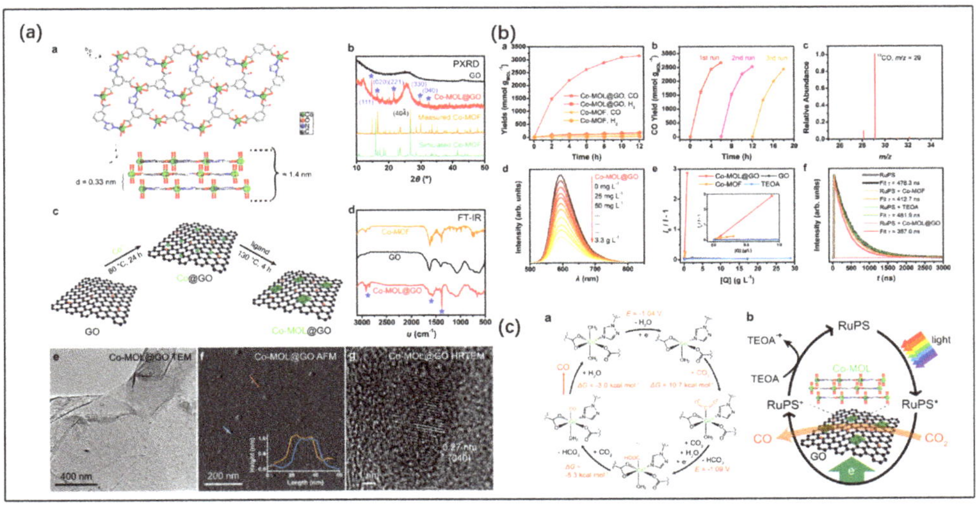


**Figure 1.** (**a**) Synthesis and characterization of samples, (**b**) photocatalytic $CO_2$ reduction, and (**c**) proposed mechanisms [110].

### 2.2.2. Carbon Nanotubes

The discovery of fullerenes, carbon nanotubes (CNTs), and graphene materials has strongly stimulated active research across multiple disciplines in the fields of nanotechnology, electronics, materials science and engineering, energy, and the environment in past decades, and these low-dimensional carbon nanostructured materials have aroused widespread interest among researchers and have shown great potential for a multitude of applications. Among them, Iijima [111] first reported the properties of carbon nanotubes in detail in 1991. Microscopically, CNTs are helical microtubules with one-dimensional (1D) graphitic carbon, with diameters down to a few nanometers. CNTs can be viewed as single-walled or multiple concentric layers of seamless multi-walled cylindrical tubes formed by convolutions of a honeycomb network of carbon hexagons. The presence of strong covalent C-C bonds along the tube axes gives CNTs ultrahigh stiffness (up to 1 TPa) and tensile strength (theoretically close to 100 GPa), whereas the presence of sp2 hybridization endows CNTs with impressive electrochemical properties (metal or semiconductor electronic behavior depending on the tube diameter and helicity). This unique combination of outstanding thermodynamic, electrical, and mechanical properties, along with a high degree of structural flexibility, distinguishes carbon nanotubes (CNTs) from most traditional bulk or micro- or nanomaterials composed of metals, alloys, polymers, ceramics, or other compounds. More typically, carbon nanotubes are used as ballistic transport quantum wires [112], the finest and toughest fatigue-resistant strings [113,114], and the strongest fiber materials ever made [115], as examples.

Depending on the number of layers of carbon atoms, carbon nanotubes can be classified as single-walled carbon nanotubes (SWCNT) or multi-walled carbon nanotubes (MWCNT), consisting of multilayered graphene sheets, which can be tailored to their specific properties for their application or use. Unlike graphene and many 2D materials that can be fabricated via mechanical or liquid exfoliation, carbon nanotubes are typically synthesized by using high-temperature chemical vapor deposition (CVD) processes to synthesize hydrocarbon and metal catalyst precursors, where problems such as excessive tube curvature are overcome by precipitating carbon from the catalyst nanoparticles to generate the associated strain energy. This has important implications for the macroscopic bottom-up construction of carbon nanotube functional materials (CNTFMs) and the pursuit of practical applications. In a sense, high-temperature environments (700–1100 °C) induce the formation of high-quality single-, double-, or multi-walled carbon nanotubes capable of maintaining excellent mechanical, electrical, and other properties, as well as structural stability in carbon nanotubes. In addition, due to van der Waals interactions between the intertwined high specific surface area carbon nanotubes, the preparatively grown carbon nanotubes are strongly orientated for 1D, 2D, or 3D self-assembly with well-aligned configurations and robust built-in interconnected structures.

Carbon nanotube fibers exhibit excellent thermal, mechanical, and electrical properties. In terms of conductivity, doping can broaden the intertube electron jump channels, potentially exceeding the conductivity limits of metal conductors, making them advantageous for lightweight wires. SWCNTs can grow up to several centimeters in length, exhibiting metallic or semiconducting properties, which have led to significant breakthroughs in integrated circuits and computer development. Moreover, the thermal conductivity of CNT–metal composites can greatly enhance the current-carrying capacity of conductors, potentially replacing traditional metal conductors in future ultrahigh-current applications. The unique assembly properties of CNTs also make thermal radiation on their surfaces particularly significant, which can be optimized by improving the phonon transport between nanotubes to regulate both the apparent and actual thermal conductivity. The electronic properties of SWCNTs, for example, depend on the chiral index (i.e., the way SWCNTs curl along their axes) formed during growth. Controlling the intrinsic factors of SWCNT selective growth contributes to the reproducibility and stability of chiral SWCNT preparation and scale-up applications. Figure 2 shows the work of Christophe Bichara's team [116] at the University of Aix-Marseille, France, who constructed a thermodynamic model linking the energy and temperature of the tube–catalyst interface to the chirality of carbon nanotubes. This model reveals that the chirality of SWCNTs originates from entropy-driven processes at the nanoscale boundaries, providing theoretical guidance for the selective growth of chiral SWCNTs.

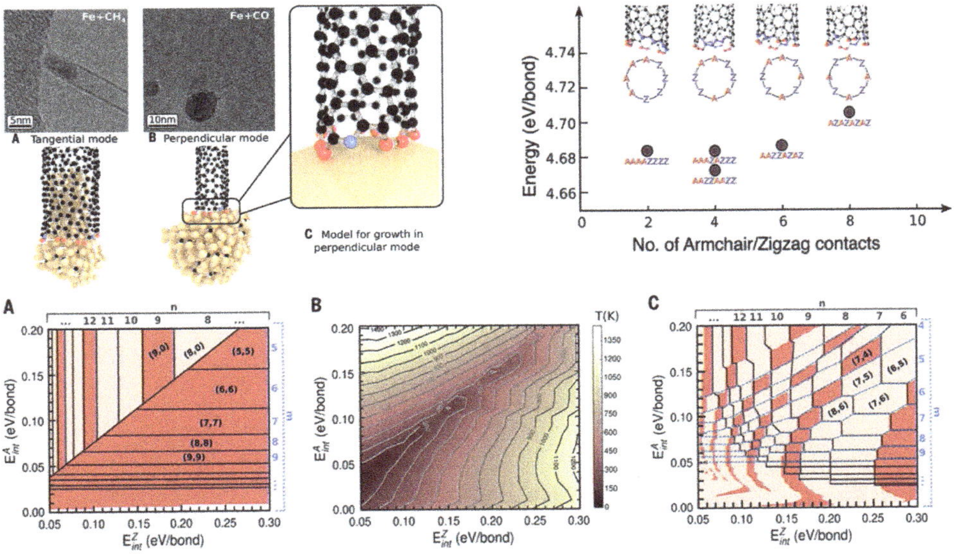


**Figure 2.** (**A**) From experiments to a model. (**B**) Key elements of the model and (**C**) structural maps [116].

In terms of mechanical properties, significant enhancements in fiber fracture strength and elastic modulus can be achieved through solvent densification, mechanical densification, stepwise drafting, the introduction of polymer network structures within the fibers, and induced covalent linkages between tubes. The rich interfacial structure within the fibers brings about diverse energy dissipation processes, giving carbon nanotube fibers (as well as thin films and composites) dynamic mechanical properties such as damping and creep, which are not found in traditional carbon fibers. This results in a bifunctional combination of rigidity and flexibility. Additionally, the yarn structure and unique flexibility of CNT fibers show unique advantages in areas such as rotary drives and bioelectrodes. Developing composite structural systems that fully utilize the unique properties of carbon nanotube bundle fiber materials is an important challenge for future carbon nanotube engineering.

Carbon nanotube fiber functional materials are an important research area for translating nanoscience and nanotechnology into practical applications, with a wide range of potential impacts in science, technology, and engineering. CNTs with entangled networks, low redox potentials, and high specific surface areas are ideal hosts and carriers for semiconducting composite photocatalyst materials, providing sufficient space for the dispersion and effective migration of photogenerated carriers during photo-redox processes. In particular, semiconducting single-walled carbon nanotubes (s-SWCNTs) with diameters of about 1.0–1.5 nm have band gaps comparable to silicon, and their inner surfaces have suitable curvature and large contact areas, making them attractive for host–guest chemical reactions triggered by electron transfer. This makes them ideal materials for photoelectrocatalytic applications. Therefore, the research is focused on deriving and composing macroscopic composite or layered materials with the different characteristics of carbon nanotube fibers, combined with their compatible structures, properties, and prospective application areas, such as mechanically robust conductive skeletons, high-performance flexible electrodes, advanced energy conversion and storage systems, and environmental applications. The development of high-performance, low-cost carbon-nanotube-based composite carrier catalyst materials with multifunctional catalytic properties will promote the future development of this field. Chen et al. [117] first constructed a novel bifunctional hybrid device comprising CoSnS@CNTs, using carbon nanotubes as a carrier through a simple, surfactant-free one-pot hydrothermal method. As shown in Figure 3, CoSnS@CNTs are composed of ultrathin CoSnS uniformly and tightly immobilized on a highly conductive CNT backbone with a unique hybridized three-dimensional (3D) porous nanostructure, which not only provides abundant catalytic sites for ion diffusion but also accelerates electron mobility. With these advantages, CoSnS@CNTs have become bifunctional catalysts with enhanced electrocatalytic and photocatalytic properties, providing strong oxygen evolution reaction (OER) performance and excellent stability at a low overpotential of 330 mV and a current density of 10 mA/cm$^2$, and a photocatalytic degradation rate for rhodamine B dye of up to 91.72%, which is twice that of pure CoSnS. The introduction of carbon nanotube carriers effectively enhanced the photoelectrocatalytic activity of the non-precious metal-based bifunctional catalysts.

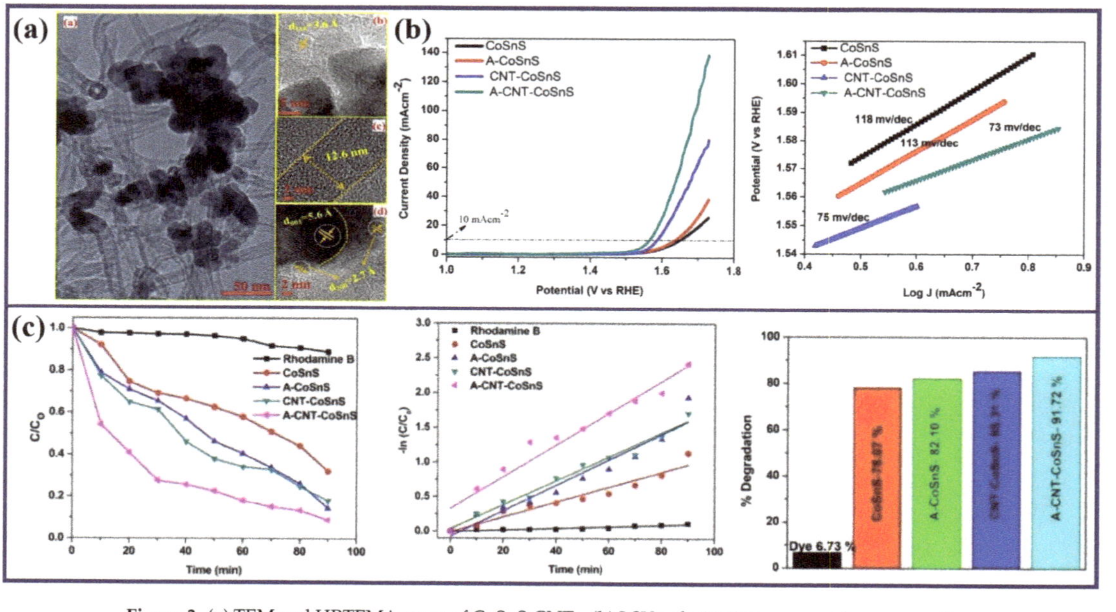


**Figure 3.** (**a**) TEM and HRTEM images of CoSnS-CNTs. (**b**) LSV polarization curve and corresponding Tafel plots. (**c**) RhB dye photodegradation efficiency [117].

### 2.2.3. Activated Carbon

Activated carbon is widely used in water purification processes due to its low cost, large specific surface area, and numerous micropores. The raw materials for activated carbon can be any carbon-rich organic matter, including coal, asphalt, ore, wood, and fruit shells (such as coconut shells, walnut shells, and apricot shells). These characteristics make activated carbon versatile. Rice-husk-based activated carbon exhibits the capability to adsorb various dyes, including methyl orange, methylene blue, and rhodamine B, as well as heavy metal ions such as chromium (Cr), cadmium (Cd), iron (Fe), and manganese (Mn). Additionally, it has applications in supercapacitors, dye-sensitized batteries, and hydrogen storage systems. Activated carbon from Chinese parasol leaves can adsorb antibiotics, phenols, and other pharmaceuticals, while coconut shell activated carbon (AC5) shows excellent adsorption and desorption properties for carbon dioxide.

### *2.3. $C_3N_4$ Carrier Material*

#### 2.3.1. One-Dimensional Carbon Nitride

One-dimensional (1D) carbon nitride photocatalysts, including nanorods, nanotubes, nanowires, and nanoribbons, have attracted much attention because of their excellent optical and chemical properties and ability for enhance photocatalytic activity by tuning the diameter, length, and aspect ratio. Li et al. [118] synthesized g-$C_3N_4$ nanorods with an average diameter of 260 nm through the thermal polymerization of nitrile in the nanochannel of an anodized alumina film template. In addition, Li's team [119] used SBA-15 nanorods as a hard template to fabricate mesoporous g-$C_3N_4$ nanorods with a length of about 650 nm and a diameter of 100 nm. Zheng et al. [120] synthesized helical g-$C_3N_4$ nanorods using mesoporous silica as a template.

#### 2.3.2. Two-Dimensional Sheet Carbon Nitride

Bulk-phase g-$C_3N_4$ is formed by the aggregation of sheet g-$C_3N_4$, which reduces the specific surface area and carrier migration rate. However, sheet g-$C_3N_4$ has unique photocatalytic properties, such as a larger surface area, more active sites, longer charge carrier lifetime, and higher photoexcited electron reduction potential. Additionally, flaky g-$C_3N_4$ can effectively disperse particles, making it a subject of wide attention. Zhang et al. [121] synthesized ultrathin g-$C_3N_4$ nanosheets using a simple liquid-phase stripping method. This experiment demonstrated that a single g-$C_3N_4$ nanosheet can remain stably suspended for up to a week under both acidic and alkaline conditions. Lin et al. [122] studied a mixed solvent's effect on the stripping of bulk-phase g-$C_3N_4$, obtaining ultrathin g-$C_3N_4$ nanosheets. Liu et al. [123] treated the precursor dicyandiamide with ammonium chloride, drying the aqueous solution and calcining the mixture at 823 K for 4 h to obtain sheet g-$C_3N_4$. These methods provide an important basis for the preparation of g-$C_3N_4$ nanosheets with high specific surface areas, which are widely used.

Graphite-like semiconductor materials with two-dimensional structures present a new opportunity to construct carrier composite photocatalysts, fully utilizing their carrier effects. Two-dimensional layered graphitic carbon nitride (g-$C_3N_4$) is a high-quality non-metallic semiconductor material that has shown excellent performance in catalytic chemistry, materials chemistry, and photoelectron chemistry [124,125]. Compared to bulk-phase materials and other general carrier materials, the two-dimensional structure of g-$C_3N_4$ offers advantages such as smaller thickness and larger specific surface area values, providing numerous surface atoms as catalytic active sites to improve catalytic processes and for enhanced catalytic activity. By further reducing the thickness of 2D g-$C_3N_4$, it is easier to prepare ultrathin 2D g-$C_3N_4$ materials for surface modification, which can enrich the structure of the carrier catalyst model and optimize the catalytic performance. Ultrathin layered g-$C_3N_4$, as a representative two-dimensional structure of graphitic-phase carbon and nitrogen materials, has a higher specific surface area compared to bulk-phase g-$C_3N_4$ due to its low nano-size thickness, two-dimensional anisotropy, and quantum confinement effect, effectively improving the electron mobility efficiency and prolonging the lifetime of photo-excited carriers. The design of composite catalyst systems supported by ultrathin two-dimensional g-$C_3N_4$ exposes a significant number of coordination-unsaturated surface atoms, thereby enriching the active sites for catalytic reactions. This enhancement makes the material widely applicable in various fields, including photocatalysis, electrocatalysis, photoelectrocatalysis, organocatalysis, and other catalytic applications. Therefore, it is suitable for g-$C_3N_4$ to act as a carrier and composite with other materials to further enhance its carrier effect, significantly improving the photocatalytic efficiency and increasing the conversion and reduction efficiency of $CO_2$ [126]. Various types of two-dimensional g-$C_3N_4$ materials have been successfully prepared via stripping, liquid-phase synthesis, and chemical oxidation. However, pristine g-$C_3N_4$ still faces challenges due to its small specific surface area, poor long-wavelength light absorption, easy recombination of photogenerated electron–hole pairs, and low quantum efficiency, which limit its photocatalytic performance and application. Utilizing doping modification, the photoelectric properties and band structure of catalytic materials can be optimized, and the chemical composition of catalysts can be adjusted to improve the photocatalytic performance, making it one of the most commonly used modification methods for developing high-efficiency two-dimensional composite g-$C_3N_4$ photocatalysts. Wang's group developed a series of metal-doped g-$C_3N_4$ catalysts, which exhibited photocatalytic performance several times greater than that of pure g-$C_3N_4$ in the degradation of dye wastewater, such as rhodamine B, and in the decomposition of water for hydrogen production [127]. Ion doping changes the electronic structure of g-$C_3N_4$ semiconductors, adjusting the semiconductor valence band to a more suitable potential energy position in the reaction direction, enhancing the electronic band structure and redox ability of the catalyst and improving the visible-light absorption and separation of photogenerated electron–hole pairs, resulting in a significant increase in photocatalytic activity [128,129].

However, from the perspective of carrier function, ordinary 2D g-$C_3N_4$ materials have a relatively single-layer structure, and their interfacial behavior is relatively weak when compounded with other semiconductors or metal or non-metal materials compared to traditional 2D carrier materials. This hinders the research application of 2D ultrathin carbon nitride materials in the field of composite photocatalysts. Therefore, it is necessary to design and develop novel ultrathin 2D g-$C_3N_4$ materials to fully utilize their 2D structural advantages through surface modification, regulate the electronic structure more precisely, optimize the catalytic performance, and expand the types of composite 2D g-$C_3N_4$ photocatalytic materials. Among the numerous modification strategies, point defect engineering (i.e., tunable vacancy and dopant introduction) can exploit the excellent structural, optical, and electronic properties of g-$C_3N_4$ to improve the photocatalytic activity. While heteroelements are doped, structural defects are inevitably introduced. For example, Zhang et al. [130] found defect energy levels in the energy band structure of P-doped g-$C_3N_4$ catalysts; some researchers found that O doping of g-$C_3N_4$ was accompanied by the introduction of new defect energy levels [131], and these defect energy levels play a crucial role in improving the photocatalytic activity. Liu et al. [132] prepared nitrogen-deficient g-$C_3N_4$, where the introduction of nitrogen defects formed defect energy levels below the conduction band, narrowing the semiconductor's forbidden bandwidth, effectively broadening the light absorption range of g-$C_3N_4$, and improving the visible photocatalytic activity for hydrogen production. Surface defect studies provide new research ideas for developing novel loaded ultrathin two-dimensional g-$C_3N_4$ composite photocatalysts and a solid theoretical basis for further enhancing the catalytic activity of defective two-dimensional g-$C_3N_4$ materials (Figure 4).

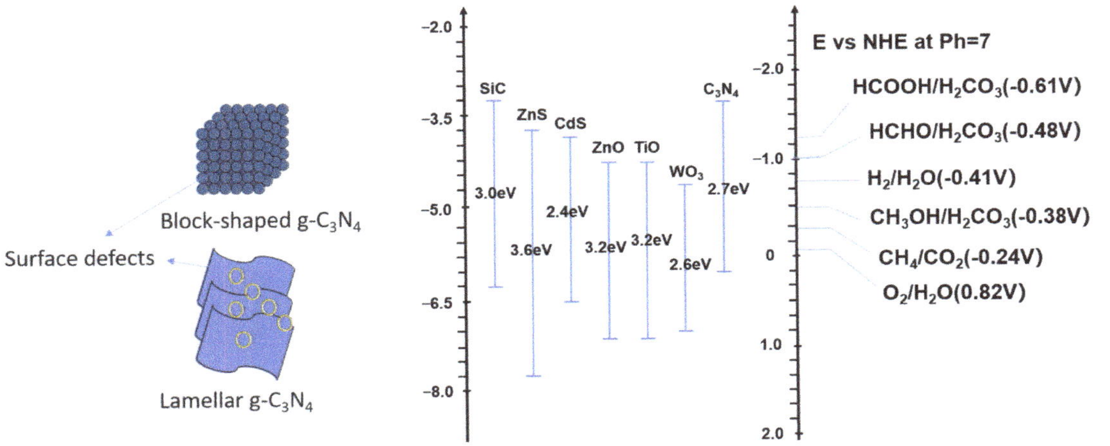


**Figure 4.** Schematic of 2D g-$C_3N_4$ surface defect active sites and band energy values for different semiconductors.

Given the booming development in this field, it is timely to review and study the recent advances in g-$C_3N_4$ point defect engineering, particularly the rational utilization of two-dimensional g-$C_3N_4$, an ideal artificial photosynthesis non-metallic photocatalyst, for photocatalytic energy storage and conversion. To address the inherent drawbacks of g-$C_3N_4$, Yu et al. [133] detailed the role, synthesis, characterization, and systematic control of point defects, as well as the wide range of applications of defective g-$C_3N_4$-based nanomaterials for photocatalytic water decomposition, carbon dioxide reduction, and nitrogen fixation. A systematic review of g-$C_3N_4$ point defect engineering is presented, including the central role of point defects in g-$C_3N_4$, synthesis strategies, characterization, and the application of defective g-$C_3N_4$-based nanomaterials in many photo-redox energy processes. It is evident from the large number of research activities that the introduction of point defects, either as cavities or dopants, plays a powerful role in aiding the surface modification and optical and electrical properties of g-$C_3N_4$, thereby improving its photocatalytic performance in water decomposition, carbon dioxide reduction, and nitrogen fixation. Overall, g-$C_3N_4$ nanomaterials undergo remarkable changes upon the introduction of point defects, such as tunable band gaps, defect-induced mid-gap states, increased surface areas, the inhibition of photogenerated electron–hole recombination, and the improved adsorption and activation of reactant molecules. The work on point defect engineering in g-$C_3N_4$ is still underdeveloped, and there is a lack of reliable methods to accurately and uniformly engineer defects. Notably, a common synthetic problem with g-$C_3N_4$ is its tendency to aggregate, which requires a layering process to achieve the uniform introduction of defects. Additionally, the synthesis methods should be low-cost and scalable to allow the controlled large-scale production of defective catalysts, which remains a significant challenge. Another obstacle is the limitations of the existing material characterization methods in accurately identifying and quantifying g-$C_3N_4$ defects. Revealing the stability of defects could play an important role in elucidating the effective mechanism of defects on catalytic activity. The relationship between defects and the effects of different morphologies of g-$C_3N_4$ on photocatalytic activity require further exploration, both experimentally and computationally, in terms of their analysis. Further studies of g-$C_3N_4$ defects could be focused on the design of defect-integrated composite structures. In addition, the combination of photodriving with other electrochemical or biocatalytic reactions is an innovative and advanced concept (Figure 5).

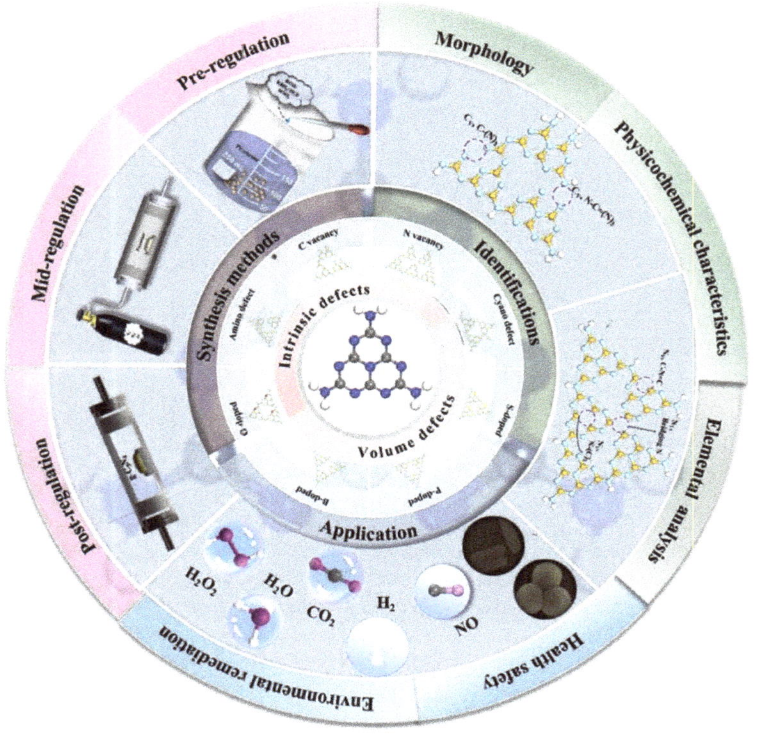


**Figure 5.** The applications of defective g-$C_3N_4$ and various health and safety aspects, along with the characterization of defect types and the synthesis methods of various defects [133].

Photocatalytic total water splitting can be achieved using Z-type heterojunction catalysts by connecting different semiconductors in series to mimic natural photosynthesis. However, it is still challenging to prepare catalyst materials that catalyze the production of both $H_2$ and $O_2$ efficiently. Shen et al. [134] addressed this issue by designing boron-doped, nitrogen-deficient two-dimensional (2D) $C_3N_4$ nanosheets based on boron-doped, nitrogen-deficient $C_3N_4$ nanosheets for the photocatalytic total water splitting study. As shown in Figure 6, the authors prepared ultrathin carbon nitride nanosheets with varying degrees of boron doping and nitrogen defects, allowing them to act as photocatalysts for $H_2$ or $O_2$ generation. Using an electrostatic self-assembly strategy, the nanosheets were coupled to obtain 2D/2D polymer heterostructures. Due to their ultrathin nanostructures, strong interfacial interactions, and staggered energy band arrangement, an efficient Z-type heterojunction pathway for charge–carrier separation and transfer can be achieved. In the presence of Pt and $Co(OH)_2$ co-catalysts, the resulting heterostructures achieve stoichiometric $H_2$ and $O_2$ release. The hydrogen production efficiency reached 1.16% under solar illumination.

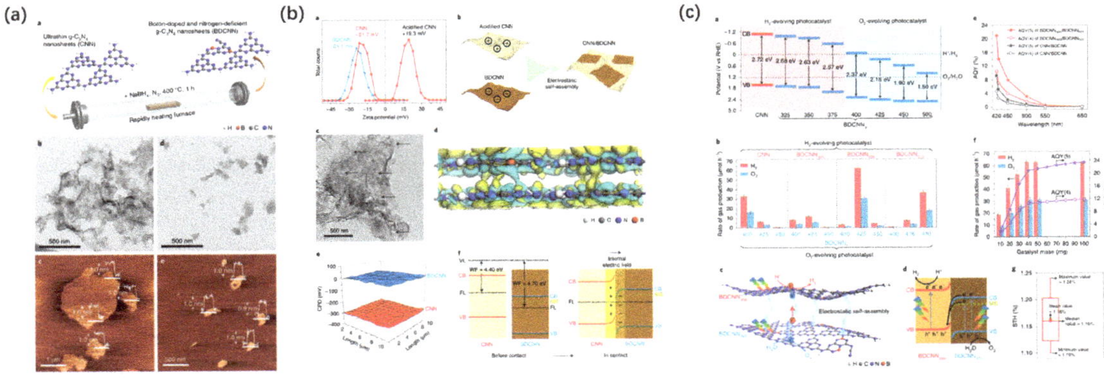


**Figure 6.** (**a**) Catalyst synthesis and microscopy. (**b**) Construction of Z-scheme heterostructures. (**c**) Band structure modulation of the self-based Z-scheme heterostructures [134].

#### 2.3.3. Three-Dimensional Carbon Nitride

The development of ordered and macro-sized 3D g-$C_3N_4$ components is of great significance for the practical application of photocatalysis. Liang et al. [135] demonstrated a template method by using a melamine sponge as a support frame, filling the sponge voids with urea, and then obtaining a 3D porous g-$C_3N_4$ material through thermal polymerization. Wang et al. [136] reported the synthesis of porous g-$C_3N_4$ using calcium carbonate ($CaCO_3$) particles as a hard template, providing a green pathway instead of using toxic and expensive fluorine-containing etchers. Additionally, Wang et al. [137] synthesized mesoporous g-$C_3N_4$ through the thermal polymerization of dicyandiamide with various soft templates, such as non-ionic surfactants, amphiphilic block polymers, and ionic surfactants.

### *2.4. Other Carrier Materials*

In addition, there are many other common inorganic semiconductor carriers with photocatalytic activity such as metal oxides, metal sulfides, and metal frame organic compounds, while semiconductor photocatalyst carriers also include alumina, silica gel, ceramics, and polymers. Alumina is one of the most widely used catalyst carriers. Alumina, with its large specific surface area. has a developed pore structure, which can make the active components of the supported catalyst highly dispersed into particles. Silica gel is a porous material and is one of the commonly used catalyst carriers. Ceramic is also a kind of porous material, which shows good adhesion to photocatalyst particles, good acid and alkali resistance, and high temperature resistance, and can also be used as a catalyst carrier. Polymers, such as saturated carbon chain polymer and fluoropolymer, have strong antioxidant capacity rates, so they can also be used in the study of supported photocatalysts. However, due to the strong oxidation of $\cdot OH^-$ and $\cdot O_2^-$, these polymer carriers can only be used in the short term. Glass carriers such as glass sheets, glass fiber nets (cloth), and hollow glass beads, due to their strong chemical stability and good light transmission, can be processed into various shapes of reactors according to one's needs, are also widely used as carriers of semiconductor photocatalysts.

When investigating how natural mineral carrier materials (e.g., hydrotalcite, bumpy clay, montmorillonite), carbon-based materials (e.g., carbon nanotubes, graphene), and carbon-nitride materials can specifically enhance the photocatalytic activity at the electronic level, the charge transfer kinetics, modulation of the energy–band structure, and interfacial effects have to be considered together. These materials work synergistically through different mechanisms to enhance the overall performance; natural mineral materials optimize the charge separation and transport pathways by modulating the metal composition of the laminate or via intercalation modification (e.g., hydrotalcite improves the charge separation efficiency by adjusting the Mg/Al ratio, while bumpy clay and montmorillonite improve the charge transfer capability via surface modification or as composites with other materials). Carbon-based materials significantly enhance the charge separation and transfer efficiency by virtue of their excellent electrical conductivity and tunable energy band structure. Graphene and carbon nanotubes further optimize the charge separation through doping or functionalization and form heterojunctions with semiconductor materials to promote charge transfer at the interface. Carbon nitride materials (g-$C_3N_4$) modulate the energy band structure by introducing defect states or compositing with other materials to extend their light absorption range and improve their charge separation efficiency. In addition, the interfacial interactions between these materials form an effective heterojunction, which not only improves the charge separation efficiency but also reduces the chance of electron–hole complexation. Taken together, the overall performance of the photocatalysts can be significantly enhanced by rationally designing the material compositions, structures, and interfacial interactions, thereby causing efficient and stable photocatalytic reactions. This multidimensional optimization strategy provides a solid theoretical foundation and technical guidance for the development of high-performance photocatalysts (Table 1).

**Table 1.** Key parameters of different photocatalysts.

| Material | Efficiency (Relative) | Light Absorption Range (nm) | Charge Separation Capability | Practical Application |
|---|---|---|---|---|
| Hydrotalcite | Moderate | UV | Ordinary | Environmental purification |
| Attapulgite | Slightly below | UV | Ordinary | Absorbent |
| Montmorillonite | Slightly below | UV | Ordinary | Absorbent |
| Carbon nanotubes | High | visible area | Large | Composite photocatalysts |
| Graphene carrier materials | High | Visible region and near infrared | Large | Composite photocatalysts |
| g-$C_3N_4$ | Medium–high | visible area | Slightly higher | Environmental purification |

The following table shows the photocatalytic efficiency of some different carrier materials. For natural mineral materials, when used as photocatalysts alone, they often face problems such as low utilization rates of visible light and easy compounding of the electron–hole pairs. They usually have to be combined with other semiconductors, and the photocatalytic efficiency is low. Carbon-based materials such as graphene and carbon nanotubes themselves have excellent conductive properties and are easy to modify for various operations. They have excellent mechanical properties and their charge separation ability is excellent, and often carbon-based materials will show good photocatalytic efficiency. Regardless of the dimensionality of $C_3N_4$ materials, they show unique advantages in the field of photocatalysis. One-dimensional materials excel in directional transport, two-dimensional materials possess excellent optical and electrical properties, and three-dimensional materials stand out with their porous structure and good stability. Therefore, the photocatalytic efficiency is also high.

## 3. Preparation of Carrier-Based Composite Photocatalytic Materials

### 3.1. Solvent (Water) Thermal Method

The hydrothermal method involves placing reactants in a sealed reaction vessel with an aqueous solution or other solvents as the reaction medium. By heating the reaction vessel (autoclave), the reactants are subjected to high temperature and pressure levels, which typically results in the dissolution of insoluble reactants and crystal growth. The hydrothermal method offers several advantages: (1) The high-temperature and high-pressure environment accelerates the thermodynamic reaction rate compared to room-temperature conditions [138]. (2) By coordinating reaction conditions such as the pH, reaction time, temperature, concentration, and reactant ratios, products with different structures, shapes, and particle sizes can be obtained [139]. (3) It enables the production of highly homogeneous nanocrystalline particles [140]. (4) With advancements in technology, the hydrothermal method can be combined with other techniques, such as microwave hydrothermal and microemulsion-assisted hydrothermal methods, expanding its application scope [141]. For carrier-based composite photocatalysts, the hydrothermal method not only enhances the stability of the composite but also optimizes and improves carrier materials by leveraging the high-temperature and high-pressure synthesis environment.

### 3.2. Precipitation Method

The precipitation method is a widely used technique for synthesizing nanomaterials. It involves adding a precipitating agent to a mixed solution to precipitate nanoparticles of different components, followed by heating, ultrasound, or other processing methods to obtain the desired product. Common precipitating agents include alkaline substances such as ammonia, urea, sodium bicarbonate, sodium carbonate, and sodium hydroxide. This method allows control over the crystalline phase, shape, and size of the resulting product by adjusting the amount of precipitating agent, the rate of precipitation, and temperature. It also includes efficient variations such as co-precipitation. The nano-precipitation method, used in biomedicine, catalysis, and electrochemistry, offers simplicity, uniform dispersion, and stable properties, making it particularly suitable for preparing loaded composites.

### 3.3. Sol–Gel Method

The sol–gel method involves using chemically active compounds as precursors. These precursors are uniformly mixed in a liquid phase, where hydrolysis and condensation reactions occur, forming a stable transparent sol. As the sol ages, gel particles slowly polymerize, creating a three-dimensional network structure filled with solvent. After drying, the solvent is removed, leaving a porous dry gel, which is then sintered to produce molecular or nanostructured materials. Sol–gel technology is a mature process with several advantages, including simplicity, low equipment costs, energy efficiency, and the ability to dope a wide range of materials. It allows for homogeneous mixing at the atomic and molecular levels, making it ideal for producing uniform materials and doping. This method is widely used in anti-corrosion, biomedical applications, and electronic devices. For carrier photocatalysts, the sol–gel method offers unique benefits. The precursor is dispersed in a solvent to form a low-viscosity solution, enabling molecular-level homogeneity in a short time. During gel formation, the reactants can be uniformly mixed with the carrier at the molecular level, resulting in more homogeneous loading and higher material purity. Additionally, the sol–gel process operates at lower temperatures, allowing better control over the material composition, especially for multi-component materials. It also enables control over the porosity of the carrier, enhancing the performance and stability of composite photocatalytic materials.

### 3.4. Electrospinning Technology

Electrospinning technology produces nanofibers with a large specific surface area, high porosity level, diverse composition, and uniform diameter distribution. These nanofibers can be easily functionalized (e.g., surface coating or modification) and have significant applications in biomedicine, environmental engineering, and textiles. For instance, electrospun nanofibers are used in optical and resistance sensors, water treatment, protective clothing, masks, catalysts, wound dressings, tissue engineering scaffolds, and drug carriers. For photocatalyst carrier materials, electrospinning is an excellent method for preparing composites. It effectively distributes nanofibers on the carrier surface, allows flexible adjustment of the nanomaterial size for different photocatalytic applications, and facilitates functional processing. By combining electrospinning with photocatalytic technology, the unique advantages of both methods can be leveraged to maximize their applications in environmental energy and other fields.

### 3.5. Vapor Deposition

Chemical vapor deposition (CVD) involves gaseous or vapor substances reacting in the gas phase or at the gas–solid interface to form solid deposits. Two or more gas sources are introduced into a reaction chamber, where they react to produce new materials, such as synthetic coatings or nanomaterials, which are deposited on a substrate. CVD is widely used in the semiconductor industry and includes reactions such as thermal decomposition, chemical synthesis, and chemical transport. CVD offers several advantages, including the ability to coat complex shapes and deep holes uniformly; deposit a wide range of materials (metallic, non-metallic, and multi-component alloys; ceramics; and chemical compounds); and produce films with high purity, good densification, low residual stress, and complete crystallization. By adjusting the deposition parameters, the chemical composition, morphology, crystal structure, and grain size of the coating can be controlled. CVD is also used in the preparation and optimization of high-performance graphene films and carbon nanotubes, making it valuable in designing carrier-based semiconductor photocatalytic systems.

### 3.6. Surface Modification

The surface modification of carrier materials can significantly enhance the catalytic efficiency of composite photocatalysts. For semiconductor carriers such as graphitic carbon nitride (g-$C_3N_4$), surface modification leverages the interfacial coordination effect between quantum dots and the carrier matrix. Carbon dot (CD)-modified multiphase nanocomposites offer broad spectral absorption, suitable electronic band structures, fast carrier mobility, abundant reserves, excellent chemical stability, and easy synthesis, making them promising for photocatalytic solar fuel production and pollutant decomposition. The process of Cr(VI) removal from an aqueous solution using hydrophobic Schiff base modified $TiO_2$ nanoparticles (HSB@$TiO_2$) under metal halide lamp irradiation was carried out as in the study by Mohamed et al. [142]. HSB@$TiO_2$ composites were synthesized, which greatly improved the degradation rate and mitigated the Cr(VI) contamination in light water sources.

### 3.7. Ball Milling

The mechanochemical method utilizes the strong impact, shear, and friction effects of the grinding media (usually hard spheres) on the raw materials in a high-speed rotating ball mill, which leads to particle refinement, lattice defect formation, and local temperature increases, conditions that can promote chemical reactions. Compared to traditional methods, mechanochemical methods do not require large amounts of organic solvents, are environmentally friendly, involve low energy consumption rates, and are easy to operate, making them suitable for large-scale production. However, it is sometimes difficult to precisely control the particle size distribution of the products, which may affect the consistency and stability of the photocatalytic performance (Table 2).

Table 2. Advantages and disadvantages of different synthesis methods.

| Synthesis Method | Advantages | Disadvantages |
|---|---|---|
| Solvent (water) thermal method | High crystallinity and good particle size control | Higher energy consumption |
| Precipitation method | Higher energy consumption | Low product purity |
| Sol–gel method | Uniformity of composition and variety of forms | High cost and complex drying process |
| Electrospinning technology | Controlled fiber morphology | Limited production |
| Vapor deposition | Continuous production with high purity | Process is complex |
| Surface modification | Flexible and long life | Flexible and long life |
| Ball milling | Environmentally friendly, good nano-sizing effect | Many crystal defects |

## 4. Application of Carrier-Based Composite Photocatalytic Materials in Environmental Photocatalysis

### 4.1. Photocatalytic Degradation

The morphology and dimensions of carrier materials directly affect their specific surface area, which influences the interfacial loading behavior of composites and their contact area with reactants or pollutants. The characteristics and adsorption capacity of the carrier are crucial for the performance of composite photocatalysts in liquid environments. Whether using natural clay mineral carriers, carbon-based carriers, or biomass carriers, different carrier materials exhibit unique features in the photocatalytic degradation of organic pollutants.

A key challenge is identifying suitable carrier materials and modifying them to enhance the performance of composite photocatalysts. For example, the presence of antibiotics in water bodies has become a major environmental concern due to their toxic effects and the risk of spreading drug-resistant genes. The use of photocatalytic technology, capable of generating highly oxidative radicals, is recognized as an efficient and sustainable method for degrading such pollutants. Qu et al. [143] optimized the degradation of lincomycin using a g-$C_3N_4$-based, metal-free composite photocatalytic system. Through doping with oxygen and modification with carbon quantum dots (CDs) loaded on reduced graphene oxide (rGO), they achieved a 10-fold increase in degradation compared to intrinsic g-$C_3N_4$. The study also elucidated the catalytic mechanism, highlighting the roles of intermediate $H_2O_2$, rGO as an active site, and the peroxidase-like activity of CDs.

### 4.2. Photocatalytic Reduction of $CO_2$

The search for efficient and stable photocatalytic materials for $CO_2$ reduction has led to the development of noble metal-free molecular catalysts hybridized with carbon-based light-absorbing carriers. For example, Song et al. [144] constructed a gold-modified CN-based donor-acceptor system coupled with a dual photothermal effect for the effective photoreduction of carbon dioxide.

Although challenges remain, such as understanding the electronic coupling between the complex and the carrier, this hybrid material represents an ideal system for studying catalytic reactions, combining the stability of solid materials with the high activity and selectivity of molecular catalysts.

### 4.3. Photocatalytic $H_2(O_2)$ Products

Photocatalytic water splitting is a promising method for producing renewable hydrogen. Carrier-based composite photocatalysts are widely studied for this purpose. For instance, Professor Zhu Yongfa's team [145] constructed a g-$C_3N_4$/rGO/PDIP Z-scheme heterojunction with high charge transfer efficiency and photocatalytic water decomposition activity levels. The composite exhibited hydrogen and oxygen evolution rates 12.1 times higher than pure g-$C_3N_4$, demonstrating the potential of carrier materials for enhancing the photocatalytic performance.

### 4.4. Nitrogen Fixation

Ammonia ($NH_3$) is a vital industrial chemical but its conventional production via the Haber–Bosch process is energy-intensive and polluting. Photocatalytic nitrogen fixation offers a sustainable alternative under mild conditions. However, developing high-performance catalysts remains challenging.

Efficient photocatalytic nitrogen fixation using boron-doped graphene quantum dot photocatalysts was achieved by Liu et al. [146]. This catalyst exhibited ultrahigh catalytic activity for $NH_3$ synthesis, with excellent stability and selectivity. The study demonstrated the potential of GDY-based catalysts in photocatalytic nitrogen fixation, offering a new approach to sustainable ammonia production.

### 4.5. Photocatalytic Antibacterial Agents

Microbial contamination, including from drug-resistant strains, is a growing concern. Silver-based nanomaterials are effective antibacterial agents due to their cytotoxicity and broad-spectrum activity. For example, the bioactive carbon-modified magnesium oxide particles developed by Muhammad et al. [147] have excellent photocatalytic antimicrobial properties. Similarly, the surface plasmon resonance technique designed by Wang et al. [148] produces excellent antimicrobial properties for enhancing graphene heterojunction materials.

## 5. Conclusions and Prospects

This paper reviews the construction of supported photocatalysts with different types of carrier materials, systematically introducing a series of composite photocatalyst materials with different functional carrier effects and their applications in the photocatalytic degradation of organic pollutants, photoreduction of $CO_2$, photocatalytic production of hydrogen and oxygen, nitrogen fixation, and enhancement of antibacterial activity. However, there are still shortcomings in the construction of carrier composite photocatalytic material systems, material characterization, the application of carrier effects, and the practical application of photocatalysis. Further suggestions for constructing composite photocatalytic materials via the functional carrier effect are as follows:

(1) In terms of the coupling between carrier materials and other materials in the composite material system, the advantages of carrier materials should be further utilized for the compositing of different semiconductors or other modified materials, reducing rejection phenomena and enhancing the degree of compositing and synergistic effect between carrier materials and other materials, achieving the accurate optimization of the internal balance of the carrier migration in different types of carrier-based composite photocatalytic material systems.

(2) Most composite photocatalyst materials are multi-dimensional composite material systems, conducive to directional and accelerated charge transfer, although the synthesis process and preparation methods of materials in various dimensions, especially the combination of three-dimensional and low-dimensional materials, need further optimization. Ensuring the integrity of the dominant morphology, structure, and properties of monomer materials in the composite system is crucial for improving the performance and stability of carrier-based composite photocatalysts.

(3) Accurate monitoring for material band structure modulation and the intuitive characterization of surface defect states are relatively lacking, and the specific interface behavior of each material in the composite photocatalytic system is relatively vague. The carrier migration and transport behavior processes need further exploration.

Carrier-based photocatalysts show great potential for application in the fields of environmental purification and energy conversion but still face many challenges in their practical applications. Catalysts may be affected by chemical changes in specific environments (e.g., strong acids and bases), so we should select materials with high chemical stability. In photodegradation, the generation of by-products may cause secondary pollution, and parameters such as the temperature and pH should be optimized. As the activity of photocatalysts tends to decrease with time, we should study effective regeneration methods, such as high-temperature calcination and chemical cleaning. Small batches of high-performance photocatalysts prepared under laboratory conditions are often difficult to be scaled up directly to industrial-scale production, facing problems related to cost control and technology transfer. We could explore synthesis processes suitable for large-scale production, such as continuous flow reactor and spray drying methods, to ensure product quality. There is a need to establish a sound quality inspection and control system to monitor key indicators in real time and ensure batch-to-batch consistency.

**Author Contributions:** H.W.: Investigation, Writing-Original Draft. C.Y.: Investigation, Validation. M.X.: Validation, Methodology. Supervision, Conceptualization. X.S.: Writing-review & editing, Funding acquisition. All authors have read and agreed to the published version of the manuscript.

**Funding:** This research received no external funding.

**Data Availability Statement:** No new data were created or analyzed in this study.

**Conflicts of Interest:** The authors declare no conflict of interest.

## References

1. Xu, Y.; Hassan, M.; Ali, S.; Li, H.; Ou, O.; Chen, Q. Self-cleaning-mediated SERS chip coupled chemometric algorithms for detection and photocatalytic degradation of pesticides in food. *J. Agric. Food. Chem.* **2021**, *69*, 1667–1674. [CrossRef] [PubMed]
2. Alexandratos, S.; Barak, N.; Bauer, D.; Davidson, F.; Gibney, B.; Hubbard, S.; Westerhof, P. Sustaining water resources: Environmental and economic impact. *ACS Sustain. Chem. Eng.* **2019**, *7*, 2879–2888. [CrossRef]
3. Jiang, L.; Wei, W.; Liu, S.; Haruna, S.; Zareef, M.; Ahmad, W.; Hassan, M.; Chen, Q. A tailorable and recyclable $TiO_2$ NFSF/Ti@ Ag NPs SERS substrate fabricated by a facile method and its applications in prohibited fish drugs detection. *J. Food Meas. Charact.* **2022**, *16*, 2890–2898. [CrossRef]
4. Xu, B.; Zhang, M.; Bhandari, B.; Sun, J.; Gao, Z. Infusion of $CO_2$ in a solid food: A novel method to enhance the low-frequency ultrasound effect on immersion freezing process. *Innov. Food Sci. Emerg. Technol.* **2016**, *35*, 194–203. [CrossRef]
5. Zhang, W.; Wang, X.; Xu, C.; Chen, Y.; Sun, W.; Liu, Q.; Dong, Q. Modeling inhibition effects of Lactobacillus plantarum subsp. plantarum CICC 6257 on growth of Listeria monocytogenes in ground pork stored at $CO_2$-rich atmospheres. *LWT* **2018**, *97*, 811–817. [CrossRef]
6. Akhlaq, M.; Zhang, C.; Yan, H.; Ou, M.; Zhang, W.; Liang, S.; Ikram, R. Response of tomato growth to continuous elevated $CO_2$ concentration under controlled environment. *Int. J. Agri. Biol. Eng.* **2022**, *15*, 51–59. [CrossRef]
7. Zhang, C.; Akhlaq, M.; Yan, H.; Ni, Y.; Liang, S.; Zhou, J.; Li, J. Chlorophyll fluorescence parameter as a predictor of tomato growth and yield under $CO_2$ enrichment in protective cultivation. *Agric. Water Manag.* **2023**, *284*, 108333. [CrossRef]
8. Khan, K.; Tang, Y.; Cheng, P.; Song, Y.; Li, X.; Lou, J.; Iqbal, B.; Zhao, X.; Hameed, R.; Li, G.; et al. Effects of degradable and non-degradable microplastics and oxytetracycline co-exposure on soil $N_2O$ and $CO_2$ emissions. *Appl. Soil Ecol.* **2024**, *197*, 105331. [CrossRef]
9. Lu, Y.; Zheng, J.; Hu, H.; Pan, Q.; Cui, L.; Hu, Y. Determination of critical crop water stress index of tea under drought stress based on the intercellular $CO_2$ concentration. *Agronomy* **2024**, *14*, 2154. [CrossRef]
10. Hayat, A.; Sohail, M.; Taha, T.A.; Alenad, A.M.; Uddin, I.; Hayat, A.; Ali, T.; Shah, R.; Irfan, A.; Khan, W. A superficial intramolecular alignment of carbon nitride through conjugated monomer for optimized photocatalytic $CO_2$ reduction. *Catalysts* **2021**, *11*, 935. [CrossRef]

11. Ye, L.; Wang, H.; Jin, X.; Su, Y.; Wang, D.; Xie, H.; Liu, X.; Liu, X. Synthesis of olive-green few-layered BiOI for efficient photoreduction of $CO_2$ into solar fuels under visible/near-infrared light. *Sol. Energy Mater. Sol. Cells* **2016**, *144*, 732–739. [CrossRef]
12. Wang, X.; Liu, X.; Liu, G.; Zhang, C.; Liu, G.; Xu, S.; Cui, P.; Li, D. Rapid synthesis of BiOCl graded microspheres with highly exposed (110) facets and oxygen vacancies at room temperature to enhance visible light photocatalytic activity. *Catal. Commun.* **2019**, *130*, 105769. [CrossRef]
13. Pellegrino, F.; Sordello, F.; Mino, L.; Hodoroaba, V.; Martra, G.; Maurino, V. Formic acid photoreforming for hydrogen production on shape-controlled anatase $TiO_2$ nanoparticles: Assessment of the role of fluorides, {101}/{001} surfaces ratio, and platinization. *ACS Catal.* **2019**, *9*, 6692–6697. [CrossRef]
14. Wang, K.; Endo-Kimura, M.; Belchi, R.; Zhang, D.; Habert, A.; Boucle, J.; Ohtani, B.; Kowalska, E.; Herlin-Boime, N. Carbon/graphene-modified titania with enhanced photocatalytic activity under UV and vis irradiation. *Materials* **2019**, *12*, 4158. [CrossRef]
15. Wu, H.; Han, D.; Hu, B.; Liang, J.; Tang, X.; Zhu, Z.; Huo, P. Insights into enhanced photocatalytic $CO_2$ reduction on carbon nitride: A strategy of simultaneous O, S co-doping. *J. Environ. Chem. Eng.* **2025**, *13*, 116121. [CrossRef]
16. Ruiz, C.; Rincón, L.; Contreras, R.; Sidney, C.; Almarza, J. Sustainable and negative carbon footprint Solid-based NaOH technology for $CO_2$ capture. *ACS Sustain. Chem. Eng.* **2020**, *8*, 19003–19012. [CrossRef]
17. Schuur, E.; Vogel, J.; Crummer, K.; Lee, H.; Sickman, J.; Osterkamp, T. The effect of permafrost thaw on old carbon release and net carbon exchange from tundra. *Nature* **2009**, *459*, 556–559. [CrossRef] [PubMed]
18. Gil, L.; Bernardo, J. An approach to energy and climate issues aiming at carbon neutrality. *Renew. Energy Focus* **2020**, *33*, 37–42. [CrossRef]
19. Chi, C.; Khosrowabadi, K.; Sheehan, S. Progress toward commercial application of electrochemical carbon dioxide reduction. *Chem* **2018**, *4*, 2571–2586.
20. Raza, A.; Hu, Y.; Lu, Y. Improving carbon flux estimation in tea plantation ecosystems: A machine learning ensemble approach. *Eur. J. Agron.* **2024**, *160*, 127297. [CrossRef]
21. Liu, Y.; Jing, Z.; Zhang, T.; Chen, Q.; Qiu, F.; Peng, Y.; Tang, S. Fabrication of functional biomass carbon aerogels derived from sisal fibers for application in selenium extraction. *Food Bioprod. Process.* **2018**, *111*, 93–103. [CrossRef]
22. Ma, S.; Wang, M.; You, T.; Wang, K. Using magnetic multiwalled carbon nanotubes as modified QuEChERS adsorbent for simultaneous determination of multiple mycotoxins in grains by UPLC-MS/MS. *J. Agric. Food Chem.* **2019**, *67*, 8035–8044. [CrossRef]
23. Weng, B.; Qi, M.; Han, C.; Tang, Z.; Xu, Y. Photocorrosion inhibition of semiconductor-based photocatalysts: Basic principle, current development and future perspective. *ACS Catal.* **2019**, *9*, 4642–4687. [CrossRef]
24. Miao, Z.; Wang, Q.; Zhang, Y.; Meng, L.; Wang, X. In situ construction of S-scheme AgBr/BiOBr heterojunction with surface oxygen vacancy for boosting photocatalytic $CO_2$ reduction with $H_2O$. *Appl. Catal. B Environ.* **2021**, *301*, 120802. [CrossRef]
25. Gerber, I.; Serp, P. A Theory/experience description of support effects in Carbon-supported catalysts. *Chem. Rev.* **2020**, *120*, 1250–1349. [CrossRef]
26. Chang, C.; Tseng, T.; Chen, B.; Wu, Y.; Huang, C.; Chen, J. Silver nanoparticle-mediated synthesis of fluorescent thiolated gold nanoclusters. *Nanomaterials* **2021**, *11*, 2835. [CrossRef] [PubMed]
27. Pathania, D.; Katwal, R.; Sharma, G.; Naushad, M.; Khan, M.; Ala'a, H. Novel guar gum/$Al_2O_3$ nanocomposite as an effective photocatalyst for the degradation of malachite green dye. *Int. J. Biol. Macromol.* **2016**, *87*, 366–374. [CrossRef]
28. Li, C.; Sun, Z.; Ma, R.; Xue, Y.; Zheng, S. Fluorine doped anatase $TiO_2$ with exposed reactive (001) facets supported on porous diatomite for enhanced visible-light photocatalytic activity. *Microporous Mesoporous Mater* **2017**, *243*, 281–290. [CrossRef]
29. Endo-Kimura, M.; Janczarek, M.; Bielan, Z.; Zhang, D.; Wang, K.; Markowska-Szczupak, A.; Kowalska, E. Photocatalytic and antimicrobial properties of $Ag_2O/TiO_2$ heterojunction. *ChemEngineering* **2019**, *3*, 3. [CrossRef]
30. Wei, Z.; Janczarek, M.; Endo, M.; Colbeau-Justin, C.; Ohtani, B.; Kowalska, E. Silver-modified octahedral anatase particles as plasmonic photocatalyst. *Catal. Today* **2018**, *310*, 19–25. [CrossRef]
31. Rong, J.; Zhou, Z.; Wang, Y.; Han, J.; Li, C.; Zhang, W.; Ni, L. Immobilization of horseradish peroxidase on multi-armed magnetic graphene oxide composite: Improvement of loading amount and catalytic activity. *Food Technol. Biotechnol.* **2019**, *57*, 260–271. [CrossRef] [PubMed]
32. Wang, K.L.; Wei, Z.S.; Ohtani, B.; Kowalska, E. Interparticle electron transfer in methanol dehydrogenation on platinum-loaded titania particles prepared from P25. *Catal. Today* **2018**, *303*, 327–333. [CrossRef]
33. Yu, Z.; Zhan, B.; Ge, B.; Zhu, Y.; Dai, Y.; Zhou, G.; Yu, F.; Wang, P.; Huang, B.; Zhan, J. Synthesis of high efficient and stable plasmonic photocatalyst $Ag/AgNbO_3$ with specific exposed crystal-facets and intimate heterogeneous interface via combustion route. *Appl. Surf. Sci.* **2019**, *488*, 485–493. [CrossRef]

34. Wei, Z.; Janczarek, M.; Endo, M.; Wang, K.L.; Balcytis, A.; Nitta, A.; Mendez-Medrano, M.G.; Colbeau-Justin, C.; Juodkazis, S.; Ohtani, B.; et al. Noble metal-modified faceted anatase titania photocatalysts: Octahedron versus decahedron. *Appl. Catal. B Environ.* **2018**, *237*, 574–587. [CrossRef] [PubMed]
35. Endo, M.; Janczarek, M.; Wei, Z.; Wang, K.; Markowska-Szczupak, A.; Ohtani, B.; Kowalska, E. Bactericidal properties of plasmonic photocatalysts composed of noble-metal nanoparticles on faceted ana-tase titania. *J. Nanosci. Nanotechnol.* **2019**, *19*, 442–452. [CrossRef]
36. Feng, C.; Chen, Z.; Jing, J.; Hou, J. The photocatalytic phenol degradation mechanism of Ag-modified ZnO nanorods. *J. Mater. Chem. C* **2020**, *8*, 3000–3009. [CrossRef]
37. Li, C.; Liu, D.; Huang, M.; Huang, W.; Li, Y.; Feng, J. Interfacial engineering strategy to improve the stabilizing effect of curcumin-loaded nanostructured lipid carriers. *Food Hydrocoll.* **2022**, *127*, 107552. [CrossRef]
38. Wen, C.; Zhang, J.; Zhang, H.; Duan, Y. New perspective on natural plant protein-based nanocarriers for bioactive ingredients delivery. *Foods* **2022**, *11*, 1701. [CrossRef]
39. Lan, Y.; Li, Z.; Xie, W.; Li, D.; Yan, G.; Guo, S.; Pan, C.; Wu, J. In situ fabrication of I-doped $Bi_2O_2CO_3$/g-$C_3N_4$ heterojunctions for enhanced photodegradation activity under visible light. *J. Hazard. Mater.* **2020**, *385*, 121622. [CrossRef]
40. Deng, X.; Zhang, Q.; Wang, L.; Han, J.; Wu, Y.; Sun, Z.; Li, W.; Li, X.; Xu, L.; Feng, C. Fabrication of Ag-modified porous ZnMgO nanorods with enhanced photocatalytic performance. *J. Mater. Sci. Mater. Electron.* **2018**, *29*, 16962–16970. [CrossRef]
41. Zhang, S.; Guo, X. Experimental and theoretical studies of natural mineral attapulgite supported iron-based oxygen carriers in chemical looping hydrogen production. *Fuel* **2023**, *347*, 128439. [CrossRef]
42. Shang, A.; Liu, H.; Luo, M.; Xia, Y.; Yang, X.; Li, H.; Gan, R. Sweet tea (*Lithocarpus polystachyus* rehd.) as a new natural source of bioactive dihydrochalcones with multiple health benefits. *Crit. Rev. Food Sci. Nutr.* **2022**, *62*, 917–934. [CrossRef]
43. Qiu, L.; Zhou, Z.; Ma, M.; Li, P.; Lu, J.; Hou, Y.; Duo, S. Enhanced visible-light photocatalytic performance of sapo-5-based g-$C_3N_4$ composite for rhodamine B (RhB) degradation. *Materials* **2019**, *12*, 3948. [CrossRef]
44. Azam, K.; Akhtar, S.; Gong, Y.; Routledge, M.; Ismail, A.; Oliveira, C.; Ali, H. Evaluation of the impact of activated carbon-based filtration system on the concentration of aflatoxins and selected heavy metals in roasted coffee. *Food Control* **2021**, *121*, 107583. [CrossRef]
45. Shao, Z.; Xing, C.; Xue, M.; Fang, Y.; Li, P. Selective removal of Pb (II) from yellow rice wine using magnetic carbon-based adsorbent. *J. Sci. Food Agri.* **2023**, *103*, 6929–6939. [CrossRef]
46. Liu, F.; Xu, Y.; Li, R.; Deng, C.; Gao, L.; Deng, H.; Jing, Y.; Jiang, D.; Yao, Q.; Zhong, C.; et al. Dielectric screening modulating charge carrier recombination in high-performance of $CsPbI_2Br$ carbon-based perovskite solar cells. *Chem. Eng. J.* **2024**, *500*, 157370. [CrossRef]
47. Ma, S.; Pan, L.G.; You, T.; Wang, K. g-$C_3N_4$/$Fe_3O_4$ nanocomposites as adsorbents analyzed by UPLC-MS/MS for highly sensitive simultaneous determination of 27 mycotoxins in maize: Aiming at increasing purification efficiency and reducing time. *J. Agri. Food Chem.* **2021**, *69*, 4874–4882. [CrossRef] [PubMed]
48. Liu, Z.; Wang, L.; Liu, P.; Zhao, K.; Ye, S.; Liang, G. Rapid, ultrasensitive and non-enzyme electrochemiluminescence detection of hydrogen peroxide in food based on the ssDNA/g-$C_3N_4$ nanosheets hybrid. *Food Chem.* **2021**, *357*, 129753. [CrossRef]
49. Tang, Z.; Wang, C.; He, W.; Wei, Y.; Zhao, Z.; Liu, J. The Z-scheme g-$C_3N_4$/3DOM-$WO_3$ photocatalysts with enhanced activity for $CO_2$ photoreduction into CO. *Chin. Chem. Lett.* **2022**, *33*, 939–942. [CrossRef]
50. Liu, H.; Li, P.; Qiu, F.; Zhang, T.; Xu, J. Controllable preparation of FeOOH/CuO@ WBC composite based on water bamboo cellulose applied for enhanced arsenic removal. *Food Bioprod. Process.* **2020**, *123*, 177–187. [CrossRef]
51. Zhang, C.; Shi, X.; Yu, F.; Quan, Y. Preparation of dummy molecularly imprinted polymers based on dextran-modified magnetic nanoparticles $Fe_3O_4$ for the selective detection of acrylamide in potato chips. *Food Chem.* **2020**, *317*, 126431. [CrossRef] [PubMed]
52. Barimah, A.; Guo, Z.; Agyekum, A.; Guo, C.; Chen, P.; El-Seedi, H.; Hesham, R.; Zou, X.; Chen, Q. Sensitive label-free $Cu_2O$/Ag fused chemometrics SERS sensor for rapid detection of total arsenic in tea. *Food Control* **2021**, *130*, 108341. [CrossRef]
53. Liu, L.; Dai, K.; Zhang, J.; Li, L. Plasmonic Bi-enhanced ammoniated alpha-MnS/$Bi_2MoO_6$ S-scheme heterostructure for visible-light-driven $CO_2$ reduction. *J. Colloid Interface Sci.* **2021**, *604*, 844–855. [CrossRef]
54. Zhang, Q.; Wu, T.; Che, H.; Tang, C.; Liu, B.; Ao, Y. In-situ growing $Ni_2P$ on CdS@g-$C_3N_4$ composites for highly efficient synergistically photocatalytic $H_2$ evolution and antibiotic degradation. *Surf. Interfaces* **2024**, *47*, 104205. [CrossRef]
55. Liu, W.; Wang, P.; Ao, Y.; Chen, J.; Gao, X.; Jia, B.; Ma, T. Directing charge transfer in a chemical-bonded $BaTiO_3$@$ReS_2$ Schottky heterojunction for piezoelectric enhanced photocatalysis. *Adv. Mater.* **2022**, *34*, 2202508. [CrossRef] [PubMed]
56. Gan, Z.; Hu, X.; Xu, X.; Zhang, W.; Zou, X.; Shi, J.; Zheng, K.; Arslan, M. A portable test strip based on fluorescent europium-based metal–organic framework for rapid and visual detection of tetracycline in food samples. *Food Chem.* **2021**, *354*, 129501. [CrossRef]
57. Marimuthu, M.; Arumugam, S.; Sabarinathan, D.; Li, H.; Chen, Q. Metal organic framework based fluorescence sensor for detection of antibiotics. *Trends Food Sci. Technol.* **2021**, *116*, 1002–1028. [CrossRef]

58. Wang, X.; Xu, Y.; Li, Y.; Li, Y.; Li, Z.; Zhang, W.; Zou, X.; Shi, J.; Huang, X.; Liu, C.; et al. Rapid detection of cadmium ions in meat by a multi-walled carbon nanotubes enhanced metal-organic framework modified electrochemical sensor. *Food Chem.* **2021**, *357*, 129762. [CrossRef]
59. Zhang, Z.; Jia, J.; Zhi, Y.; Ma, S.; Liu, X. Porous organic polymers for light-driven organic transformations. *Chem. Soc. Rev.* **2022**, *51*, 2444–2490. [CrossRef]
60. Corrigan, N.; Shanmugam, S.; Xu, J.; Boyer, C. Photocatalysis in organic and polymer synthesis. *Chem. Soc. Rev.* **2016**, *45*, 6165–6212. [CrossRef]
61. Yang, D.; Tao, Y.; Ding, X.; Han, B. Porous organic polymers for electrocatalysis. *Chem. Soc. Rev.* **2022**, *51*, 761–791. [CrossRef] [PubMed]
62. Jiang, J.; Wang, X.; Xu, Q.; Mei, Z.; Duan, L.; Guo, H. Understanding dual-vacancy heterojunction for boosting photocatalytic $CO_2$ reduction with highly selective conversion to $CH_4$. *Appl. Catal. B Environ.* **2022**, *316*, 121679. [CrossRef]
63. Yan, C.; Xu, M.; Cao, W.; Chen, Q.; Song, X.; Huo, P.; Zhou, W.; Wang, H. Fabricated S-scheme $BiOBr/Cu_2O$ heterojunction photocatalyst for adjusting conversion of $CO_2$ to $CH_4$. *J. Environ. Chem. Eng.* **2023**, *11*, 111479. [CrossRef]
64. Yin, S.; Zhao, X.; Jiang, E.; Yan, Y.; Zhou, P.; Huo, P. Boosting water decomposition by sulfur vacancies for efficient $CO_2$ photo-reduction. *Energy Environ. Sci.* **2022**, *15*, 1556–1562. [CrossRef]
65. Wang, L.; Zhang, H.; Wang, J.; Wang, J.; Zhang, Y. Long-term fertilization with high nitrogen rates decreased diversity and stability of diazotroph communities in soils of sweet potato. *Appl. Soil Ecol.* **2022**, *170*, 104266. [CrossRef]
66. Dzah, C.; Duan, Y.; Zhang, H.; Wen, C.; Zhang, J.; Chen, G.; Ma, H. The effects of ultrasound assisted extraction on yield, antioxidant, anticancer and antimicrobial activity of polyphenol extracts: A review. *Food Biosci.* **2020**, *35*, 100547. [CrossRef]
67. Cui, H.; Ma, C.; Li, C.; Lin, L. Enhancing the antibacterial activity of thyme oil against Salmonella on eggshell by plasma-assisted process. *Food Control* **2016**, *70*, 183–190. [CrossRef]
68. Cui, H.; Li, W.; Lin, L. Antibacterial activity of liposome containing curry plant essential oil against Bacillus cereusin rice. *J. Food Saf.* **2017**, *37*, e12302. [CrossRef]
69. Zhang, J.; Liu, T.; Chen, R.; Liu, X. Vermiculite as a natural silicate crystal for hydrogen generation from photocatalytic splitting of water under visible light. *RSC Adv.* **2014**, *4*, 406–408. [CrossRef]
70. Kumar, P.; Boukherroub, R.; Shankar, K. Sunlight-driven water-splitting using two dimensional carbon based semiconductors. *J. Mater. Chem. A* **2018**, *6*, 12876–12931. [CrossRef]
71. Yamashita, H.; Mori, K.; Kuwahara, Y.; Kamegawa, T.; Wen, M.; Verma, P.; Che, M. Single-site and nano-confined photocatalysts designed in porous materials for environmental uses and solar fuels. *Chem. Soc. Rev.* **2018**, *47*, 8072–8096. [CrossRef] [PubMed]
72. Chen, D.; Cheng, Y.; Zhou, N.; Chen, P.; Wang, Y.; Li, K.; Ruan, R. Photocatalytic degradation of organic pollutants using $TiO_2$-based photocatalysts: A review. *J. Clean. Prod.* **2020**, *268*, 121725. [CrossRef]
73. Petronella, F.; Truppi, A.; Ingrosso, C.; Placido, T.; Striccoli, M.; Curri, M.; Comparelli, R. Nanocomposite materials for photocatalytic degradation of pollutants. *Catal. Today* **2017**, *281*, 85–100. [CrossRef]
74. Prishchenko, D.; Zenkov, E.; Mazurenko, V.; Fakhrullin, R.; Lvov, Y.; Mazurenko, V. Molecular dynamics of the halloysite nanotubes. *Phys. Chem. Chem. Phys.* **2018**, *20*, 5841–5849. [CrossRef] [PubMed]
75. Zhang, X.; Wang, H.; Yan, C. Facile Preparation of Attapulgite-Supported Ag-AgCl Composite Photocatalysts for Enhanced Degradation of Tetracycline. *Catalysts* **2024**, *14*, 464. [CrossRef]
76. González, B.; Trujillano, R.; Vicente, M.; Rives, V.; Korili, S.; Gil, A. Photocatalytic degradation of trimethoprim on doped Ti-pillared montmorillonite. *Appl. Clay Sci.* **2019**, *167*, 43–49. [CrossRef]
77. Yu, X.; Wang, B.; Wang, C.; Zhuang, C.; Yao, Y.; Li, Z.; Wu, C.; Feng, J.; Zou, Z. 2D high-entropy hydrotalcites. *Small* **2021**, *17*, 2103412. [CrossRef] [PubMed]
78. Lisuzzo, L.; Cavallaro, G.; Parisi, F.; Milioto, S.; Lazzara, G. Colloidal stability of halloysite clay nanotubes. *Ceram. Int.* **2019**, *45*, 2858–2865. [CrossRef]
79. Yuan, P.; Tan, D.; Annabi-Bergaya, F. Properties and applications of halloysite nanotubes: Recent research advances and future prospects. *Appl. Clay Sci.* **2015**, *112*, 75–93. [CrossRef]
80. Lim, S.; Park, S.; Sohn, D. Modification of halloysite nanotubes for enhancement of gas-adsorption capacity. *Clays. Clay. Miner.* **2020**, *68*, 189–196. [CrossRef]
81. Xu, J.; Zhang, B.; Jia, L.; Bi, N.; Zhao, T. Metal-enhanced fluorescence detection and degradation of tetracycline by silver nanoparticle-encapsulated halloysite nano-lumen. *J. Hazard. Mater.* **2020**, *386*, 121630. [CrossRef] [PubMed]
82. Massaro, M.; Colletti, C.; Lazzara, G.; Milioto, S.; Noto, R.; Riela, S. Halloysite nanotubes as support for metal-based catalysts. *J. Mater. Chem. A.* **2017**, *5*, 13276–13293. [CrossRef]
83. Ortiz-Quiñonez, J.; Vega-Verduga, C.; Díaz, D.; Zumeta-Dubé, I. Transformation of Bismuth and beta-$Bi_2O_3$ nanoparticles into $(BiO)_2CO_3$ and $(BiO)_4(OH)_2CO_3$ by capturing $CO_2$: The role of halloysite nanotubes and "Sunlight" on the crystal shape and size. *Cryst. Growth Des.* **2018**, *18*, 4334–4346. [CrossRef]

84. Wang, Y.; Zhao, J.; Wang, T.; Li, Y.; Li, X.; Yin, J.; Wang, C. $CO_2$ photoreduction with $H_2O$ vapor on highly dispersed $CeO_2/TiO_2$ catalysts: Surface species and their reactivity. *J. Catal.* **2016**, *337*, 293–302. [CrossRef]
85. Si, S.; Shou, H.; Mao, Y.; Bao, X.; Zhai, G.; Song, K.; Wang, Z.; Wang, P.; Liu, Y.; Zheng, Z.; et al. Low-Coordination single Au atoms on ultrathin $ZnIn_2S_4$ nanosheets for selective photocatalytic $CO_2$ reduction towards $CH_4$. *Angew. Chem. Int. Ed.* **2022**, *134*, 202209446. [CrossRef]
86. Lin, S.; Zhang, Y.; You, Y.; Zeng, C.; Xiao, X.; Ma, T.; Huang, H. Bifunctional hydrogen production and storage on 0D-1D heterojunction of $Cd_{0.5}Zn_{0.5}S$@halloysites. *Adv. Funct. Mater.* **2019**, *29*, 1903825. [CrossRef]
87. Xu, Z.; Luo, Y.; Zhang, D.; Wang, H.; Sun, X.; Li, Z. Iridium complex-linked porous organic polymers for recyclable, broad-scope photocatalysis of organic transformations. *Green Chem.* **2020**, *22*, 136–143. [CrossRef]
88. Chen, Y.; Shen, C.; Wang, J.; Xiao, G.; Luo, G. Green synthesis of Ag-$TiO_2$ supported on porous glass with enhanced photocatalytic performance for oxidative desulfurization and removal of dyes under visible light. ACS Sustainable. *Chem. Eng.* **2018**, *6*, 13276–13286. [CrossRef]
89. Zhou, Z.; Li, B.; Liu, X.; Li, Z.; Zhu, S.; Liang, Y.; Wu, S. Recent progress in photocatalytic antibacterial. *ACS Appl. Bio Mater.* **2021**, *4*, 3909–3936. [CrossRef]
90. Yu, K.; Jing, S.; Tong, Z. Photo-activated graphene oxide to enhance photocatalytic reduction of $CO_2$. *ACS Appl. Mater. Interfaces* **2020**, *12*, 3580–3591.
91. Zhu, C.; Zhu, M.; Sun, Y.; Zhou, Y.; Gao, J.; Huang, H.; Kang, Z. Carbon-supported oxygen vacancy-rich $Co_3O_4$ for robust photocatalytic $H_2O_2$ production via coupled water oxidation and oxygen reduction reaction. *ACS Appl. Energy Mater.* **2019**, *2*, 8737–8746. [CrossRef]
92. Li, C.; Zhu, N.; Yang, S.; He, X.; Zheng, S.; Sun, Z.; Dionysiou, D. A review of clay based photocatalysts: Role of phyllosilicate mineral in interfacial assembly, microstructure control and performance regulation. *Chemosphere* **2021**, *273*, 129723–129735. [CrossRef] [PubMed]
93. Bailey, S.; Alietti, A.; Brindley, G. Summary of recommendations of AIPEA nomenclature committee. *Clays Clay Miner.* **1980**, *28*, 73–78. [CrossRef]
94. Ma, J.; Zou, J.; Li, L.; Yao, C.; Kong, Y.; Cui, B.; Li, D. Nanocomposite of attapulgite-$Ag_3PO_4$ for Orange II photodegradation. *Appl. Catal. B Environ. Energy* **2014**, *144*, 36–40. [CrossRef]
95. Guo, R.; Wang, G.; Liu, W. Clever use of natural clay materials in the synthesis of non-symmetric carbonates by utilizing $CO_2$ as a feedstock: Ag/attapulgite nano-catalyst. *Dalton Trans.* **2020**, *49*, 10232–10239. [CrossRef]
96. Wang, W.; Fan, M.; Zhang, Z.; Wu, L.; Liu, Y.; Cao, Y.; Peng, W. In-situ nanofication of Cu-doped $BiVO_4$ on montmorillonite nanosheets with enhanced photocatalytic efficiency for phenolic degradation. *Sep. Purif. Technol.* **2025**, *361*, 131341. [CrossRef]
97. Fatimah, I.; Syu'aib, Y.; Ramanda, G.; Kooli, F.; Sagadevan, S.; Oh, W. Facile synthesis of highly active and reusable NiO/montmorillonite photocatalyst for tetracycline removal by photocatalytic oxidation. *Inorg. Chem. Commun.* **2025**, *172*, 113731. [CrossRef]
98. Wang, J.; Chen, Y.; Liu, G.; Cao, Y. Synthesis, characterization and photocatalytic activity of inexpensive and non-toxic $Fe_2O_3-Fe_3O_4$ nano-composites supported by montmorillonite and modified by graphene. *Compos. Part B Eng.* **2017**, *114*, 211–222. [CrossRef]
99. Zeng, L.; Sun, H.; Peng, T.; Lv, X. Comparison of the phase transition and degradation of methylene blue of $TiO_2$, $TiO_2$/montmorillonite mixture and $TiO_2$/montmorillonite composite. *Front. Chem.* **2019**, *7*, 538. [CrossRef]
100. Liu, K.; Wang, L.; Fu, T.; Zhang, H.; Lu, C.; Tong, Z.; Yang, Y.; Peng, Y. Oxygen-functionalized $Ti_3C_2$ MXene/exfoliated montmorillonite supported S-scheme $BiOBr/Bi_2MoO_6$ heterostructures for efficient photocatalytic quinolone antibiotics degradation. *Chem. Eng. J.* **2023**, *457*, 141271. [CrossRef]
101. Mohamed, I.; Wu, X.; Zhu, J.; Abd El-Lateef, H.; Mousa, H.; Xing, F. Microstructure and interface analyses of novel external anode mortar incorporated calcined hydrotalcite nanoparticles towards an enhanced impressed current cathodic protection. *J. Taiwan Inst. Chem. Eng.* **2023**, *145*, 104803. [CrossRef]
102. Zheng, C.; Jiang, G.; Jin, Z. ZnCdS/NiAl hydrotalcite S-scheme heterojunction for efficient photocatalytic hydrogen evolution. *Int. J. Hydrogen Energy* **2022**, *47*, 292–304. [CrossRef]
103. Tao, L.; Wang, Y.; Zou, Y.; Zhang, N.; Zhang, Y.; Wu, Y.; Wang, S. Charge transfer modulated activity of carbon-based electrocatalysts. *Adv. Energy Mater.* **2020**, *10*, 1901227. [CrossRef]
104. Wu, Q.; Yang, L.; Wang, X.; Hu, Z. Carbon-based nanocages: A new platform for advanced energy storage and conversion. *Adv. Mater.* **2020**, *32*, 1904177. [CrossRef] [PubMed]
105. Xin, Q.; Shah, H.; Nawaz, A.; Xie, W.; Akram, M.; Batool, A.; Gong, J. Antibacterial carbon-based nanomaterials. *Adv. Mater.* **2019**, *31*, 1804838. [CrossRef]
106. Zhu, Y.; Sokolowski, J.; Song, X.; He, Y.; Mei, Y.; Wu, G. Engineering local coordination environments of atomically dispersed and heteroatom-coordinated single metal site electrocatalysts for clean energy-conversion. *Adv. Energy Mater.* **2020**, *10*, 1902844. [CrossRef]

107. Wu, J.; Sharifi, T.; Gao, Y.; Zhang, T.; Ajayan, P. Emerging carbon-based heterogeneous catalysts for electrochemical reduction of carbon dioxide into value-added chemicals. *Adv. Mater.* **2019**, *31*, 1804257. [CrossRef]
108. Ma, B.; Blanco, M.; Calvillo, L.; Chen, L.; Chen, G.; Lau, T.; Granozzi, G. Hybridization of molecular and graphene materials for $CO_2$ photocatalytic reduction with selectivity control. *J. Am. Chem. Soc.* **2021**, *143*, 8414–8425. [CrossRef]
109. Bie, C.; Yu, H.; Cheng, B.; Ho, W.; Fan, J.; Yu, J. Design, fabrication, and mechanism of nitrogen-doped graphene-based photocatalyst. *Adv. Mater.* **2021**, *33*, 2003521. [CrossRef]
110. Wang, J.; Qiao, L.; Nie, H.; Huang, H.; Li, Y.; Yao, S.; Lu, T. Facile electron delivery from graphene template to ultrathin metal-organic layers for boosting $CO_2$ photoreduction. *Nat. Commun.* **2021**, *12*, 21084–21089. [CrossRef]
111. Iijima, S. Helical microtubules of graphitic carbon. *Nature* **1991**, *354*, 56–58. [CrossRef]
112. Javey, A.; Guo, J.; Wang, Q.; Lundstrom, M.; Dai, H. Ballistic carbon nanotube field-effect transistors. *Nature* **2003**, *424*, 654–657. [CrossRef] [PubMed]
113. Zeng, K.; Wei, W.; Jiang, L.; Zhu, F.; Du, D. Use of carbon nanotubes as a solid support to establish quantitative (centrifugation) and qualitative (filtration) immunoassays to detect gentamicin contamination in commercial milk. *J. Agric. Food Chem.* **2016**, *64*, 7874–7881. [CrossRef]
114. Bai, Y.; Yue, H.; Wang, J.; Shen, B.; Sun, S.; Wang, S.; Wei, F. Super-durable ultralong carbon nanotubes. *Science* **2020**, *369*, 1104–1106. [CrossRef]
115. Bai, Y.; Zhang, R.; Ye, X.; Zhu, Z.; Xie, H.; Shen, B.; Wei, F. Carbon nanotube bundles with tensile strength over 80 GPa. *Nature Nanotechnol.* **2018**, *13*, 589–595. [CrossRef] [PubMed]
116. Magnin, Y.; Amara, H.; Ducastelle, F.; Loiseau, A.; Bichara, C. Entropy driven stability of chiral single-walled carbon nanotubes. *Science* **2018**, *362*, 212–215. [CrossRef]
117. Jeyagopal, R.; Chen, Y.; Ramadoss, M.; Marimuthu, K.; Wang, B.; Li, W.; Zhang, X. A three-dimensional porous CoSnS@CNT nanoarchitecture as a highly efficient bifunctional catalyst for boosted OER performance and photocatalytic degradation. *Nanoscale* **2020**, *12*, 3879–3887. [CrossRef] [PubMed]
118. Li, X.; Zhang, J.; Chen, X.; Fischer, A.; Thomas, A.; Antonietti, M.; Wang, X. Condensed graphitic carbon nitride nanorods by nanoconfinement: Promotion of crystallinity on photocatalytic conversion. *Chem. Mater.* **2011**, *23*, 4344–4348. [CrossRef]
119. Li, X.; Wang, X.; Antonietti, M. Mesoporous g-$C_3N_4$ nanorods as multifunctional supports of ultrafine metal nanoparticles: Hydrogen generation from water and reduction of nitrophenol with tandem catalysis in one step. *Chem. Sci.* **2012**, *3*, 2170. [CrossRef]
120. Zheng, Y.; Lin, L.; Ye, X.; Guo, F.; Wang, X. Helical graphitic carbon nitrides with photocatalytic and optical activities. *Angew. Chem. Int. Ed. Engl.* **2014**, *53*, 11926–11930. [CrossRef]
121. Zhang, X.; Meng, Z.; Rao, D.; Wang, Y.; Shi, Q.; Liu, Y.; Wu, H.; Deng, K.; Liu, H.; Lu, R. Efficient band structure tuning, charge separation, and visible-light response in $ZrS_2$-based van der waals heterostructures. *Energy Environ. Sci.* **2016**, *9*, 841–849. [CrossRef]
122. Lin, Q.; Li, L.; Liang, S.; Liu, M.; Bi, J.; Wu, L. Efficient synthesis of monolayer carbon nitride 2D nanosheet with tunable concentration and enhanced visible-light photocatalytic activities. *Appl. Catal. B Environ.* **2015**, *163*, 135–142. [CrossRef]
123. Liu, Q.; Chen, T.; Guo, Y.; Zhang, Z.; Fang, X. Ultrathin g-$C_3N_4$ nanosheets coupled with carbon nanodots as 2D/0D composites for efficient photocatalytic $H_2$ evolution. *Appl. Catal. B Environ.* **2016**, *193*, 248–258. [CrossRef]
124. Cui, L.; Ding, X.; Wang, Y.; Shi, H.; Huang, L.; Zuo, Y.; Kang, S. Facile preparation of Z-scheme $WO_3$/g-$C_3N_4$ composite photocatalyst with enhanced photocatalytic performance under visible light. *Appl. Surf. Sci.* **2017**, *391*, 202–210. [CrossRef]
125. Yu, W.; Xu, D.; Peng, T. Enhanced photocatalytic activity of g-$C_3N_4$ for selective $CO_2$ reduction to $CH_3OH$ via facile coupling of ZnO: A direct Z-scheme mechanism. *J. Mater. Chem. A* **2015**, *3*, 19936–19947. [CrossRef]
126. Jung, H.; Pham, T.T.; Shin, E.W. Interactions between ZnO nanoparticles and amorphous g-$C_3N_4$ nanosheets in thermal formation of g-$C_3N_4$/ZnO composite materials: The annealing temperature effect. *Appl. Surf. Sci.* **2018**, *458*, 369–381. [CrossRef]
127. Yang, P.; Guo, S.; Yu, X.; Zhang, F.; Yu, B.; Zhang, H.; Liu, Z. Photocatalytic reduction of carbon dioxide over quinacridone nanoparticles supported on reduced graphene oxide. *Ind. Eng. Chem. Res.* **2019**, *58*, 9636–9643. [CrossRef]
128. Lin, P.; Shen, J.; Yu, X.; Liu, Q.; Li, D.; Tang, H. Construction of $Ti_3C_2$ MXene/O-doped g-$C_3N_4$ 2D-2D schottky-junction for enhanced photocatalytic hydrogen evolution. *Ceram. Int.* **2019**, *45*, 24656–24663. [CrossRef]
129. Zhu, Z.; Tang, X.; Fan, W.; Liu, Z.; Huo, P.; Wang, T.; Li, C. Studying of Co-doped g-$C_3N_4$ and modified with $Fe_3O_4$ quantum dots on removing tetracycline. *J. Alloys Compd.* **2019**, *775*, 248–258. [CrossRef]
130. Liu, B.; Ye, L.; Wang, R.; Yang, J.; Zhang, Y.; Guan, R.; Chen, X. Phosphorus-doped graphitic carbon nitride nanotubes with amino-rich surface for efficient $CO_2$ capture, enhanced photocatalytic activity, and product selectivity. *ACS Appl. Mater. Interfaces* **2018**, *10*, 4001–4009. [CrossRef]
131. Tran, D.; Pham, C.; Ngoc, T.; Phi, H.; Ta, Q.; Truong, D.; Vo, V. One-step synthesis of oxygen doped g-$C_3N_4$ for enhanced visible-light photodegradation of Rhodamine B. *J. Phys. Chem. Solids* **2021**, *151*, 109900. [CrossRef]

132. Liu, X.; Wang, P.; Zhai, H.; Zhang, Q.; Huang, B.; Wang, Z.; Zhang, X. Synthesis of synergetic phosphorus and cyano groups (CN) modified g-$C_3N_4$ for enhanced photocatalytic $H_2$ production and $CO_2$ reduction under visible light irradiation. *Appl. Catal. B Environ.* **2018**, *232*, 521–530. [CrossRef]
133. Chen, M.; Sun, M.; Cao, X.; Wang, H.; Xia, L.; Jiang, W.; Huang, M.; He, L.; Zhao, X.; Zhou, Y. Progress in preparation, identification and photocatalytic application of defective g-$C_3N_4$. *Coord. Chem. Rev.* **2024**, *510*, 215849. [CrossRef]
134. Zhao, D.; Wang, Y.; Dong, C.; Huang, Y.; Chen, J.; Xue, F.; Guo, L. Boron-doped nitrogen-deficient carbon nitride-based Z-scheme heterostructures for photocatalytic overall water splitting. *Nat. Energy* **2021**, *6*, 388–397. [CrossRef]
135. Liang, Q.; Li, Z.; Yu, X.; Huang, Z.; Kang, F.; Yang, Q. Macroscopic 3D porous graphitic carbon nitride monolith for enhanced photocatalytic hydrogene evolution. *Adv. Mater.* **2015**, *27*, 4634–4639. [CrossRef]
136. Wang, J.; Zhang, C.; Shen, Y.; Zhou, Z.; Yu, J.; Li, Y.; Wei, W.; Liu, S.; Zhang, Y. Environment-friendly preparation of porous graphite-phase polymeric carbon nitride using calcium carbonate as templates, and enhanced photoelectrochemical activity. *J. Mater. Chem. A* **2015**, *3*, 5126–5131. [CrossRef]
137. Wang, Y.; Zhang, J.; Wang, X.; Antonietti, M.; Li, H. Boron- and fluorine-ontaining mesoporous carbon nitride polymers: Metal-free catalysts for cyclohexane oxidation. *Angew. Chem.-Int. Edit.* **2010**, *49*, 3356–3359. [CrossRef]
138. Li, N.; Gao, X.; Su, J.; Gao, Y.; Ge, L. Metallic $WO_2$-decorated g-$C_3N_4$ nanosheets as noble-metal-free photocatalysts for efficient photocatalysis. *Chin. J. Catal.* **2023**, *47*, 161–170. [CrossRef]
139. Zeng, J.; Han, C.; Wang, B.; Cao, G.; Yao, C.; Li, X. Construction of plasmonic CuS/attapulgite nanocomposites toward photothermal reforming of biomass for hydrogen production. *J. Alloys Compd.* **2024**, *985*, 174038. [CrossRef]
140. Chen, C.; Zhang, X.; Liu, E.; Xu, J.; Sun, J.; Shi, H. Biochar decorated $Bi_4O_5Br_2$/g-$C_3N_4$ S-scheme heterojunction with enhanced photocatalytic activity for Norfloxacin degradation. *J. Mater. Sci. Technol.* **2024**, *198*, 1–11. [CrossRef]
141. Cao, G.; Xing, H.; Gui, H.; Yao, C.; Chen, Y.; Chen, Y.; Li, X. Plasmonic quantum dots modulated nano-mineral toward photothermal reduction of $CO_2$ coupled with biomass conversion. *Nano Res.* **2024**, *17*, 5061–5072. [CrossRef]
142. Mohamed, I.; Hossam, G.; Shaker, A.; Nassr, L. Surface functionalization of $TiO_2$ nanoparticles with hydrophobic tridentate schiff base for a promising photocatalytic reduction of toxic Cr (VI) in aqueous solutions. *Ceram. Int.* **2025**, *in press*. [CrossRef]
143. Deng, Y.; Liu, J.; Huang, Y.; Ma, M.; Liu, K.; Dou, X.; Wang, Z. Engineering the photocatalytic behaviors of g-$C_3N_4$ based metal-free materials for degradation of a representative antibiotic. *Adv. Funct. Mater.* **2020**, *30*, 2002353. [CrossRef]
144. Song, X.; Li, G.; Zhou, W.; Wu, Y.; Liu, X.; Zhu, Z.; Huo, P.; Wang, M. Construction of Au-modified CN-based donor-acceptor system coupled with dual photothermal effects for efficient photoreduction of carbon dioxide. *J. Colloid Interface Sci.* **2024**, *664*, 868–881. [CrossRef]
145. Chen, X.; Wang, J.; Chai, Y.; Zhang, Z.; Zhu, Y. Efficient photocatalytic overall water splitting induced by a giant internal electric field of g-$C_3N_4$/rGO/PDIP Z-scheme heterojunction. *Adv. Mater.* **2021**, *33*, 2007479. [CrossRef]
146. Liu, Z.; Luo, M.; Yuan, S.; Meng, L.; Ding, W.; Su, S.; Cao, Y.; Wang, Y.; Li, X. Boron-doped graphene quantum dot/bismuth molybdate composite photocatalysts for efficient photocatalytic nitrogen fixation reactions. *J. Colloid Interface Sci.* **2023**, *650*, 1301–1311. [CrossRef] [PubMed]
147. Tahir, M.; Sillanpaa, M.; Almutairi, T.; Mohammed, A.; Ali, S. Excellent photocatalytic and antibacterial activities of bio-activated carbon decorated magnesium oxide nanoparticles. *Chemosphere* **2023**, *312*, 137327. [CrossRef]
148. Wang, Z.; Liu, G.; Chen, W.; Zhang, L.; Qi, Z.; Bai, G.; Fan, Y.; Liu, C.; Xiao, C.; Li, W.; et al. Contribution of surface plasmonic resonance to enhanced photocatalytic antibacterial performance of graphene-based two-dimensional heterojunction. *Chem. Eng. J.* **2023**, *460*, 141720. [CrossRef]

catalysts

MDPI

Review

# Agricultural Plastic Mulch: A Brief Review of Development, Composition and Catalytic Upcycling Strategies

Yang Wan [1], Yangyang Yang [1] and Weiqiang Zhou [1,2,*]

1 Institute of Green Chemistry and Chemical Technology, School of Chemistry & Chemical Engineering, Jiangsu University, Zhenjiang 212013, China; m15051390946@163.com (Y.W.); yyyang@ujs.edu.cn (Y.Y.)
2 Foshan (Southern China) Institute for New Materials, Foshan 528200, China
* Correspondence: wqzhou@ujs.edu.cn

**Abstract:** Agricultural plastic mulch film, valued for its superior heat insulation and moisture retention, is widely used globally but has led to significant microplastic accumulation in soils, threatening agricultural ecosystems. This paper reviews its development and environmental impact, focusing on recycling and upcycling technologies, particularly catalytic recovery methods (with nearly 100% conversion efficiency) such as photocatalysis, thermocatalysis, and photothermal catalysis. It analyzes technical challenges and future directions in upcycling, emphasizing the role of catalysis in converting waste plastic mulch into hydrocarbon resources. This paper also evaluates the progress and challenges of biodegradable alternatives. By offering scientific insights and innovative approaches, it aims to reduce plastic mulch pollution, enhance resource utilization, and promote sustainable agriculture.

**Keywords:** agricultural plastic mulch; white pollution; recycling technology; catalytic upgrading

## 1. Introduction

In the process of agricultural modernization, the wide application of mulch is undoubtedly a revolutionary initiative. Since the 1950s, mulch technology has come into use, acting like a layer of magic on the Earth's 'coat' to create a more favorable environment for crop growth [1–3]. Plastic mulch can significantly improve the micro-environment of arable fields by virtue of its excellent heat-insulating and moisture-retaining properties. By raising the temperature of the tillage layer by 3–5 °C and reducing water evaporation by 30–50 per cent, it creates an ideal growing environment for crop root development. Mulch technology creates a physical barrier that reduces weed germination by more than 60%, blocks the spread of soil-borne diseases, and improves the utilization of light and heat resources by about 20%, ultimately increasing crop yields by 15–25% and significantly increasing commercial yields [4,5]. Zhang and his co-workers compared two maize–soybean intercropping systems (with and without mulch) with monocropping of maize and soybean. Their findings revealed that mulch cover significantly increased root length or mass in the interspecific overlap area, improved maize root morphology, and balanced soil resource utilization and above-ground development. As a result, these improvements led to an increase in grain yield within the intercropping system [6].

However, with the continuous growth in the usage volume of plastic mulch, the issue of plastic mulch pollution has become increasingly prominent. According to the UN Food and Agriculture Organization (FAO), approximately 2 million tons of plastic mulch are utilized annually worldwide, covering an estimated 30 million hectares of agricultural land each year. However, the recovery rate of this material remains alarmingly low, averaging


https://doi.org/10.3390/catal15040310

just 30%. Consequently, significant quantities of plastic mulch accumulate in croplands, with estimates suggesting an average buildup of 50 to 100 kg per hectare. This persistent presence poses a growing environmental challenge for agricultural ecosystems globally [7]. Most plastic mulches are fabricated from high-molecular polymers such as polyethylene (PE) and polyvinyl chloride (PVC). These materials possess stable chemical properties and are highly resistant to degradation in the natural environment. The plastic mulch residues in the soil gradually fragment into small pieces, thereby giving rise to "white pollution" [8]. The long-term accumulation of plastic mulch fragments can disrupt the soil structure, impede the transfer of soil moisture and nutrients, and have an impact on the growth and development of crop roots, ultimately resulting in a decline in crop yields. Research indicates that when the residual amount of plastic mulch in the soil reaches 3.9 kg/mu, the yields of crops like corn and wheat may experience a reduction ranging from 9% to 23% [9,10]. Furthermore, plastic mulch fragments may enter water bodies along with surface runoff, inflicting damage on the aquatic ecosystem. Once ingested by animals, they can pose a threat to their health and even their lives [11]. Gan et al. investigated the influence of environmentally simulated microplastics (EMPs) on the bioavailability of Cd and Cr in soil and their phytotoxicity through pot experiments. The results demonstrated that a 5% dosage of PE and EMP significantly decreased the soil pH value, water-holding capacity, and organic carbon content. The decrease in the pH value enhanced the mobility and bioavailability of Cd and Cr. Notably, PE and EMP increased the bioavailability of Cr in the 15 cm deep soil layer [12].

Given such a severe situation, it is extremely urgent to seek effective treatment methods for agricultural plastic mulch films. In recent years, scholars at home and abroad have carried out extensive research, mainly focusing on aspects such as the recycling and upcycling technologies of waste mulch films, the research and development of degradable mulch films, and source reduction measures. Technologies such as mechanical recycling and chemical recycling (incineration, pyrolysis, and catalytic upgrading) provide various solutions for the treatment of waste mulch films, but they still face challenges such as high costs and low efficiency. Meanwhile, the research and development of new degradable mulch films (such as starch-based materials, polylactic acid, and polyhydroxyalkanoates) has provided a new direction for reducing mulch film pollution, yet their costs, performances, and degradation rates still need further optimization [13,14].

This brief review begins with an overview of the development history and fundamental composition of agricultural plastic mulch films. Subsequently, it examines the adverse environmental impacts of these plastic mulch films. The focus then shifts to the recycling and upcycling strategies for waste mulch films, encompassing traditional recycling methods as well as catalytic conversion techniques. Following this, the development and application of novel agricultural mulch films are discussed. Finally, the challenges faced by current recycling technologies and potential future directions are explored, including the integration of multiple catalytic approaches and the design of advanced catalysts, providing forward-looking recommendations for future research. Through this comprehensive review, this paper aims to offer theoretical foundations and technical insights for the sustainable utilization and recycling of agricultural plastic mulch films, particularly through catalytic upcycling, thereby promoting the green transition of agricultural production.

## 2. Introduction to Plastic Mulch Films and Their Hazards

### 2.1. *Development History of Plastic Mulch Films*

Over the course of several hundred years, depending on the climatic conditions in different regions of the world, various natural materials (such as straw and volcanic ash) have been used for covering crops [15,16]. In the 1950s, the concept of agricultural plastic mulch films was first proposed and applied to cover horticultural crops (such as strawberries and vegetables) to increase yields and improve quality. In the 1960s, Japan and the United States took the lead in promoting the plastic mulch film technology. Subsequently, polyethylene (PE) mulch films were widely adopted on a large scale globally, which significantly enhanced agricultural production efficiency [15]. However, the long-term use of mulch films has led to the accumulation of microplastics in the soil. Research by Li shows that after 32 consecutive years of using plastic mulch films, the number of microplastics in the surface soil (0–10 cm) reached 7183–10,586 pieces/kg, and, in the deep subsoil (80–100 cm), it reached 2268–3529 pieces/kg, seriously polluting both the surface and subsurface soils [17]. The photo-degradable and oxygen-degradable plastics introduced since the 1980s have failed to effectively solve the pollution problem. Instead, they have exacerbated microplastic pollution [18].

In the 1990s, plastic mulch technology underwent significant optimization, marked by the introduction of various functional films (such as black, silver, and dual-color mulches) and advancements in thickness and strength to cater to the diverse needs of different crops and climatic conditions. In recent years, driven by heightened environmental awareness, the focus of agricultural plastic mulch has shifted toward the development of biodegradable materials (such as polylactic acid (PLA) and polyhydroxyalkanoates (PHA)) and efficient recycling technologies. However, developing countries encounter substantial hurdles in adopting biodegradable materials and lag behind developed nations due to technical, infrastructural, and economic constraints. The lack of production expertise and inadequate composting and recycling facilities hinder material degradation, while high costs and limited funding render biodegradable products less competitive. Overcoming these challenges requires technology transfer, robust policy support, and international collaboration to enable sustainable large-scale applications.

Notably, catalytic upgrading offers a promising method to convert waste mulch films into high-value products like fuels, chemicals, or catalyst carriers, reducing their environmental impact [19,20]. These topics will be explored in greater detail in subsequent sections. These innovations aim to foster a circular economy for plastic mulch films and promote the harmonious coexistence of agriculture and the environment. Looking ahead, through technological advancements (such as multi-catalytic collaborative systems and novel catalyst designs) and robust policy support, agricultural plastic mulch is expected to become more environmentally sustainable and efficient. This evolution will not only bolster sustainable agricultural development but also offer innovative solutions to address the pressing issue of plastic pollution.

### 2.2. Composition of Plastic Mulch Films

At present, the composition of agricultural mulch films primarily includes the base material, additives, and functional components, each of which plays a critical role in optimizing performance and achieving desired functionalities. The base material, as the primary component, is typically composed of plastic polymers, with common materials including polyolefin plastics such as polyethylene (PE), polypropylene (PP), and polyvinyl chloride (PVC). Among these, polyethylene (PE) has emerged as the most widely used material in mulch film production due to its exceptional properties. Depending on the polymerization process and density, polyethylene can be categorized into low-density polyethylene (LDPE), linear low-density polyethylene (LLDPE), and high-density polyethylene (HDPE). LDPE is characterized by its excellent flexibility and transparency, making it easy to install and operate. Mulch films made from LDPE provide effective thermal insulation and moisture retention environment for crops, and thus are extensively utilized in agricultural production [10,21]. LLDPE is renowned for its superior tensile strength and puncture resistance, which significantly extend the service life of mulch films and reduce damage caused by external forces, making it ideal for planting scenarios requiring high mechanical strength. HDPE, with its higher density and hardness, produces more durable mulch films, often employed in environments demanding greater strength and longevity.

In addition to polyethylene, polypropylene (PP), and polyvinyl chloride (PVC) are also commonly used plastic polymers in mulch film production. Polypropylene is distinguished by its excellent strength and heat resistance, making it particularly suitable for high-temperature planting environments. It maintains shape stability and resists chemical erosion, thereby offering a longer service life. However, its flexibility is inferior to polyethylene, limiting its application in scenarios requiring frequent bending or high operational flexibility. Polyvinyl chloride, on the other hand, excels in weather resistance and mechanical properties, effectively withstanding ultraviolet rays and harsh climatic conditions while remaining resistant to aging even under prolonged sun exposure. By incorporating different plasticizers and other additives, its softness and other properties can be flexibly adjusted to meet diverse agricultural needs. Nevertheless, PVC may generate harmful substances such as hydrogen chloride and dioxins during production, and its resistance to degradation after disposal poses significant environmental challenges, which restricts its widespread application. Table 1 presents a concise summary of the aforementioned content.

**Table 1.** Comparison of three types of agricultural plastic mulch films.

| Plastic | Types | Structural Formula | Cost (RMB/m$^2$) | Degradability | Features | Environmental Impact |
|---|---|---|---|---|---|---|
| Polyethylene (PE) | low-density polyethylene (LDPE) | -(-$CH_2$-$CH_2$-)$_n$ | 0.3~3.5 | It is not degradable and takes hundreds of years to decompose | Excellent flexibility and transparency | The relative influence among the three is small, but it can pollute water, soil, crops, climate, etc. |
| | high-density polyethylene (HDPE) | | 1.7~13.5 | | High flexibility and high light transmittance, medium weather resistance | |
| Polyvinyl chloride (PVC) | / | -(-$CH_2$-CHCl-)$_n$ | 8~30 | | High strength, tear resistance, high weather resistance, UV resistance, medium flexibility and light transmittance | The most toxic plastic; it releases highly toxic gases when burned and may cause heavy metal contamination. |
| Polypropylene (PP) | / | -[-$CH_2$-CH($CH_3$)-]$_n$ | 0.3~15 | | High strength, tear resistance, high weather resistance, UV resistance, medium flexibility and light transmittance | It can pollute water, soil, crops, climate, etc. At the same time, harmful substances may be released during the production process, but toxicity is low. |

Note: The cost will be affected by crude oil price, supply and demand, specifications, etc.

In the production of mulch films, the incorporation of additives is essential, as they primarily enhance the physical and chemical properties of the films and extend their service life [22–24]. For instance, antioxidants are employed to prevent oxidative degradation, thereby prolonging the films' durability, while plasticizers improve flexibility and processability, thus facilitating easier installation and handling. Additionally, to address the specific requirements of diverse planting environments, functional components are often integrated into mulch films [25]. For example, water-retaining agents such as polyacrylamide (PAM) can significantly enhance the films' water retention capacity, reduce soil water evaporation, and optimize crop water-use efficiency [26,27]. Insect-repellent components, including plant extracts or chemical repellents, can effectively deter pests, reduce the reliance on pesticides, and mitigate environmental pollution. Furthermore, given the potential electrostatic properties of certain base materials or additives, antistatic agents are also incorporated to minimize dust adsorption, improve the films' light transmittance, and optimize crop photosynthesis efficiency [28].

In recent years, with the continuous development of catalytic technology, the functional scope of plastic film has been further expanded. Researchers have developed novel mulch films with photocatalytic or biocatalytic functions by introducing catalytic components into mulch materials. These catalytic mulch films not only retain the advantages of traditional mulch films, but can also degrade pesticide residues and organic pollutants in soil through photocatalysis, and promote soil microbial activities through biocatalysis, thereby further improving soil environment.

In summary, the composition design of mulch films optimizes multiple properties, including thermal insulation, moisture retention, anti-aging, and puncture resistance, through the synergistic effects of base materials, additives, and functional components (such as catalysts and moisturizers), providing crucial technical support for agricultural production. However, with the increasing awareness of environmental protection, the selection of mulch film materials and the use of additives must place greater emphasis on environmental friendliness and sustainability to promote the harmonious development of agriculture and the environment.

### 2.3. Hazards of Plastic Mulch Films

While the properties of plastic mulch films offer substantial benefits to agricultural production, they also present a series of significant challenges. Primarily composed of high-molecular-weight polymers such as polyethylene (PE) and polyvinyl chloride (PVC), mulch films exhibit stable chemical properties that render them resistant to natural degradation. Over time, these films can fragment into microplastics through processes such as ultraviolet radiation, weathering, and microbial activity. Furthermore, the increasing use of mulch films has led to the accumulation of residual film fragments in the soil, resulting in severe "white pollution." This phenomenon poses considerable threats to soil health, crop growth, and broader ecosystems [29–32].

The environmental impacts of plastic mulch films are multifaceted, affecting soil, water, and the atmosphere [33,34]. In soil, plastic mulch films alter key properties such as bulk density, pH, structure, water evaporation rates, and enzyme activity. These changes reduce soil aeration and water retention capacity, disrupt soil physical structure, and ultimately impair microbial activity and nutrient cycling, leading to a decline in soil fertility. Moreover, the degradation of plastic mulch films into microplastics through processes such as light exposure and weathering further exacerbates soil degradation, posing risks to plant and animal health [30,35–37]. For instance, Lahive and co-workers found that when the concentration of polyamide (PA) microplastics in soil exceeded 90 g/kg, the reproduction of Enchytraeus crypticus was significantly inhibited [38] (Figure 1a). Similarly, Dong conducted microscopic soil experiments to investigate the effects of polystyrene (PS) and polyphenylene sulfide (PPS) microplastics on sulfur reduction–oxidation (REDOX) processes in black, grassland, and rice soils. The results demonstrated that PS and PPS microplastics interfered with sulfur reduction in soil by affecting the activities of adenosine-5′-phosphate sulfate reductase and sulfite reductase. Additionally, PS and PPS microplastics directly altered the structure of soil enzymes, leading to changes in their activity [39].

Plastic mulch can contaminate rivers, lakes, and groundwater systems when carried into water bodies through stormwater runoff or agricultural drainage. Microplastics, which are resistant to degradation in water, are readily ingested by aquatic organisms, impairing their growth and reproduction. These microplastics infiltrate plants and aquatic species (e.g., fish and shrimp) via soil and water, and upon human consumption, they may induce inflammation, cause cellular damage, disrupt endocrine function, compromise gut health, and potentially exhibit carcinogenic effects. Comprehensive research is crucial to fully evaluate their long-term impacts on the immune system, hormonal balance, and overall health [40–42] (Figure 1b). Simultaneously, harmful chemicals released during plastic decomposition, such as plasticizers and flame retardants, further exacerbate water pollution. Lee observed the bioaccumulation of plastic particles in the gills, intestines, and liver of Pseudobagrus fulvidraco, which may enter human tissues through the food chain [43] (Figure 1c). In addition, Wang's research further explores the translocation and accumulation of microplastics from insects to mammals. The findings reveal that micro/nano-plastics (M/NPs) are transferred across trophic levels in the food chain. The study highlights their ability to cross biological barriers, accumulate in organisms, and potentially impact embryonic development through trophic transfer [44].

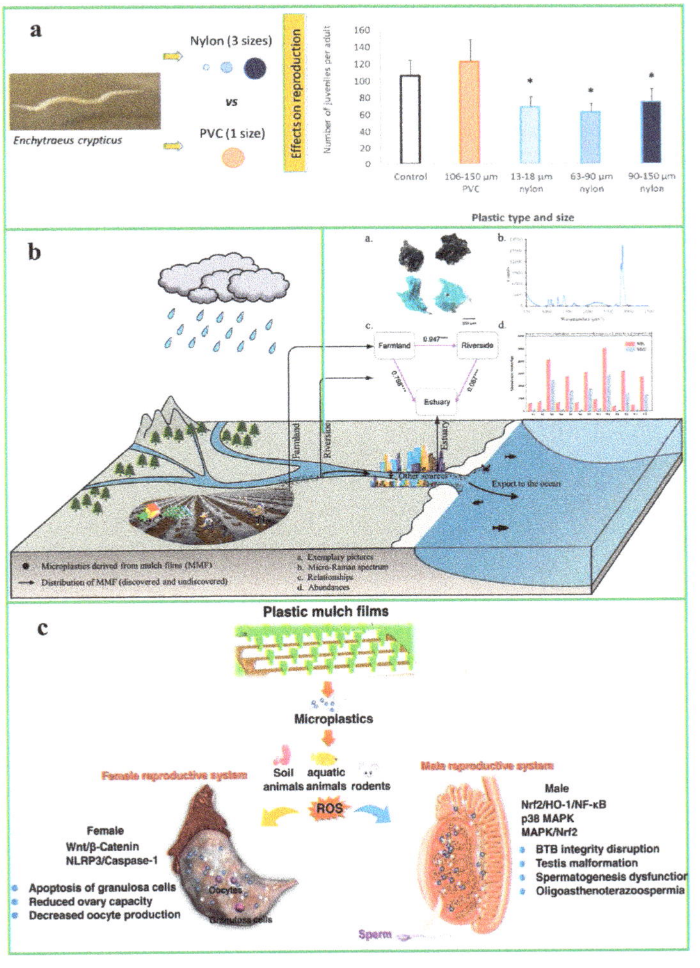


**Figure 1.** Waste plastic mulch. (**a**) Impact on soil. Reproduced with permission from reference [38]. Copyright 2019 Elsevier. (**b**) Transmission route along the water cycle. Reproduced with permission from reference [40]. Copyright 2022 Elsevier. (**c**) Impact on living organisms. Reproduced with permission from reference [43]. Copyright 2022 Elsevier.

During incineration, plastic mulch films release toxic gases such as dioxins and hydrogen chloride, polluting the atmosphere and posing risks to human respiratory health, potentially causing respiratory diseases and cancer [45]. Given the severe ecological, plant, and animal hazards posed by plastic mulch films, effective management of agricultural plastic waste has become a pressing global issue.

## 3. Treatment Technologies for Waste Agricultural Plastic Mulch

At present, the treatment technologies for waste agricultural plastic mulch encompass methods such as melting regeneration, high-temperature hydrolysis, and direct incineration for energy recovery. Advanced treatment technologies, however, focus on converting waste plastic mulch into high-value chemicals or hydrocarbon fuels through catalytic conversion. These innovative approaches offer promising solutions for mitigating environmental pollution and advancing the development of a plastic recycling economy. Below is a brief overview of the various processing techniques (Figure 2).

**Figure 2.** Recycling mode of discarded mulch.

### *3.1. Traditional Treatment Techniques*

#### 3.1.1. Mechanical Recovery

Plastic film mechanical recycling involves collection (such as manual collection or mechanical recycling), cleaning (removal of soil and weeds), crushing, washing (cold or hot water washing), drying, and granulation steps. The recovery effect is affected by the film material (such as PE or PVC), residual amount, impurity content (such as pesticide residue) and technical advancements. The recovered products can be used to make recycled plastic pots, incinerate electricity, or extract chemical raw materials [46,47]. However, current recycling technologies often cause degradation of plastic material performance, reducing their recyclability. The degradation of polymer chains during processing significantly undermines the material's strength. This recycling method, known as "downcycling" or "downward cycling", contrasts with catalytic conversion technology, which represents an "upcycling" approach. Despite breakthroughs in PET plastic recycling, downcycling cannot resolve the fundamental issue of progressive material deterioration [48].

Mechanical recycling of plastic mulch film reduces environmental pollution and protects land and water resources. However, it faces challenges such as high labor, material, and transportation costs, especially in rural areas. The process involves removing soil and impurities, which consumes significant amounts of water and energy, and requires costly specialized equipment for crushing, granulation, and maintenance.

It is crucial to emphasize that a significant gap exists between the idealized plastic recycling models developed in laboratories and the complexities of real-world applications. Studies reveal that this disjunction hinders the effective implementation of mechanical recycling processes in large-scale commercial waste plastic treatment [49]. This fundamental contradiction has led to a steady decline in the global adoption of mechanical recycling, despite its current dominant position. In contrast, emerging recycling systems, such as pyrolysis technology and chemical decomposition methods (such as efficient catalytic conversion), are rapidly advancing. These innovative technologies hold the potential to redefine the landscape of the plastic circular economy, offering more sustainable and efficient solutions for plastic waste management.

#### 3.1.2. Melting Regeneration

Melting regeneration technology is currently one of the mainstream methods for recycling and reusing agricultural films. The core process involves cleaning, crushing, and drying collected agricultural films, followed by heating and melting under specific conditions to enable re-plasticization and shaping into new products. During this process, high temperatures facilitate the reorganization of the molecular structure of the used agricultural films, restoring their plasticity. The molten plastic is then extruded through an extruder and molded using dies to produce various plastic products, such as pellets, pipes, and sheets [50,51]. These recycled plastics have broad applications in agriculture, industry, and daily life. The key advantages of melting regeneration technology include its straightforward process, relatively low equipment investment, and cost-effectiveness, enabling large-scale recycling of used agricultural films. However, this technology also has limitations. Due to environmental factors during long-term use, the performance of used agricultural films degrades, resulting in recycled plastic products that are typically inferior in quality to virgin plastics and are primarily used in applications with lower performance requirements [52–54]. Additionally, the molten plastic can be converted into bio-oil, which serves as an asphalt binder regeneration agent, reactivating aged asphalt and even enhancing some of its properties [55,56].

Pyrolysis converts plastic mulch film into fuel oil and gas, enabling energy recovery and reducing environmental pollution of land and water. However, it requires high temperatures and significant energy, often from costly fossil fuels. The equipment is complex, with high initial investment and maintenance costs. Gases and residues need further treatment, adding expenses. Additionally, the technology's complexity demands professional operator training, increasing labor costs. While beneficial, pyrolysis involves substantial financial and technical challenges. In addition, this method may also release harmful gases, causing environmental pollution, and equipment investment is large, and the economy is poor.

#### 3.1.3. Pyrolysis

Pyrolysis technology thermally decomposes waste plastic films at high temperatures in an oxygen-free or low-oxygen environment (without catalysts), breaking molecular chains into smaller gaseous, liquid, and solid molecules, typically without catalysts [57,58]. These resulting products possess significant economic value and can be effectively utilized as energy sources or chemical raw materials. During the pyrolysis process, high-molecular-weight polymers, such as polyethylene found in waste plastic films, undergo thermal degradation, producing combustible gases (primarily methane and ethylene), liquid fuels (e.g., diesel and gasoline), and solid carbon black [59].

For instance, Gracida-Alvarez and his co-workers conducted secondary degradation of pyrolysis vapors from high-density polyethylene (HDPE) waste in a two-stage micro-pyrolysis reactor (TSMR) by altering temperature and vapor residence time (VRT). At low temperatures (625 °C) and short VRT (1.4 s), various gases and liquid products were produced, while at high temperatures (675 °C) and long VRT (5.6 s), mostly hydrocarbon gases and single and multi-ring aromatics were generated [60] (Figure 3a). Some researchers have also explored alternative methods by co-pyrolyzing waste plastic films with agricultural waste to produce biochar for enhancing soil fertility. For example, as shown in Figure 3b, Han prepared the optimal pyrolysis and activation conditions of MPS–char by mixing plastic film (polyethylene) and corn stover (MPS) at a ratio of 1:5 without adding a catalyst, and experimentally clarified that the pyrolysis of biochar from farmland plastic film waste could promote the growth of maize and other crops [61].

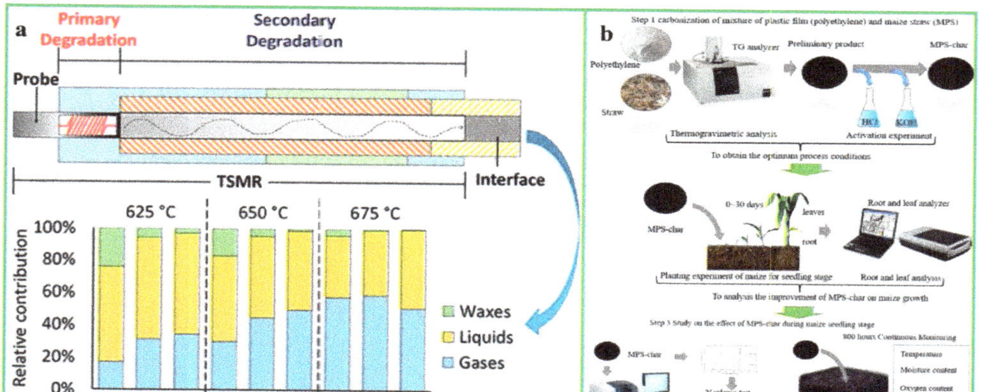


Figure 3. (a) Schematic diagram of the two-stage micro-pyrolysis reactor and the performance of direct pyrolysis of polyethylene plastics. Reproduced with permission from reference [60]. Copyright 2018 American Chemical Society. (b) Experimental process and steps of biochar. Reproduced with permission from reference [61]. Copyright 2023 Elsevier.

In addition, Yan and co-workers used agricultural mulch as an organic fertilizer precursor to produce thermochemical degradation products (TDPs) by percolating PEMF in nitric acid ($HNO_3$) using a micro-glass reactor under mild conditions. The introduction of nitrogen and oxygen-containing functional groups in TDP suggests its potential as an organic soil fertilizer [62].

Pyrolysis technology enables deep resource utilization of waste plastic film, converting it into high-value energy and chemical products while minimizing environmental pollution. However, this technology also faces some challenges. The thermal cracking process consumes a large amount of energy, which results in high equipment requirements and high investment costs. So, reducing the reaction temperature by combining with other technologies (such as adding catalysts) is the way to go. In addition, some toxic and harmful gases, such as hydrogen chloride and dioxins, may be produced during the pyrolysis process, so it is necessary to be equipped with perfect exhaust gas treatment devices to ensure environmental safety.

#### 3.1.4. Energy Recovery

Energy recovery technology can reuse resources by incinerating waste plastic film and converting its chemical energy into heat or electricity. In a specialized incinerator, waste plastic film is fully combusted with air, producing high-temperature flue gas that heats boiler water via a heat exchanger, generating steam to power turbines for electricity or directly providing heating. This technology can quickly reduce the volume of plastic film and recover energy, but the incineration process will emit carbon dioxide, sulfur dioxide and other pollutants, which need to be strictly controlled. If the combustion is not sufficient, it may also produce highly toxic substances such as dioxins, which endanger the environment and health [63].

### *3.2. Catalytic Transformation Treatment*

Catalytic conversion technology, a cornerstone of modern chemical engineering, has emerged as a pivotal tool across diverse fields owing to its exceptional efficiency, selectivity, and environmentally benign characteristics. Through catalytic oxidation, reduction, and other processes, harmful substances are transformed into non-toxic or low-toxicity compounds, thereby contributing to the purification of air, water, and soil [64,65]. Furthermore, this technology facilitates the efficient utilization of fossil fuels and the advancement of renewable energy, thereby supporting the optimization and sustainable development of energy structures [66,67]. In this context, catalytic conversion has recently been developed as an innovative approach for the treatment of plastic mulch films [68,69]. By leveraging the action of catalysts, this method promotes the conversion and degradation of waste mulch films, enhancing the quality and added value of the resulting products. As such, it represents a promising strategy for achieving sustainable development and fostering a circular economy [70].

In the catalytic transformation process, it is crucial to choose the right catalyst. Different types of catalysts have significant effects on the conversion efficiency and product distribution of waste plastic film. Metal oxide and metal sulfide catalysts can promote the degradation and transformation of waste plastics to a certain extent [71–73]. Zeolite catalysts have unique pore structure and an acidic center, which can improve product selectivity and conversion. The efficiency of catalytic conversion and product selectivity can be further improved by optimizing the composition and structure of the catalyst [74,75].

#### 3.2.1. Photocatalytic Treatment

The upcycling of waste plastic mulch films, typically polyolefins, into value-added chemicals using solar-powered photocatalysis is a promising strategy to achieve sustainable development. Xu used $ZnS/Ga_2O_3$ Z-type heterojunction material as a photocatalyst to efficiently convert polyethylene plastic into a syngas mixture ($CO + H_2$). The heterojunction agent achieved a $H_2$ formation rate of 50.15 $\mu mol\ g^{-1}\ h^{-1}$ and a CO formation rate of 27.32 $\mu mol\ g^{-1}\ h^{-1}$ (Figure 4a,b). A series of characterization and comparative experiments demonstrated that the formation of the heterojunction between ZnS and $Ga_2O_3$ promoted the efficient photocatalytic conversion of plastic waste to syngas and explained the mechanism of syngas production [76] (Figure 4c). This provides a reference for the photocatalytic treatment of waste polyolefin films. In addition, some people combine plasma treatment technology with photocatalytic technology to greatly reduce the photocatalytic on-demand time. For example, Jiang and his colleagues treated discarded polyolefin plastic with plasma, and then used $TiO_2$ loaded with Pt as a photocatalyst, which can efficiently upgrade polyolefin plastic to hydrogen and gas fuel in only 4 h. More than 100 μmol $g^{-1}$ $h^{-1}$ of hydrogen was produced [77].

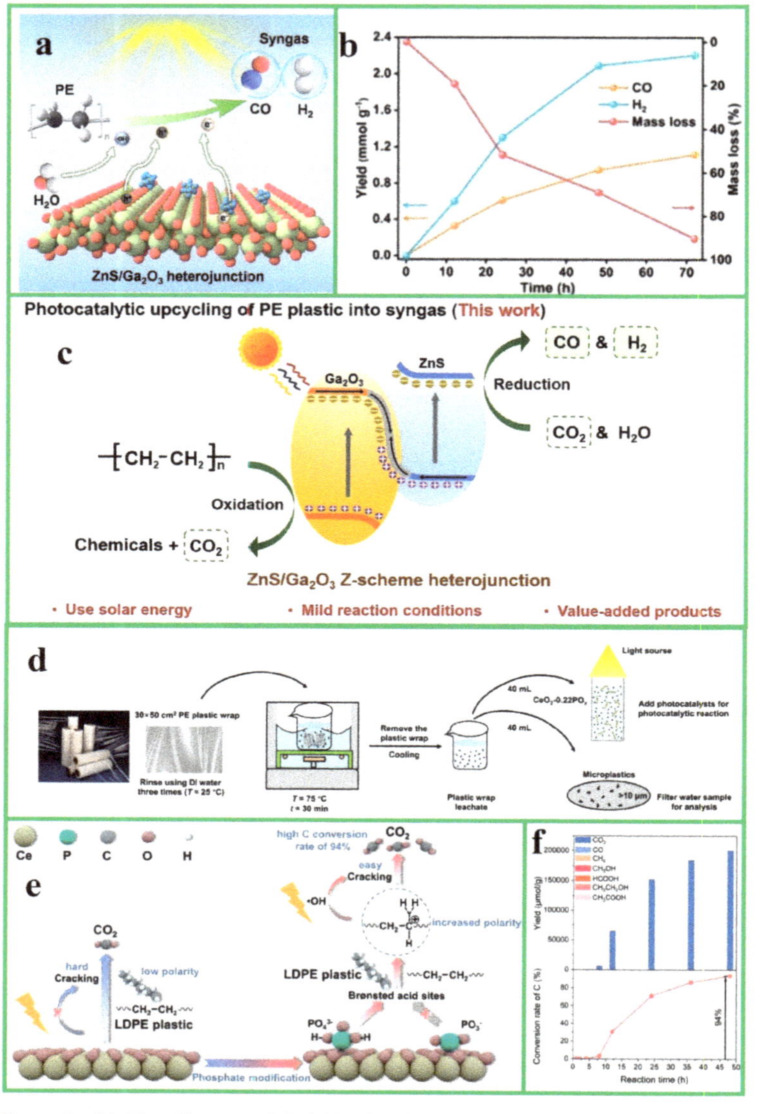


**Figure 4.** (**a**) Idea diagram of $ZnS/Ga_2O_3$ photocatalytic upgrading of polyethylene plastics. (**b**) Performance chart. (**c**) Mechanism diagram. Reproduced with permission from reference [76]. Copyright 2024 John Wiley and Sons. (**d**) Photocatalytic cracking of microparticles released from LDPE plastic wrap. (**e**) Photocatalytic product yields and the carbon conversion efficiency of $CeO_2$-0.22$PO_x$. Reproduced with permission from reference [78]. (**f**) Performance of $CeO_2$-0.22POx Copyright 2024 Elsevier.

In addition, some researchers have photocatalytic degradation of waste polyolefin plastic mulch films to $CO_2$. For example, Wen and colleagues have developed a phospho $CeO_2$ catalyst based on organophosphate precursors for efficient photocatalytic degradation of low-density polyethylene (LDPE) without acid–base pretreatment (Figure 4d). Compared with pristine $CeO_2$, surface phosphorylation introduces Brønsted acid sites to form carbocations on LDPE through protonation, while optimizing the band structure and enhancing light absorption and reactive oxygen generation capacity, thereby promoting C-C bond breakage of LDPE (Figure 4e). The catalyst achieved >94% carbon conversion and >99% $CO_2$ selectivity after exposure to 50 mW/cm$^2$ simulated sunlight for 48 h (Figure 4f). This method effectively reduces the production of microplastics and harmful gases. Moreover, $CO_2$ can be further converted into useful chemicals or fuels by other technologies, such as carbon capture and utilization (CCU), to realize the recycling of carbon resources [78].

### 3.2.2. Thermocatalytic Treatment

Thermal catalysis can improve the reaction rate, control product selectivity, and enable large-scale production based on photocatalysis, which is more advantageous in some aspects [79,80]. Xu and his co-workers used HZSM-5 as the optimal catalyst for the thermocatalytic conversion of waste polyethylene film. Using a combined platform of a cone calorimeter and a tube pyrolysis furner–mass spectrometer (MS), they achieved the conversion of PE into many alkenes and aromatic hydrocarbons (AH), such as benzene, toluene, and xylene (BTX) products, thereby reducing PE plastic by 59.7% [81].

Kong explored an innovative decomposition thermocatalytic pathway based on in situ self-assembled metal nanoparticles (NPs) co-doped with microporous carbonaceous catalysts (M/PCC) for upcycling of waste plastic film. When M/PCCs with different Zn and Ni co-additions were tested, $Zn_{50}$-$Ni_{20}$/CSB1 ($ZnCl_2$:50%; $NiCl_2$:20%) (Zn-Ni/PSB) was found to promote the formation of liquid hydrocarbons and the production of high-purity $H_2$ in the discarded plastic mulch (Figure 5a). The Zn-Ni/PSB catalyst with the highest specific surface area (671 m$^2$/g) and microporosity (76.05%) showed the highest liquid yield (51.67 wt.%), as well as the highest carbon yield of C5-C12 hydrocarbons (45.82 C%) and C8-C16 hydrocarbons (36.90 C%) [82] (Figure 5b).

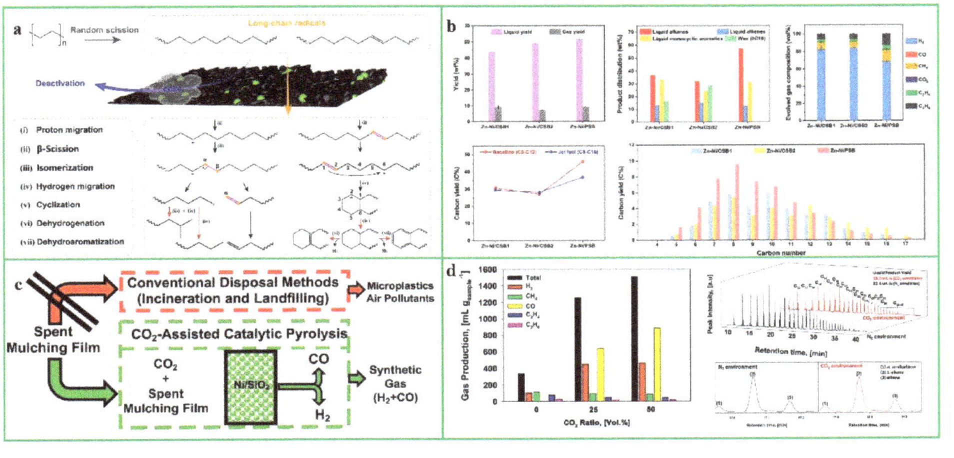


**Figure 5.** (**a**) Reaction mechanism of Zn-Ni/PSB thermocatalytic conversion of waste polyethylene mulch film. (**b**) Performance chart. Reproduced with permission from reference [82]. Copyright 2025 Elsevier. (**c**) Idea diagram of thermocatalytic conversion of waste plastic mulch film under $CO_2$ atmosphere. (**d**) Performance graph. Reproduced with permission from reference [83]. Copyright 2022 Elsevier.

In addition to the simple thermocatalytic conversion of waste plastic mulch, some researchers have combined $CO_2$ utilization with catalytic conversion of waste plastic mulch to synergistically improve the catalytic performance. Jung et al. used the reaction substrate, $Ni/SiO_2$, as a thermal catalyst for spent plastic mulch film (SMF). Under the condition of $CO_2$, the total concentration of gas obtained by catalytic pyrolysis was increased by 11.3 times compared with that by non-catalytic pyrolysis. At the same time, the conversion of hydrocarbons to $H_2$ is also improved (Figure 5c). When $CO_2$ is introduced as a flow gas during catalytic pyrolysis, substantial CO production is observed because $CO_2$ acts as an oxidant and converts long-chain hydrocarbons and $CO_2$ to CO. The formation of $H_2$ and CO quantity is proportional to the flow of $CO_2$ concentration, confirming that $CO_2$ helps to improve the process of catalytic pyrolysis SMF syngas production [83] (Figure 5d).

#### 3.2.3. Photothermal Catalytic Treatment

As is well known, photocatalytic technology integrates photocatalysis and thermal catalysis, promoting charge transfer and separation to fully leverage the advantages of thermal catalytic carriers. This effectively overcomes the limitations of single-photocatalytic processes, synergistically enhancing the restructuring performance of plastic waste [84,85]. For example, Miao and co-workers constructed an LDPE-$Ru/TiO_2$ photothermal catalytic system, and the whole system can be heated to 200–300 °C by photothermal heating. The system efficiently cuts C-C and C-H bonds in the polymer (Figure 6a). At the same time, LDPE conversion was as high as 95.0%, far exceeding photoreduction alone. By optimizing temperature and pressure, it was converted to liquid fuel in only 3 h (86% C5-C21). The method also allows for the production of a variety of products, including high-purity methane and different liquid-phase fuels [86] (Figure 6b). Under the illumination of a full-spectrum Xe lamp or concentrated sunlight, the $Ru/TiO_2$ catalyst undergoes localized photothermal heating, which facilitates the melting of the polymer. Ultraviolet light ($\lambda < 365$ nm) serves to activate the LDPE chains, thereby generating reactive sites conducive to the cleavage of Ru nanoparticles on the $Ru/TiO_2$ catalyst. This elucidated mechanism underpins the effective photothermal conversion of LDPE into diminutive gaseous and liquid/waxy hydrocarbon derivatives, achieved by the harmonious exploitation of UV, visible, and near-infrared light spectra.

Han introduced a novel photothermal dechlorination-carbonization two-step reaction to convert waste PVC plastics into valuable carbon materials (Figure 6c). Outdoor experiments have substantiated the practical applicability of photothermal carbonization in the valorization of waste polyvinyl chloride plastics (Figure 6d). The experiment showed that the carbon retention rate of the two-step thermal method reached 48.9%, which was significantly higher than 13.8% of the direct carbonization method (Figure 6e). The resulting hard carbon material can be used as a potential anode material for sodium-ion batteries [87]. Through detailed techno-economic assessment (TEA) modeling for the recovery of 96,000 tons of plastic, they demonstrate that photothermal conversion using solar energy saves approximately $2.34 \times 10^{12}$ kJ of electricity and reduces the carbon footprint by 261,912.2 tons compared to conventional thermal methods, offering substantial environmental benefits. This innovative technology not only enables the efficient use of energy, but also significantly improves cost-effectiveness, while bringing considerable environmental benefits, providing a greener and efficient solution for the sustainable recycling of PVC.

In addition, Chen successfully developed a copper–ruthenium superfine particle photothermal catalyst. Under a light intensity of 4.0 $W/cm^2$, the photothermal catalyst rapidly reached temperatures exceeding 350 °C within 60 s. These properties were effectively utilized for the photothermal recovery of waste polyethylene and polyethylene terephthalate. Hydrogenolysis of LDPE was performed using engineered Cu3Ru1 superstructures dispersed on alumina nanosheets (with a ruthenium content of 2 wt.%). After 6 h of irradiation under a xenon lamp (4 W/cm$^2$), C8-C35 liquid fuels were obtained with a yield of 87.6%. Additionally, through scaled-up experiments, complete conversion of 10 g of waste PET film was achieved, recovering approximately 9.26 g of high-value BHET with a yield of about 70%, demonstrating the robustness and scalability of the catalytic system. Furthermore, cost analysis confirmed the practical universality of the copper–ruthenium superfine particle photothermal catalytic system for converting waste plastics [88].

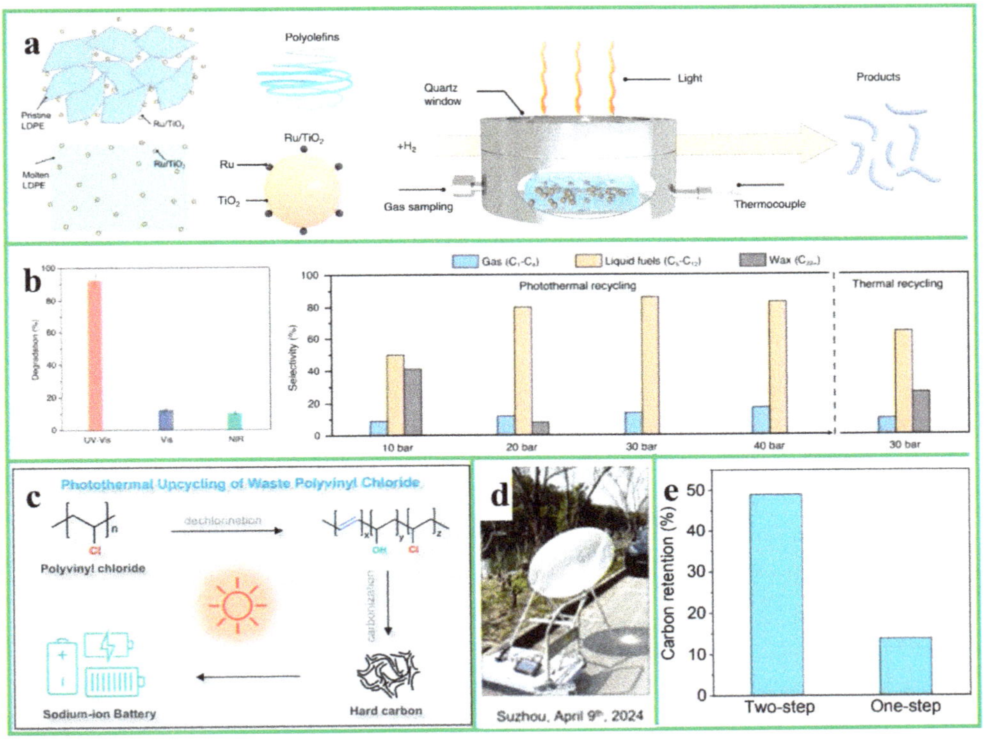


**Figure 6.** (**a**) Schematic diagram of the contact between Ru/TiO$_2$ catalyst and LDPE and plastic photothermal recovery system. (**b**) LDPE conversion performance graph. Reproduced with permission from reference [86]. Copyright 2023 Springer International Publishing. (**c**) Schematic illustration of photothermal catalytic dechlorination of PVC to hard carbon material. (**d**) Photothermal catalytic outdoor device. (**e**) Comparison of carbon retention performance of the two methods. Reproduced with permission from reference [87]. Copyright 2024 American Chemical Society.

### 3.2.4. Electric/Photoelectric Catalytic Treatment

In addition to the aforementioned recycling methods, electrocatalysis and photocatalysis involve chemical reactions driven by electricity or solar energy, respectively, promoting the transfer of electrons through the electrode/electrolyte interface. These methods enable precise control of chemical reactions by using electrical potential, current density, or light to produce valuable chemical products [89,90]. However, these methods are generally used for the upcycling of polyester plastics. Although polyester plastic is rarely used in agricultural mulch film (which can be used in agricultural greenhouses), its electric/photoelectric catalytic upgrading method can provide a new perspective for the recycling of polyolefin plastic mulch films. Qi used cation vacancy engineering cobalt selenide as a bifunctional catalyst to effectively convert waste polyethylene terphenyl ester plastic into high-value succinic acid through an electrocatalytic upcycling route, achieving a Faraday efficiency of 94.0% [91]. This highlights the potential of electro/photocatalytic degradation to break plastics down into small molecules (e.g., $H_2O$, $CO_2$), which can be reduced to produce value-added chemicals.

We can see from the comparison in Table 2 that catalytic conversion treatments have some unique advantages over traditional regenerative technologies. They can realize the transformation of waste plastic film under relatively mild conditions, reducing energy consumption and equipment wear. By selecting appropriate catalysts and reaction conditions, the composition and structure of the products can be precisely controlled, allowing for the production of materials with specific properties and uses, which improves the added value of the waste plastic film.

**Table 2.** Comparison of the three recycling treatments.

| Method of Treatment | Mechanical Recovery | Direct Pyrolysis | Catalytic Upgrading |
|---|---|---|---|
| Factor of cost | Collection and transportation; cleaning and sorting; crushing and granulation. | Energy cost; equipment investment; by-product treatment; technical threshold. | The cost is high and benefits are limited. |
| Advantages and disadvantages | Reuse of resources, reduces waste plastic film pollution, but it is costly and time-consuming. | Energy recycling, reduces waste plastic mulch pollution, but may cause gas pollution. | Quickly and effectively reduces environmental pollution, improves energy efficiency, improves product quality, reduces by-products, and yields high value-added products. |
| Comprehensive evaluation | Catalyst cost, equipment modification, and operation complexity | The cost is the highest, but benefits are significant. | The cost is high, but the long-term benefits are most significant. |
| The latest technology | Dissolution/precipitation process. | Plasma pyrolysis. | Co-catalysis with other wastes; co-catalysis (collaboration of multiple catalytic systems). |

However, this approach faces significant challenges. Durability is hindered by complex inactivation mechanisms (e.g., chemical poisoning, physical wear, and thermal degradation) and harsh operating conditions, while recyclability is constrained by limited regeneration efficiency, high costs, environmental impacts, and finite regeneration cycles. These issues increase economic burdens and undermine sustainability. Moving forward, breakthroughs in new materials, regeneration technologies, smart catalyst design, and circular economy models are essential to enhance catalyst performance and sustainability.

Finally, we believe that in the future, low-cost, green catalytic conversion technology for plastic waste can alleviate the global plastic pollution problem and promote the development of a circular economy.

## 4. The Development of New Agricultural Plastic Film

Biodegradable plastic mulch films (BDMs) are becoming increasingly popular in agriculture and are emerging as an alternative to conventional polyethylene (PE) films. These degradable plastic films can gradually degrade in the natural environment after use and do not persist for a long time like traditional plastic films, which effectively reduces soil pollution and provides strong support for sustainable agricultural development [92–94] (Figure 7a).

In recent years, a series of technological breakthroughs have been made in the research and development of degradable plastic films. In terms of materials, new degradable polymers are emerging, such as polylactic acid (PLA), polyhydroxyfatty acid ester (PHA), and polyvinyl alcohol (PVA). These materials have good biodegradability and mechanical properties, meeting the basic needs of agricultural production [95–97]. For example, Yoshinakad and colleagues synthesized bio-based and water-degradable polyester amides (PEAs) from N, N′-bis (2-hydroxyethyl) brassicamide (BHEBA) with excellent degradability in water (seawater, deionized water, etc.). The results confirm the great potential of brassylated polylamides to be an environmentally friendly alternative to certain petroleum-derived commodity plastics, especially polyethylene [98]. Zhu used waste milk to produce a biodegradable mulch film (BMF), which was used as the main substrate and mixed with gelatin, tributyl citrate, ZnO, and $(NH_4)_2SO_4$ to prepare a multi-nutrient slow-release biodegradable mulch film (HPBMF). The prepared HPBMF exhibits good water resistance as well as mechanical properties such as stiffness and flexibility at a low cost. At the same time, HPBMF can be degraded within 3–4 months during application, which aligns well with the crop season [99] (Figure 7b).

In the preparation process, by improving the processing technology, the uniformity and stability of the degradable plastic film were improved, and the production cost was reduced. Gao et al. used a green and simple process to synthesize lignin-based polyurethane (LPU), which was coated on low-quantity cellulose paper (CP) to obtain an economical and green composite LPU/CP mulch film. The physical strength of the film was improved, and it exhibited excellent water resistance and degradability [100] (Figure 7c). Similarly, Wang et al. developed a novel biodegradable and recyclable bio-based plastic preparation method. In this method, carboxymethyl cellulose (CMC) and cationic polymer ionic liquid (PILCl) were used as the main raw materials and were prepared by a complexation reaction with the assistance of $KNO_3$. This innovative process not only utilizes renewable biomass resources, but also enables material degradability and recyclability, providing a new technological approach to the development of environmentally friendly plastics [101] (Figure 7d).

These innovative degradable polymers exhibit excellent biodegradability and mechanical properties, effectively addressing the fundamental requirements of agricultural production. For instance, PLA mulch can fully degrade within just a few months under specific conditions, whereas traditional PE mulch may persist for hundreds of years without complete degradation. Moreover, the optimization of catalytic technology has significantly enhanced the synthesis efficiency and degradation rate of these materials, thereby offering robust support for the advancement of sustainable agriculture.

However, the application of degradable plastic films still faces many challenges. First, the performance of biodegradable films compared with traditional films still has gaps in aspects such as the tensile strength and weather resistance, which remains to be further improved to adapt to different environments of agricultural production [102,103]. Secondly, the high cost is an important factor restricting the promotion of degradable plastic film. In fact, degradable mulch film (about 20–40 yuan/kg) typically costs 2–3 times as much as conventional mulch film. In addition, it is difficult to precisely control the degradation conditions and rate of degradable plastic films, which may degrade too quickly or too slowly in some areas, affecting the growth and yield of crops. At the same time, challenges such as inadequate market supply, insufficient policy incentives, and deeply ingrained traditional practices further reduce farmers' willingness to embrace this technology. Additionally, weak environmental consciousness and a lack of technical training create practical difficulties for farmers, ultimately impeding the widespread application of biodegradable plastic films.

Last but not least, the lack of uniform quality standards and testing methods for degradable plastic films, and the uneven quality of products in the market, also bring difficulties to their promotion.

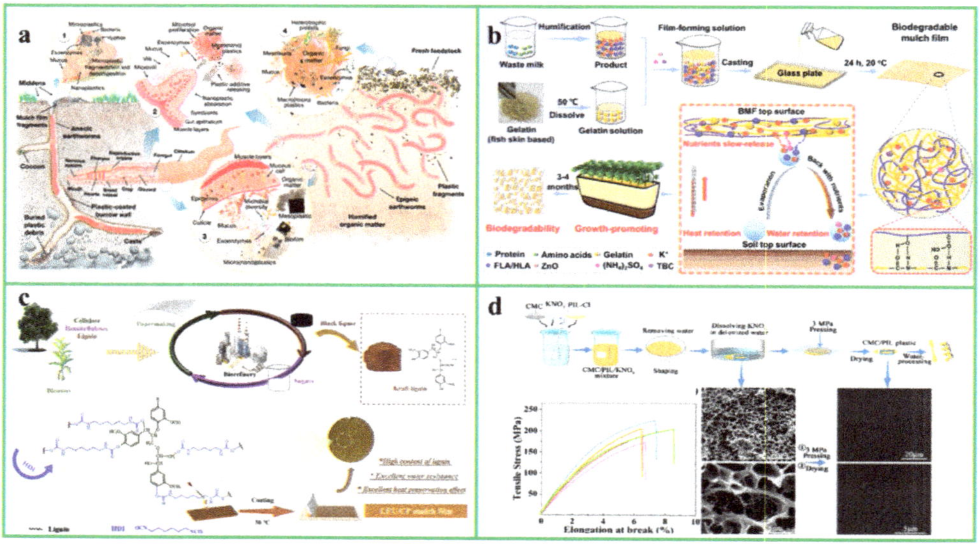

**Figure 7.** (**a**) Degradation process of degradable plastic mulch. Reproduced with permission from reference [92]. Copyright 2020 American Chemical Society. (**b**) Preparation process and properties of HPBMF. Reproduced with permission from reference [99]. Copyright 2024 Elsevier. (**c**) Preparation process of composite LPU/CP film. Reproduced with permission from reference [100]. Copyright 2024 Elsevier. (**d**) Complexation reaction preparation of degradable plastic mulch film process. Reproduced with permission from reference [101]. Copyright 2025 Elsevier.

## 5. Conclusions and Prospect

The wide application of plastic mulch film in agricultural production has caused serious environmental pollution and resource waste problems, and its recycling technology has become a key breakthrough in controlling agricultural non-point source pollution and promoting sustainable agricultural development. However, the traditional treatment technologies of polyolefin plastic film have significant defects in terms of environmental friendliness, which need to be improved through technological innovation. Therefore, the development of new environmentally friendly agricultural plastic films or advanced treatment technologies has become a priority. At present, widely used recycling technologies, such as melting regeneration, thermal cracking, and energy recovery, struggle to maximize the value of waste plastic mulch film resources. In this context, catalyst-based catalytic conversion technology has shown significant advantages, efficiently converting waste plastic mulch into fuels or other high value-added chemicals under relatively mild conditions, providing a new approach for the resource utilization of plastic mulch. However, there are still many challenges in catalytic transformation and upgrading, including the high cost of catalysts and insufficient selectivity of target products. The future development of waste plastic mulch treatment technology will be discussed in depth.

Agricultural plastic films are predominantly used in soil environments abundant with microorganisms. In contrast to photocatalysis, which requires light sources and catalysts, or thermal catalysis, which operates under high-temperature conditions, biocatalysis (or enzymatic catalysis) emerges as a more promising and sustainable approach. This method does not require stringent conditions but instead relies on the development of efficient and cost-effective enzymatic catalysts capable of harnessing the metabolic potential of enzymes or microorganisms to convert plastic mulch films into reusable monomers, intermediates, or high-value-added products [104]. For instance, advancements in polyethylene oxidase research hold the potential to achieve the efficient biodegradation of polyethylene mulches, transforming them into organic acids, alcohols, or fuels. This process not only mitigates plastic pollution in soil and water but also contributes to the restoration of ecological balance.

The development of low-cost, high-performance catalysts is crucial for efficiently converting plastic waste into valuable products with high selectivity. Building on the success of thermal catalysis, it is important to leverage its advantages to design efficient photothermal catalysts. These catalysts can utilize light energy to generate higher temperatures, enhancing the transformation of plastic waste and advancing sustainable resource utilization.

The advancement of efficient pretreatment technologies plays a pivotal role in the transformation and upgrading of waste plastic mulch films. Implementing suitable pretreatment methods can substantially enhance catalytic conversion efficiency, elevate product quality, lower conversion costs, and minimize environmental impact. This underscores the critical importance of pretreatment as a foundational step in optimizing the overall process of plastic waste valorization.

Since plastics are primarily composed of hydrocarbon elements, the conventional transformation pathway for waste plastic mulch has focused on conversion into fuel. However, we posit that repurposing waste plastic mulch into carbon-based adsorption materials, catalyst supports, conductive materials, and other functional products holds substantial promise for addressing environmental challenges while unlocking economic potential. Such diversified valorization strategies not only mitigate plastic pollution but also contribute to the development of sustainable materials for advanced industrial applications, thereby aligning with the principles of a circular economy.

Addressing plastic pollution demands a unified approach involving the public, governments, NGOs, research institutions, media, and global organizations. Effective management practices and rigorous legal frameworks are crucial to curbing the release of plastic waste into the environment.

Policy advocacy is essential: we must establish clear standards and certification systems to ensure material biodegradability; introduce tax incentives or subsidies to reduce production costs and boost market competitiveness; and strengthen international cooperation to create a unified global regulatory framework for addressing cross-border pollution.

In short, the development of new degradable plastics or green and low-cost catalytic recycling methods for plastic mulch film is expected to achieve efficient degradation and high value-added transformation of plastic mulch film in the future. These technologies will not only help solve the problem of plastic pollution, but also promote the development of the circular economy and green chemistry, providing important support for the sustainable development goals.

**Author Contributions:** Conceptualization, methodology, resource investigation, writing—original draft, Y.W.; funding acquisition, writing—review and editing, Y.Y.; visualization, funding acquisition, project administration, writing—review and editing, supervision, W.Z. All authors have read and agreed to the published version of the manuscript.

**Funding:** This research was funded by the Young Talent Support Fund from Jiangsu University (5501310013), Guangdong Basic and Applied Basic Research Foundation (No. 2022A1515111017), the Natural Science Research Project of Universities in Jiangsu Province (24KJB610004), the Natural Science Foundation of Jiangsu Province (BK20240885), and the Senior Talents Fund of Jiangsu University (5501310033).

**Data Availability Statement:** No new data were created or analyzed in this study.

**Conflicts of Interest:** The authors declare that they have no known competing financial interests or personal relationships that could have appeared to influence the work reported in this paper.

## References

1. Huang, C.; Wang, L.; Wu, W.-M.; Capowiez, Y.; Qiao, Y.; Hou, D. When plastisphere and drilosphere meet: Earthworms facilitate microbiome and nutrient turnover to accelerate biodegradation of agricultural plastic films. *Environ. Int.* **2025**, *196*, 109309. [CrossRef] [PubMed]
2. Helm, L.T.; Venier-Cambron, C.; Verburg, P.H. The potential land-use impacts of bio-based plastics and plastic alternatives. *Nat. Sustain.* **2025**, *8*, 190–201. [CrossRef]
3. Sahai, H.; Aguilera del Real, A.M.; Alcayde, A.; Bueno, M.J.M.; Wang, C.; Hernando, M.D.; Alba, A.R.F. Key insights into microplastic pollution in agricultural soils: A comprehensive review of worldwide trends, sources, distribution, characteristics and analytical approaches. *Trends Anal. Chem.* **2025**, *185*, 118176. [CrossRef]
4. Wang, X.; Wang, G.; Guo, T.; Xing, Y.; Mo, F.; Wang, H.; Fan, J.; Zhang, F. Effects of plastic mulch and nitrogen fertilizer on the soil microbial community, enzymatic activity and yield performance in a dryland maize cropping system. *Eur. J. Soil Sci.* **2021**, *72*, 400–412. [CrossRef]
5. Mo, F.; Zhu, Y.; Wang, Z.-Y.; Deng, H.-L.; Li, P.-F.; Sun, S.-K.; Xiong, Y.-C. Polyethylene film mulching enhances the microbial carbon-use efficiency, physical and chemical protection of straw-derived carbon in an Entisol of the Loess Plateau. *Sci. Total Environ.* **2021**, *792*, 148357. [CrossRef]
6. Zhang, W.; Li, S.; Shen, Y.; Yue, S. Film mulching affects root growth and function in dryland maize-soybean intercropping. *Field Crops Res.* **2021**, *271*, 108240. [CrossRef]
7. Li, Y.; Yang, R.; Guo, L.; Gao, W.; Su, F.; Xu, Z.; Xiao, H.; Ma, Z.; Liu, X.; Gao, P.; et al. The composition, biotic network, and assembly of plastisphere protistan taxonomic and functional communities in plastic-mulching croplands. *J. Hazard. Mater.* **2022**, *430*, 128390. [CrossRef]
8. Ellis, L.D.; Rorrer, N.A.; Sullivan, K.P.; Otto, M.; McGeehan, J.E.; Román-Leshkov, Y.; Wierckx, N.; Beckham, G.T. Chemical and biological catalysis for plastics recycling and upcycling. *Nat. Catal.* **2021**, *4*, 539–556. [CrossRef]
9. Cui, H.; Yuan, L.; Li, C.; Lin, L. Control of Staphylococcus aureus on soya bean products by D-amino acids/nutmeg essential oil-co-loaded nanofilms. *Int. J. Food Sci. Technol.* **2017**, *52*, 2393–2403. [CrossRef]
10. Yan, H.; Ma, J.; Zhang, J.; Wang, G.; Zhang, C.; Akhlaq, M.; Huang, S.; Yu, J. Effects of film mulching on the physiological and morphological parameters and yield of cucumber under insufficient drip irrigation. *Irrig. Drain.* **2022**, *71*, 897–911. [CrossRef]
11. Shah, S.; Ilyas, M.; Liu, Z.-Y.; Zhang, D.-J.; Wu, G.; Huang, Q.-Y.; Elgizawy, K.K.; Yang, F.-L. Interaction of micro and nanoplastics (MNPs) with agricultural stored products and their pests. *Sci. Total Environ.* **2025**, *963*, 178534.
12. Gan, C.-D.; Liao, Y.-L.; Liu, H.-B.; Yang, J.-Y.; Nikitin, A. Microplastic-induced changes in Cd and Cr behavior in the agricultural soil-wheat system: Insights into metal bioavailability and phytotoxicity. *J. Hazard. Mater.* **2025**, *482*, 136592. [PubMed]
13. Shi, Z.; Zhang, X.; Liu, X.; Kang, M.; Yao, J.; Guo, L. Analysis and Test of the Tillage Layer Roll-Type Residual Film Recovery Mechanism. *Appl. Sci.* **2023**, *13*, 7598. [CrossRef]
14. Suzuki, G.; Uchida, N.; Tanaka, K.; Higashi, O.; Takahashi, Y.; Kuramochi, H.; Yamaguchi, N.; Osako, M. Global discharge of microplastics from mechanical recycling of plastic waste. *Environ. Pollut.* **2024**, *348*, 123855. [PubMed]
15. Mansoor, Z.; Tchuenbou-Magaia, F.; Kowalczuk, M.; Adamus, G.; Manning, G.; Parati, M.; Radecka, I.; Khan, H. Polymers Use as Mulch Films in Agriculture—A Review of History, Problems and Current Trends. *Polymers* **2022**, *14*, 5062. [CrossRef]
16. Khan, I.; Iqbal, B.; Khan, A.A.; Inamullah; Rehman, A.; Fayyaz, A.; Shakoor, A.; Farooq, T.H.; Wang, L.-X. The Interactive Impact of Straw Mulch and Biochar Application Positively Enhanced the Growth Indexes of Maize (*Zea mays* L.) Crop. *Agronomy* **2022**, *12*, 2584. [CrossRef]
17. Li, S.; Ding, F.; Flury, M.; Wang, Z.; Xu, L.; Li, S.; Jones, D.L.; Wang, J. Macro- and microplastic accumulation in soil after 32 years of plastic film mulching. *Environ. Pollut.* **2022**, *300*, 118945.
18. Zhang, Y.; Gao, Q.; Gao, Q.; Xu, M.; Fang, N.; Mu, L.; Han, X.; Yu, H.; Zhang, S.; Li, Y.; et al. Microplastics and nanoplastics increase major adverse cardiac events in patients with myocardial infarction. *J. Hazard. Mater.* **2025**, *489*, 137624. [CrossRef]
19. Campanale, C.; Galafassi, S.; Di Pippo, F.; Pojar, I.; Massarelli, C.; Uricchio, V.F. A critical review of biodegradable plastic mulch films in agriculture: Definitions, scientific background and potential impacts. *Trends Anal. Chem.* **2024**, *170*, 117391.
20. Li, X.; Zheng, G.; Li, Z.; Fu, P. Formulation, performance and environmental/agricultural benefit analysis of biomass-based biodegradable mulch films: A review. *Eur. Polym. J.* **2024**, *203*, 112663.
21. Cui, H.; Ma, C.; Lin, L. Synergetic antibacterial efficacy of cold nitrogen plasma and clove oil against Escherichia coli $O_{157}$:$H_7$ biofilms on lettuce. *Food Control* **2016**, *66*, 8–16. [CrossRef]
22. Tang, S.; Sun, P.; Ma, S.; Jin, W.; Zhao, Y. The interfacial behaviors of different arsenic species on polyethylene mulch film microplastics: Roles of the plastic additives. *J. Hazard. Mater.* **2023**, *442*, 130037. [CrossRef]

23. Iqbal, B.; Zhao, T.; Yin, W.; Zhao, X.; Xie, Q.; Khan, K.Y.; Zhao, X.; Nazar, M.; Li, G.; Du, D. Impacts of soil microplastics on crops: A review. *Appl. Soil Ecol.* **2023**, *181*, 104680. [CrossRef]
24. Xu, Z.; Zhang, J.; Qi, R.; Liu, Q.; Cao, H.; Wen, F.; Liao, Y.; Shih, K.; Tang, Y. Complex Release Dynamics of Microplastic Additives: An Interplay of Additive Degradation and Microplastic Aging. *J. Hazard. Mater.* **2025**, *490*, 137711. [CrossRef]
25. Wang, Y.; Wu, J.; Hartzell, E.J.; Hu, W.; Mahle, R.; Li, X.; Chen, Y.; Sahoo, J.K.; Chan, C.; Longo, B.N.; et al. Living plastics from plasticizer-assisted thermal molding of silk protein. *Nat. Commun.* **2025**, *16*, 52. [CrossRef]
26. Pettinelli, N.; Sabando, C.; Rodríguez-Llamazares, S.; Bouza, R.; Castaño, J.; Valverde, J.C.; Rubilar, R.; Frizzo, M.; Recio-Sánchez, G. Sodium alginate-g-polyacrylamide hydrogel for water retention and plant growth promotion in water-deficient soils. *Ind. Crops Prod.* **2024**, *222*, 119759. [CrossRef]
27. Bu, F.; Liu, J.; Song, Z.; Sun, M.; Wang, Z.; Jing, M.; Qian, W. Enhancement of water-retaining property and cracking resistance of soil under wetting-drying cycles: Synergistic effect of sisal fiber and polyacrylamide. *Constr. Build. Mater.* **2024**, *435*, 136841. [CrossRef]
28. Maes, T.; Preston-Whyte, F.; Lavelle, S.; Gomiero, A.; Booth, A.M.; Belzunce-Segarra, M.J.; Bellas, J.; Brooks, S.; Bakir, A.; Devriese, L.I.; et al. A recipe for plastic: Expert insights on plastic additives in the marine environment. *Mar. Pollut. Bull.* **2023**, *196*, 115633. [CrossRef]
29. Lakhiar, I.A.; Yan, H.; Zhang, J.; Wang, G.; Deng, S.; Bao, R.; Zhang, C.; Syed, T.N.; Wang, B.; Zhou, R.; et al. Plastic Pollution in Agriculture as a Threat to Food Security, the Ecosystem, and the Environment: An Overview. *Agronomy* **2024**, *14*, 548. [CrossRef]
30. Zhang, T.; Luo, X.-S.; Xu, J.; Yao, X.; Fan, J.; Mao, Y.; Song, Y.; Yang, J.; Pan, J.; Khattak, W.A. Dry–wet cycle changes the influence of microplastics (MPs) on the antioxidant activity of lettuce and the rhizospheric bacterial community. *J. Soils Sediments* **2023**, *23*, 2189–2201. [CrossRef]
31. Nihart, A.J.; Garcia, M.A.; El Hayek, E.; Liu, R.; Olewine, M.; Kingston, J.D.; Castillo, E.F.; Gullapalli, R.R.; Howard, T.; Bleske, B.; et al. Bioaccumulation of microplastics in decedent human brains. *Nat. Med.* **2025**. [CrossRef]
32. Cao, J.; Li, C.; Gao, X.; Cai, Y.; Song, X.; Siddique, K.H.M.; Zhao, X. Agricultural soil plastic as a hidden carbon source stimulates microbial activity and increases carbon dioxide emissions. *Resour. Conserv. Recycl.* **2023**, *198*, 107151. [CrossRef]
33. Khan, I.; Tariq, M.; Alabbosh, K.F.; Rehman, A.; Jalal, A.; Khan, A.A.; Farooq, M.; Li, G.; Iqbal, B.; Ahmad, N.; et al. Soil microplastics: Impacts on greenhouse gasses emissions, carbon cycling, microbial diversity, and soil characteristics. *Appl. Soil Ecol.* **2024**, *197*, 105343.
34. Khan, K.Y.; Tang, Y.; Cheng, P.; Song, Y.; Li, X.; Lou, J.; Iqbal, B.; Zhao, X.; Hameed, R.; Li, G.; et al. Effects of degradable and non-degradable microplastics and oxytetracycline co-exposure on soil $N_2O$ and $CO_2$ emissions. *Appl. Soil Ecol.* **2024**, *197*, 105331. [CrossRef]
35. Zhang, Y.-F.; Huang, Z.-Y.; Li, Y.-F.; Lu, X.-L.; Li, G.-R.; Qi, S.-S.; Khan, I.U.; Li, G.-L.; Dai, Z.-C.; Du, D.-L. The Degradability of Microplastics May Not Necessarily Equate to Environmental Friendliness: A Case Study of Cucumber Seedlings with Disturbed Photosynthesis. *Agriculture* **2024**, *14*, 53.
36. Zhi, L.; Li, Z.; Li, R.; Su, Z.; Chen, F.; Chen, G.; Huang, X.; Wang, J. Micro/nanoplastics in aquaponic systems: Sources, features and their effects on the immune system of fish. *Trends Anal. Chem.* **2025**, *184*, 118158.
37. Hoang, V.-H.; Nguyen, M.-K.; Hoang, T.-D.; Rangel-Buitrago, N.; Lin, C.; Pham, M.-T.; Ha, M.C.; Nguyen, T.P.; Shaaban, M.; Chang, S.W.; et al. Microplastic characteristics, transport, risks, and remediation in groundwater: A review. *Environ. Chem. Lett.* **2025**. [CrossRef]
38. Lahive, E.; Walton, A.; Horton, A.A.; Spurgeon, D.J.; Svendsen, C. Microplastic particles reduce reproduction in the terrestrial worm *Enchytraeus crypticus* in a soil exposure. *Environ. Pollut.* **2019**, *255*, 113174.
39. Dong, Y.-M.; Gao, M.-L.; Cai, Q.-Q.; Qiu, W.-W.; Xiao, L.; Chen, Z.-M.; Peng, H.-C.; Liu, Q.-H.; Song, Z.-G. The impact of microplastics on sulfur REDOX processes in different soil types: A mechanism study. *J. Hazard. Mater.* **2024**, *465*, 133432.
40. Jiao, M.; Wang, Y.; Li, T.; Li, R.; Liu, B. Riverine microplastics derived from mulch film in Hainan Island: Occurrence, source and fate. *Environ. Pollut.* **2022**, *312*, 120093.
41. Eze, C.G.; Nwankwo, C.E.; Dey, S.; Sundaramurthy, S.; Okeke, E.S. Food chain microplastics contamination and impact on human health: A review. *Environ. Chem. Lett.* **2024**, *22*, 1889–1927. [CrossRef]
42. Li, X.A.; Ding, G.L.; Li, T.; Pu, Q.K.; Wang, Z.H.; Li, Y.; Jiang, X.; Li, X.X. Microplastics in freshwater food chains: Priority list based on identification of oxidative stress response characteristic. *J. Hazard. Mater.* **2024**, *480*, 135893. [CrossRef] [PubMed]
43. Yuan, Y.; Qin, Y.-W.; Wang, M.; Xu, W.-Q.; Chen, Y.; Zheng, L.-P.; Chen, W.; Luo, T. Microplastics from agricultural plastic mulch films: A mini-review of their impacts on the animal reproductive system. *Ecotox. Environ. Saf.* **2022**, *244*, 114030. [CrossRef]
44. Wang, Y.J.; Nan, X.Y.; Sun, H.Y.; Miao, J.X.; Li, Y.H.; Han, X.Y.; Zhang, N.; Wang, H.Q.; Ren, N.Q.; Zhao, X.; et al. From insects to mammals! Tissue accumulation and transgenerational transfer of micro/nano-plastics through the food chain. *J. Hazard. Mater.* **2024**, *480*, 136424. [CrossRef] [PubMed]
45. Abbasi, S.; Turner, A. Dry and wet deposition of microplastics in a semi-arid region (Shiraz, Iran). *Sci. Total Environ.* **2021**, *786*, 147358. [CrossRef]

46. Wang, X.; Hong, T.; Fang, W.; Chen, X. Optimized Design for Vibration Reduction in a Residual Film Recovery Machine Frame Based on Modal Analysis. *Agriculture* **2024**, *14*, 543. [CrossRef]
47. Mulakkal, M.C.; Castillo Castillo, A.; Taylor, A.C.; Blackman, B.R.K.; Balint, D.S.; Pimenta, S.; Charalambides, M.N. Advancing mechanical recycling of multilayer plastics through finite element modelling and environmental policy. *Resour. Conserv. Recycl.* **2021**, *166*, 105371. [CrossRef]
48. Wang, C.; Han, H.; Wu, Y.; Astruc, D. Nanocatalyzed upcycling of the plastic wastes for a circular economy. *Coord. Chem. Rev.* **2022**, *458*, 214422. [CrossRef]
49. Vollmer, I.; Jenks, M.J.F.; Roelands, M.C.P.; White, R.J.; van Harmelen, T.; de Wild, P.; van der Laan, G.P.; Meirer, F.; Keurentjes, J.T.F.; Weckhuysen, B.M. Beyond Mechanical Recycling: Giving New Life to Plastic Waste. *Angew. Chem. Int. Ed.* **2020**, *59*, 15402–15423. [CrossRef]
50. Khan, A.; Naveed, M.; Rabnawaz, M. Melt-reprocessing of mixed polyurethane thermosets. *Green Chem.* **2021**, *23*, 4771–4779. [CrossRef]
51. Ghosh, A. Performance modifying techniques for recycled thermoplastics. *Resour. Conserv. Recycl.* **2021**, *175*, 105887.
52. Zhang, Z.; Ju, R.; Zhou, H.; Chen, H. Migration characteristics of heavy metals during sludge pyrolysis. *Waste Manag.* **2021**, *120*, 25–32.
53. Rahimi, A.; García, J.M. Chemical recycling of waste plastics for new materials production. *Nat. Rev. Chem.* **2017**, *1*, 0046.
54. Singh, N.; Hui, D.; Singh, R.; Ahuja, I.P.S.; Feo, L.; Fraternali, F. Recycling of plastic solid waste: A state of art review and future applications. *Compos. Part B Eng.* **2017**, *115*, 409–422.
55. Xuan Lu, D.; Giustozzi, F. Recycled plastics as synthetic coarse and fine asphalt aggregate. *Int. J. Pavement Eng.* **2023**, *24*, 2068550.
56. Aldagari, S.; Karam, J.; Kazemi, M.; Kaloush, K.; Fini, E.H. Comparing the critical aging point of rubber-modified bitumen and plastic-modified bitumen. *J. Clean. Prod.* **2024**, *437*, 140540.
57. Lee, T.; Jung, S.; Lee, S.; Tsang, Y.F.; Lee, K.H.; Kwon, E.E. Production of aviation fuel via thermal cracking of plastic waste. *Energy Convers. Manag.* **2024**, *315*, 118827.
58. Kusenberg, M.; Eschenbacher, A.; Djokic, M.R.; Zayoud, A.; Ragaert, K.; De Meester, S.; Van Geem, K.M. Opportunities and challenges for the application of post-consumer plastic waste pyrolysis oils as steam cracker feedstocks: To decontaminate or not to decontaminate? *Waste Manag.* **2022**, *138*, 83–115.
59. Dai, L.; Zhou, N.; Lv, Y.; Cheng, Y.; Wang, Y.; Liu, Y.; Cobb, K.; Chen, P.; Lei, H.; Ruan, R. Pyrolysis technology for plastic waste recycling: A state-of-the-art review. *Prog. Energy Combust. Sci.* **2022**, *93*, 101021.
60. Gracida-Alvarez, U.R.; Mitchell, M.K.; Sacramento-Rivero, J.C.; Shonnard, D.R. Effect of Temperature and Vapor Residence Time on the Micropyrolysis Products of Waste High Density Polyethylene. *Ind. Eng. Chem. Res.* **2018**, *57*, 1912–1923.
61. Zixi, H.; Xu, Z.; Wuzhi, H.; Yong, H. Pyrolytic biochar from plastic film waste addition on farmland for maize growth improvement: Process and effect study. *Waste Manag.* **2023**, *157*, 210–218. [CrossRef] [PubMed]
62. Yan, Y.; Wang, C.; Liu, Q.; Liu, Z. Thermochemical degradation of agricultural mulch films for potential organic fertilizer precursors: Low-temperature pyrolysis in nitric acid. *J. Anal. Appl. Pyrol.* **2023**, *176*, 106256. [CrossRef]
63. Praveenkumar, T.R.; Sekar, M.; Pasupuleti, R.R.; Gavurová, B.; Arun Kumar, G.; Vignesh Kumar, M. Current technologies for plastic waste treatment for energy recovery, it's effects on poly aromatic hydrocarbons emission and recycling strategies. *Fuel* **2024**, *357*, 129379. [CrossRef]
64. Chen, R.; Wang, J.; Zhang, C.; Sun, Y.; Li, J.; Dong, F. Purification and Value-Added Conversion of NOx under Ambient Conditions with Photo-/Electrocatalysis Technology. *Environ. Sci. Technol.* **2025**, *59*, 1013–1033. [CrossRef]
65. Yuan, Y.; Zhang, Y.; Menzel, J.P.; Santoro, J.; Dolack, M.; Wang, H.; Batista, V.; Wang, D. High-Conversion Propane Dehydrogenation by Photocatalysis under Ambient Conditions. *ACS Catal.* **2024**, *14*, 17445–17452.
66. Shen, L.; Zhou, X.; Zhang, C.; Yin, H.; Wang, A.; Wang, C. Functional characterization of bimetallic CuPd nanoparticles in hydrothermal conversion of glycerol to lactic acid. *J. Food Biochem.* **2019**, *43*, e12931.
67. Li, M.; Jiang, H.; Zhang, L.; Yu, X.; Liu, H.; Yagoub, A.E.A.; Zhou, C. Synthesis of 5-HMF from an ultrasound-ionic liquid pretreated sugarcane bagasse by using a microwave-solid acid/ionic liquid system. *Ind. Crops Prod.* **2020**, *149*, 112361.
68. Shao, S.; Sun, T.; Li, X.; Wang, Y.; Ma, L.; Liu, Z.; Wu, S. Preparation of heavy bio-oil-based porous carbon by pyrolysis gas activation and its performance in the aldol condensation for aviation fuel as catalyst carrier. *Ind. Crops Prod.* **2024**, *218*, 118963.
69. Lee, K.; Jing, Y.; Wang, Y.; Yan, N. A unified view on catalytic conversion of biomass and waste plastics. *Nat. Rev. Chem.* **2022**, *6*, 635–652. [CrossRef]
70. Xu, S.; Tang, J.; Fu, L. Catalytic Strategies for the Upcycling of Polyolefin Plastic Waste. *Langmuir* **2024**, *40*, 3984–4000.
71. Ji, Q.; Yu, X.; Chen, L.; Yarley, O.P.N.; Zhou, C. Facile preparation of sugarcane bagasse-derived carbon supported $MoS_2$ nanosheets for hydrogen evolution reaction. *Ind. Crops Prod.* **2021**, *172*, 114064.
72. Wan, Y.; Wang, H.; Liu, J.; Liu, X.; Song, X.; Zhou, W.; Zhang, J.; Huo, P. Enhanced degradation of polyethylene terephthalate plastics by $CdS/CeO_2$ heterojunction photocatalyst activated peroxymonosulfate. *J. Hazard. Mater.* **2023**, *452*, 131375.

73. Wan, Y.; Wang, H.; Liu, J.; Lei, Y.; Ma, W.; Song, X.; Zhou, W.; Liu, X.; Zhang, J.; Li, B.; et al. Fabricated S-scheme $CoS/CeO_2$ heterojunction photocatalyst efficiently activates peroxymonosulfate for polyethylene terephthalate plastic degradation: Insight into radicals and electron transfer. *Chem. Eng. J.* **2024**, *502*, 158124. [CrossRef]
74. Liu, X.; Wang, X.; Chu, M.; Zhang, W.; Fu, J.; Li, S.; Wang, L.; Chen, J.; Zhang, Q.; Cao, M. Selective Liquid Chemical Production in Waste Polyolefin Photorefinery by Controlling Reactive Species. *J. Am. Chem. Soc.* **2025**, *147*, 5228–5237. [CrossRef]
75. Liang, J.; Li, H.; Chen, L.; Ren, M.; Fakayode, O.A.; Han, J.; Zhou, C. Efficient hydrogen evolution reaction performance using lignin-assisted chestnut shell carbon-loaded molybdenum disulfide. *Ind. Crops Prod.* **2023**, *193*, 116214.
76. Xu, T.; Shan, T.; Jiang, Y.; Xu, L.-h.; Zhang, H.; Chu, S. Photocatalytic Upcycling of Plastic Waste into Syngas by $ZnS/Ga_2O_3$ Z-Scheme Heterojunction. *ChemSusChem* **2024**, e202402310. [CrossRef]
77. Jiang, Y.; Zhang, H.; Hong, L.; Shao, J.; Zhang, B.; Yu, J.; Chu, S. An Integrated Plasma–Photocatalytic System for Upcycling of Polyolefin Plastics. *ChemSusChem* **2023**, *16*, e202300106. [PubMed]
78. Wen, Q.; Zhang, Q.; Liu, Z.; Wang, H.; Hao, S.; Zhang, F.; Zhang, L.; Han, Q.; Zheng, G. Brønsted-acid sites induced photocatalytic cracking of low-polarity polyethylene plastics. *J. Energy Chem.* **2024**, *96*, 509–515. [CrossRef]
79. Bu, Q.; Chen, K.; Morgan, H.M.; Liang, J.; Zhang, X.; Yan, L.; Mao, H. Thermal Behavior and Kinetic Study of the Effects of Zinc-Modified Biochar Catalyst on Lignin and Low-Density Polyethylene (LDPE) Co-Pyrolysis. *Trans. ASABE* **2018**, *61*, 1783–1793.
80. Li, J.; Liu, T.; Palansooriya, K.N.; Yu, D.; Wan, G.; Sun, L.; Chang, S.X.; Wang, Y. Zeolite-catalytic pyrolysis of waste plastics: Machine learning prediction, interpretation, and optimization. *Appl. Energy* **2025**, *382*, 125258. [CrossRef]
81. Xu, D.; Huang, G.; Guo, L.; Chen, Y.; Ding, C.; Liu, C. Enhancement of catalytic combustion and thermolysis for treating polyethylene plastic waste. *Adv. Compos. Hybrid Mater.* **2022**, *5*, 113–129. [CrossRef]
82. Kong, G.; Yuan, X.; Zhao, X.; Wang, J.; Zhong, W.; Qiu, R.; Song, K.; Cheng, Q.; Zhang, X.; Han, L. Solvent-free upcycling of agricultural plastic waste using in situ self-assembly metal nanoparticles co-doped microporous carbocatalyst for advanced transportation fuels. *J. Hazard. Mater.* **2025**, *486*, 137086. [CrossRef] [PubMed]
83. Jung, J.-M.; Cho, S.-H.; Jung, S.; Lin, K.-Y.A.; Chen, W.-H.; Tsang, Y.F.; Kwon, E.E. Disposal of plastic mulch film through $CO_2$-assisted catalytic pyrolysis as a strategic means for microplastic mitigation. *J. Hazard. Mater.* **2022**, *430*, 128454. [CrossRef] [PubMed]
84. Liu, Y.; Li, C.; Yu, H.; Guo, J. The "Midas touch": Photothermal catalysis enables recycling of plastic waste. *Chem* **2022**, *8*, 2586–2588. [CrossRef]
85. Xing, C.; Cai, H.; Kang, D.; Sun, W. Photothermal Catalysis: An Emerging Green Approach to Upcycling Plastic Waste. *Adv. Energy Sustain. Res.* **2023**, *4*, 2300015. [CrossRef]
86. Miao, Y.; Zhao, Y.; Waterhouse, G.I.N.; Shi, R.; Wu, L.-Z.; Zhang, T. Photothermal recycling of waste polyolefin plastics into liquid fuels with high selectivity under solvent-free conditions. *Nat. Commun.* **2023**, *14*, 4242. [CrossRef] [PubMed]
87. Han, H.; Yan, P.; Li, Q.; Zhang, S.; Jiao, B.; Wei, G.; Wang, Z.; Cao, M.; Xu, P.; Zhang, Q.; et al. Photothermal Upcycling of Waste Polyvinyl Chloride Plastics. *Environ. Sci. Technol.* **2024**, *58*, 21861–21870. [CrossRef]
88. Chu, M.; Yan, P.; Zhou, Y.; Lou, X.; Li, C.; Cao, M.; He, L.; Zhang, Q.; Chen, J. Plasmonic Copper-Ruthenium Superstructure for Efficient Photothermal Conversion and Plastic Recycling. *Adv. Funct. Mater.* **2024**, *35*, 2417644. [CrossRef]
89. Sajwan, D.; Sharma, A.; Sharma, M.; Krishnan, V. Upcycling of Plastic Waste Using Photo-, Electro-, and Photoelectrocatalytic Approaches: A Way toward Circular Economy. *ACS Catal.* **2024**, *14*, 4865–4926. [CrossRef]
90. Ma, G.; Shi, Q.; Hou, X.; Peng, Y.; Liu, Q. An electrochemical sensor for simultaneous voltammetric detection of ascorbic acid and dopamine enabled by higher electrocatalytic activity of co-modified MCM-41 mesoporous molecular sieve. *Front. Sustain. Food Syst.* **2024**, *8*, 1448421. [CrossRef]
91. Qi, J.; Xia, Y.; Meng, X.; Li, J.; Yang, S.; Zou, H.; Ma, Y.; Zhang, Y.; Du, Y.; Zhang, L.; et al. Cation-Vacancy Engineering in Cobalt Selenide Boosts Electrocatalytic Upcycling of Polyester Thermoplastics at Industrial-Level Current Density. *Adv. Mater.* **2025**, *37*, 2419058.
92. Sanchez-Hernandez, J.; Capowiez, Y.; Ro, K.-S. Potential Use of Earthworms to Enhance Decaying of Biodegradable Plastics. *ACS Sustain. Chem. Eng.* **2020**, *8*, 4292–4316.
93. Cui, H.; Surendhiran, D.; Li, C.; Lin, L. Biodegradable zein active film containing chitosan nanoparticle encapsulated with pomegranate peel extract for food packaging. *Food Packag. Shelf Life* **2020**, *24*, 100511.
94. Payanthoth, N.S.; Mut, N.N.N.; Samanta, P.; Li, G.; Jung, J. A review of biodegradation and formation of biodegradable microplastics in soil and freshwater environments. *Appl. Biol. Chem.* **2024**, *67*, 110.
95. Eisenreich, F.; Tomović, Ž. Biorenewable circularity aids sustainability of plastics. *Nat. Sustain.* **2023**, *6*, 1293–1294. [CrossRef]
96. Wang, Z.; Yang, Z.; Qiao, P.; Fang, J.; Wu, X.; Yang, Z.; Li, H. Degradable Polyvinyl Alcohol Liquid Mulch Film Synergistically Reinforced by Cellulose and Lignin. *J. Appl. Polym. Sci.* **2025**, *142*, e56586.
97. Chen, Y.; Chen, H.; Dai, Y.; Xiao, J.; Qiu, F.; Zhang, T. Honeycomb porous regenerated cellulose aerogel films with enhanced thermal dissipation for agricultural mulch application. *Food Bioprod. Process.* **2024**, *147*, 418–427.

98. Yoshinaka, Y.; Miller, S.A. Bio-oil derived polyesteramides as water-degradable replacements for polyethylene. *Green Chem.* **2025**. [CrossRef]
99. Zhu, Y.; Fu, B.; Zhu, P.; Shu, S.; Wang, D.; Xu, H.; Cai, D. Waste milk-derived biodegradable mulch film fabrication and nutrient slow-release performance evaluation. *Chem. Eng. J.* **2024**, *497*, 155008. [CrossRef]
100. Gao, S.; Zhao, J.; Yang, C.; Zhang, Y.; Fu, Y.; Zhang, A.; Yang, X.; Wang, Z.; Tian, G. Synthesis of lignin-based polyurethane (LPU) and fabrication of degradable mulch film through LPU coating on cellulose paper. *Ind. Crops Prod.* **2024**, *210*, 118099.
101. Wang, J.; Liang, Y.; Chen, Y.; Wan, H.; Jin, W.; Luo, T.; Chen, Y.; Wei, P.; Huang, S.; He, Y.; et al. Highly degradable bio-based plastic with water-assisted shaping process and exceptional mechanical properties. *Carbohydr. Polym.* **2025**, *347*, 122773. [PubMed]
102. Li, Y.; Liu, C.; Wei, H.; Gai, X.; Lei, T.; Wang, Y.; Li, Q.; Xiao, H. Recent Advances of Biodegradable Agricultural Mulches from Renewable Resources. *ACS Sustain. Chem. Eng.* **2023**, *11*, 14866–14885.
103. Su, W.; Yang, Z.; Wang, H.; Fang, J.; Li, C.; Lyu, G.; Li, H. Synergistic Effect of Sodium Alginate and Lignin on the Properties of Biodegradable Poly(vinyl alcohol) Mulch Films. *ACS Sustain. Chem. Eng.* **2022**, *10*, 11800–11814.
104. Zan, X.; Yan, Y.; Chen, G.; Sun, L.; Wang, L.; Wen, Y.; Xu, Y.; Zhang, Z.; Li, X.; Yang, Y.; et al. Recent Advances of Oxalate Decarboxylase: Biochemical Characteristics, Catalysis Mechanisms, and Gene Expression and Regulation. *J. Agric. Food Chem.* **2024**, *72*, 10163–10178.

catalysts

MDPI

*Review*

# Application of Graphene-Based Solar Driven Interfacial Evaporation-Coupled Photocatalysis in Water Treatment

**Yining Zhang [1], Huiqin Wang [2,*] and Jisheng Zhang [1,3,*]**

1 Institute of Green Chemistry and Chemical Technology, School of Chemistry & Chemical Engineering, Jiangsu University, Zhenjiang 212013, China
2 School of Energy and Power Engineering, Jiangsu University, Zhenjiang 212013, China
3 Jiangsu Agrochem Laboratory Co., Ltd., Changzhou 213022, China
* Correspondence: hqwang@ujs.edu.cn (H.W.); jszhang@ujs.edu.cn (J.Z.)

**Abstract:** The global shortage of freshwater resources and the energy crisis have propelled solar-driven interfacial evaporation (SDIE) coupled with photocatalytic technology to become a research focus in efficient and low-carbon water treatment. Graphene-based materials demonstrate unique advantages in SDIE–photocatalysis integrated systems, owing to their broadband light absorption, ultrafast thermal carrier dynamics, tunable electronic structure, and low evaporation enthalpy characteristics. This review systematically investigates the enhancement mechanisms of graphene photothermal conversion on photocatalytic processes, including (1) improving light absorption through surface morphology modulation, defect engineering, and plasmonic material compositing; (2) reducing water evaporation enthalpy via hydrophilic functional group modification and porous structure design; (3) suppressing heat loss through thermal insulation layers and 3D structural optimization; and (4) enhancing water transport efficiency via fluid channel engineering and wettability control. Furthermore, salt resistance strategies and structural optimization significantly improve system practicality and stability. In water treatment applications, graphene-based SDIE systems achieve synergistic "adsorption–catalysis–evaporation" effects, enabling efficient the degradation of organic pollutants, reduction in/fixation of heavy metal ions, and microbial inactivation. However, practical implementation still faces challenges including low steam condensation efficiency, insufficient long-term material durability, and high scaling-up costs. Future research should prioritize enhancing heat and mass transfer in condensation systems, optimizing material environmental adaptability, and developing low-cost manufacturing processes to promote widespread application of graphene-based SDIE–photocatalysis integrated systems.

**Keywords:** graphene; solar-driven interface evaporation; photothermal conversion; photocatalytic coordination; water treatment; seawater desalination

## 1. Introduction

The rapid socio-economic development and significant global population growth have exacerbated freshwater shortages and energy crises worldwide, with nearly half of the population facing water stress [1–3]. Although the Earth has abundant water resources, only 3% is freshwater, which is unsuitable for direct consumption or use in households and agriculture [4,5]. Therefore, the development of efficient and low-energy-consumption water treatment technologies has become a core challenge in the field of environmental science. Traditional water treatment methods often face bottlenecks such as high energy consumption, reliance on chemical reagents, and secondary pollution. As a result, it is necessary to explore alternative strategies to address global freshwater scarcity and complex water pollution issues [6–8].

In pursuit of sustainable development goals, the utilization of clean solar energy has rapidly advanced across various fields in forms such as photochemical synthesis, photovoltaic power generation, photothermal conversion, and photocatalysis [8–12]. As a cutting-edge approach in this field, solar-powered interfacial evaporation (SDIE) represents a novel vapor generation methodology. This technique utilizes photothermal conversion materials operating at gas-liquid interfaces, implementing a spatially confined thermal management paradigm through its distinctive "targeted heating" mechanism. SDIE technology focuses solar energy at the water–air interface, either by heating large volumes of water with high thermal inertia or by heating only a thin layer of water at the interface, which has negligible thermal inertia and, thus, responds very quickly [13]. This technology harnesses renewable solar energy to replace traditional energy sources, obtaining clean water with minimal environmental impact, addressing freshwater shortages, and achieving low-carbon, efficient, and sustainable desalination. It has become a significant research focus in the field of clean water resource acquisition [14].

However, the singular evaporation process, being fundamentally a physical phase-change separation mechanism, exhibits constrained purification efficacy due to its dependence on pollutant vapor pressure characteristics. This process inevitably leads to co-evaporation of contaminants with water vapor. For high-boiling-point, non-volatile organic compounds, while apparent removal can be achieved through interception effects, the enrichment of pollutants in concentrated solutions still poses secondary contamination risks. Furthermore, the concentration polarization layer formed at the evaporation interface accelerates inorganic salt crystallization, inducing scaling on photothermal material surfaces that causes irreversible decay in light absorption efficiency and evaporation rate degradation. In this context, the synergistic integration of interfacial evaporation with photocatalytic processes establishes a novel paradigm for multi-energy field coordination in water treatment, enabling the efficient conversion of photo-thermal-chemical energy coupled with simultaneous pollutant removal. Photocatalytic technology leverages the photoexcitation properties of semiconductor materials to generate high-energy electron-hole pairs that initiate advanced oxidation processes (AOPs) [15]. Following photoexcitation, valence band holes directly oxidize organic compounds or react with water molecules to produce ·OH radicals, while conduction band electrons reduce $O_2$ to form $\cdot O_2^-$ species, ultimately mineralizing pollutants into $CO_2$ and $H_2O$. Nevertheless, critical limitations including high charge carrier recombination rates and low energy utilization efficiency persist as key performance constraints. Therefore, it is imperative to engineer an optimized photothermal-photocatalytic synergistic system that overcomes these inherent drawbacks while achieving enhanced evaporation performance and profound contaminant degradation.

Based on the photothermal conversion mechanism, photothermal materials are primarily categorized into four types: plasmonic metal nanoparticles [16–19], semiconductors [20–22], polymers, and carbon-based materials [23–29]. Among these, carbon-based materials have been widely used across various fields due to their excellent photothermal properties, low cost, and abundant source materials [30–32]. Graphene emerges as an exceptional medium for photon-to-thermal energy conversion, exhibiting intrinsic material properties uniquely suited to this energy transformation mechanism. Compared to other two-dimensional materials—such as transition metal dichalcogenides constrained by narrow-band light absorption, MXenes with limited photothermal conversion efficiency, and graphitic carbon nitride suffering from high carrier recombination rates—graphene emerges as an ideal platform for photothermal-photochemical synergistic water treatment, owing to its broadband light absorption, ultrafast hot carrier dynamics, tunable electronic structure, and low evaporation enthalpy [33]. Additionally, as a novel low-dimensional quantum material, graphene provides a unique platform for investigating photoexcited hot carrier dynamics and their energy conversion applications. Its unique electronic structure offers three key advantages: Firstly, the strong electron–electron interactions (with Coulomb interaction strength reaching ~25 eV) facilitate the thermalization process of carriers on an ultrafast timescale (approximately 50 fs), with experiments confirming that over 50% of the energy from photoexcited carriers can be efficiently converted into thermal energy of the electronic system in the initial stage [34,35]. This unique ultrafast thermalization mechanism effectively avoids the phonon scattering energy losses commonly found in traditional semiconductors [36]. Secondly, the two-dimensional Dirac fermion nature of graphene significantly reduces the electron–phonon coupling strength (with a coupling constant $\lambda$~0.05), allowing the thermalized carriers to maintain a non-equilibrium distribution for over 1 ps, which is 2–3 orders of magnitude longer than in conventional semiconductor materials. Notably, under continuous wave (CW) excitation, by optimizing the dynamic balance between light absorption and heat dissipation, a stable non-equilibrium state can be achieved where the electron temperature (Te) significantly exceeds the lattice temperature (Tph) (Te $\gg$ Tph). Thirdly, gate modulation allows for continuous tuning of the Fermi level within the range of 0.3–1.5 eV, enabling precise control over the Schottky barrier height, which provides an important dimension for regulating the selective energy transport of thermal carriers. Leveraging these unique quantum attributes, graphene is increasingly recognized as a prime candidate for modulable thermionic carrier injection with optimized energy conversion efficiency [37]. Despite remarkable advancements in graphene-based SDIE systems, three fundamental challenges persist: (i) insufficient mechanistic understanding of the synergistic interplay between photothermal energy conversion and interfacial fluid dynamics, (ii) unverified durability under complex aqueous matrices containing organic/inorganic contaminants, and (iii) suboptimal economic viability for scalable manufacturing.

Despite significant progress in this field, graphene-based SDIE-coupled photocatalytic systems still face critical challenges: the synergistic optimization mechanisms between photothermal conversion and photocatalytic carrier transport remain ambiguous, long-term stability under complex water matrices requires urgent verification, and the cost-effectiveness ratio of scaled-up fabrication demands further improvement. This paper mainly reviews the optimization of the graphene-based SIED system and the effective application of the graphene-based SIED catalytic system in water treatment in recent years in terms of graphene optimization modification and structural adjustment, and reviews the problems and challenges it still faces in the field of seawater desalination, in order to provide a theoretical reference and technical inspiration for the development of the next generation of efficient solar water treatment systems (Figure 1).

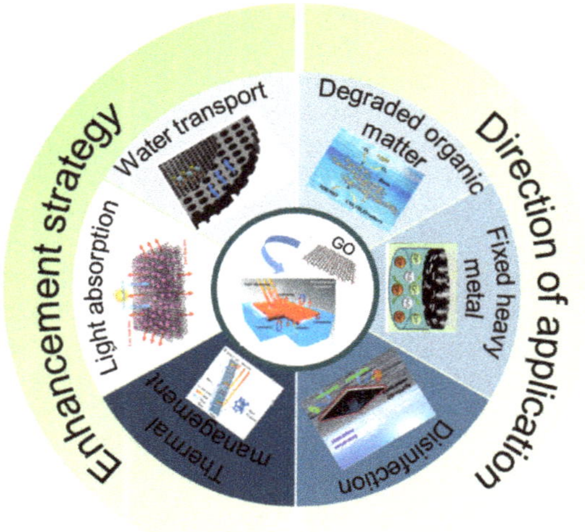


**Figure 1.** Schematic of enhancement strategies and application directions for graphene-based SDIE coupled photocatalytic systems [38–43].

## 2. Promotion of Photocatalysis by Photothermal Conversion of Graphene

Graphene is a single-layer, two-dimensional honeycomb lattice structure material composed of carbon atoms arranged in $sp^2$ hybridized orbitals [44]. This material has garnered ubiquitous research attention owing to its array of attractive properties, including excellent tunability [45–47], high fluid permeability [48,49], high thermal conductivity [50,51], abundant active sites [52,53], superior carrier mobility at room temperature [54–56], good electrical conductivity [57–59], extremely high theoretical specific surface area [60,61], excellent environmental compatibility [62], and strong adsorption capacity for both organic and inorganic molecules [63–65]. These characteristics collectively demonstrate the immense potential of graphene-based materials.

Graphene features both $\sigma$ and $\pi$ bonds, with the $\pi$ bonds formed by 2p orbitals being responsible for its unique properties. On one hand, compared to the larger energy gap between $\sigma$ and $\sigma^*$ orbitals, the $\pi$ bonds, which are generally weaker than $\sigma$ bonds, allow electrons to be excited from the $\pi$ to the $\pi^*$ orbitals with relatively low energy input. The delocalized $\pi$-electron networks induce bathochromic shifts in optical absorption profiles through conjugation-length-dependent bandgap narrowing. Enhanced $\pi$-orbital overlap progressively reduces the HOMO-LUMO energy separation, enabling photon harvesting across visible to near-infrared regimes. This structural configuration supports multi-mode $\pi$-$\pi$ electronic transitions under broad-spectrum solar excitation. Consequently, graphene exhibits strong absorption across a broad solar spectrum. Moreover, when excited electrons relax back to the ground state, they release heat. The large absorptive surface area and efficient photothermal conversion make graphene a representative photothermal material (Figure 2a). Additionally, a pristine monolayer of graphene has a Young's modulus of about 1.0 TPa, a third-order elastic stiffness of approximately 2.0 TPa, and an intrinsic strength of 130 GPa. Experimental evaluations confirm the composite's exceptional structural robustness, where graphene integration significantly improves mechanical characteristics through interfacial reinforcement mechanisms. The thermal conduction behavior in graphene arises from lattice vibrational quanta propagation dynamics, manifesting as temperature-dependent transport regimes: Disordered phonon scattering dominates at elevated temperatures, while coherent wave-like phonon transmission prevails in cryogenic conditions. State-of-the-art computational models identify defect configuration, strain field gradients, and morphological parameters as critical factors modulating thermal conductivity via selective phonon mode suppression. Although reported thermal conductivity values exhibit methodological variations, consensus measurements surpass 2000 $W{\cdot}m^{-1}{\cdot}K^{-1}$, substantiating graphene's unparalleled heat dissipation capacity for advanced thermal regulation systems [66,67].

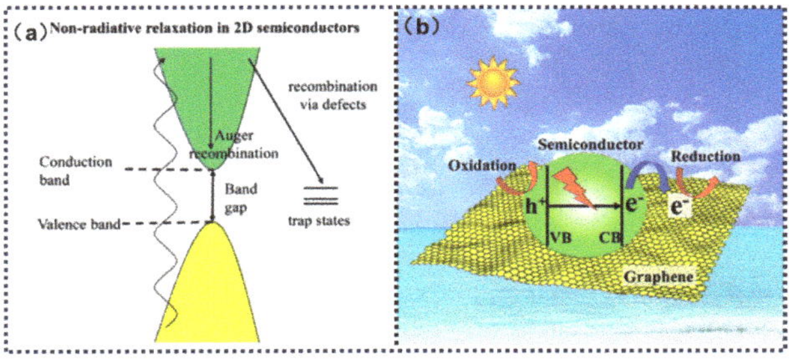


**Figure 2.** (**a**) The photothermal mechanism of graphene. Reprinted with permission from ref. [68]. Copyright 2020 John Wiley and Sons. (**b**) The photocatalytic mechanism of graphene. Reprinted with permission from ref. [69]. Copyright 2025 John Wiley and Sons.

Meanwhile, the localized temperature rise induced by the photothermal effect exerts multi-dimensional synergistic enhancements on photocatalytic processes. First, the input of thermal energy reduces the recombination probability of photogenerated carriers by enhancing their kinetic energy to overcome the Schottky barrier at heterojunction interfaces, thereby accelerating the directional separation and migration of electron–hole pairs (Figure 2b). Second, elevated temperature significantly modulates the adsorption–desorption equilibrium of reactant molecules on the catalyst surface, lowering the reaction activation energy and promoting the diffusion mass transfer of intermediate products, which optimizes the kinetic pathways of redox reactions. More importantly, the interfacial coupling effects between graphene and photocatalytic components enable cooperative regulation of energy band structures. Photothermally induced lattice vibrations (alterations in phonon modes) may facilitate the formation of dynamic charge transport channels at heterojunction interfaces, while thermal perturbations partially compensate for the bandgap limitations of semiconductor materials, allowing carrier transitions under lower-energy photon excitation. This photothermal synergy mechanism not only enhances the comprehensive utilization efficiency of solar energy but also provides theoretical foundations for designing high-efficiency photocatalytic systems capable of adapting to ambient temperature variations. Through precise regulation of the microstructure, interfacial properties, and photothermal response behavior of graphene composites, cross-scale optimization of energy transfer and material conversion in photocatalytic reactions can be achieved, paving new pathways for clean energy conversion and environmental remediation technologies.

## 3. Strategies for Enhancing System of SDIE-Coupled Photocatalytic

### *3.1. Enhancing Light Absorption*

Optical harvesting efficiency serves as a critical determinant of photothermal conversion efficacy in evaporative systems, fundamentally modulating the photonic-to-thermal energy conversion dynamics that govern vapor generation kinetics. Ideal photothermal materials should minimize the reflection and transmission of sunlight across the entire wavelength range to achieve high solar absorption and efficient photothermal conversion [70]. Graphene, with its unique electronic structure, acts as a zero-gap material with metallic-like properties and linear energy dispersion near these points. Nair and colleagues discovered that suspended graphene exhibits a broadband and relatively high light absorption rate of 2.3%. Additionally, the absorption increases linearly with the number of layers, with each additional layer increasing absorption by another 2.3% (Figure 3a) [71]. Paradoxically, the elevated infrared emissivity inherent to graphene architectures constrains their photothermal conversion efficiency, mandating hybrid integration with photon-trapping nanostructures to achieve panchromatic absorption profiles [16]. By modifying graphene-based materials through surface structure design, plasmonic material modification, defect control, and hybrid material design, researchers can develop efficient, stable, and cost-effective photothermal materials, advancing society towards more sustainable and environmentally friendly solutions.

During the interfacial solar evaporation process, the surface of the evaporator acts as the evaporation front, and its surface structure significantly influences the water evaporation behavior. Flat structures have a theoretical limit on energy efficiency due to diffuse reflection and thermal radiation. This limitation can be overcome by increasing surface roughness. Uneven structures provide a unique light-trapping space, allowing incident light to reflect multiple times, thereby significantly enhancing light absorption and photothermal conversion efficiency. Therefore, designing evaporators with high surface roughness is crucial for improving the material's light absorption properties [70,72,73]. Zhou et al. [74] proposed a novel strategy to modulate the surface morphology of graphene oxide (GO) using polyaniline (PANI) nanocone arrays. By employing electrostatic assembly and in situ polymerization, the GO surface was fully hybridized with the PANI nanocone arrays, forming a highly folded periodic structure (Figure 3b). This structure not only enhanced the material's light absorption capability but also improved the thermal localization effect, significantly boosting the energy conversion efficiency from sunlight to steam. Additionally, rational vacancy engineering can extend the wavelength range of light absorption, enhance broadband light absorption, alter the optical properties of materials, and increase carrier concentration, thereby improving the material's light absorption performance. Xiong et al. [75] developed a novel photothermal platform by preparing $CoCr_2O_4$ nanocrystals with abundant oxygen vacancies. These oxygen vacancies imparted the nanoparticles with significant localized surface plasmon resonance effects and excellent photothermal properties (Figure 3c).

Additionally, the modification with plasmonic materials can enable graphene materials to exhibit significant absorption in the visible light range. When plasmonic nanomaterials absorb incident light that matches the collective oscillation frequency of conductive carriers, the carriers are excited, leading to localized surface plasmon resonance (LSPR) [76]. This phenomenon is markedly pronounced in metallic nanostructures, wherein photoexcited carrier–lattice interactions and subsequent anharmonic lattice vibrations induce substantial surface temperature increments [77]. The resultant thermal energy propagates through interfacial thermalization processes, initiating thermal diffusion within adjacent media. Plasmonically active noble metals such as Au and Ag nanoparticles demonstrate particularly strong light-matter interactions across visible wavelengths due to their characteristic localized surface plasmon resonances. Wang et al. [39] developed a solar steam generation material composed of plasmonic TiN nanoparticles and hydrophilic semi-reduced graphene oxide (semi-rGO) (Figure 3d). As shown in Figure 3e, the incorporation of TiN nanoparticles enhanced light absorption and hydrophilicity while reducing heat loss, thanks to the layered structure and strong interfacial interactions of the TiN/semi-rGO composite.

When defects are introduced into a material, they can create new electronic states within the bandgap, known as "defect states" [78]. These defect states can effectively trap photogenerated carriers, reducing recombination rates and, thereby enhancing light absorption efficiency. Additionally, certain types of point defects, such as oxygen vacancies, can reduce the bandgap of the material, making it sensitive to a broader range of the spectrum and enabling broadband light absorption [79]. This is particularly important for photovoltaic materials and photocatalysts, which need to absorb as much energy from the solar spectrum as possible [80]. Allahbakhsh et al. [81] demonstrated that oxygen-containing functional groups and other types of defects in graphene-based nanostructures can act as phonon scattering sites. By trapping electrons in defect states and converting their energy into heat, these defects contribute to localized heat generation. This not only enhances the SDIE of water molecules but also reduces energy dissipation into the bulk water (Figure 3f).

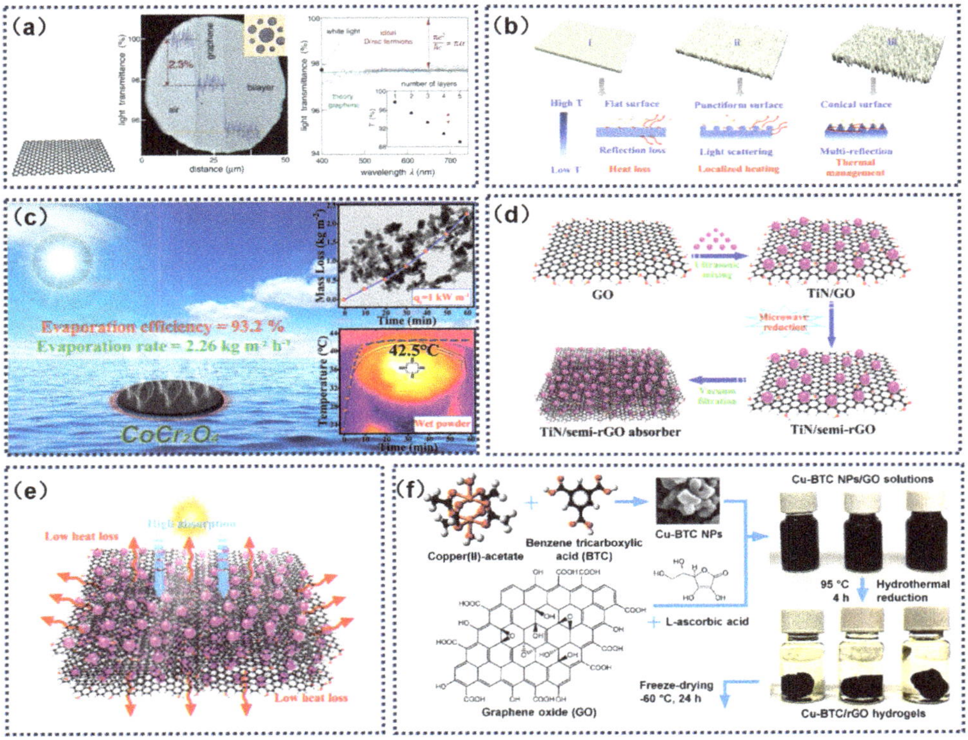


**Figure 3.** (**a**) Light absorption in two-dimensional layers. Reprinted with permission from ref. [71]. Copyright 2019 John Wiley and Sons. (**b**) Surface-modified PANI/GO composites for solar evaporation. Reprinted with permission from ref. [74]. Copyright 2022 John Wiley and Sons. (**c**) $CoCr_2O_4$ nanocrystals with abundant oxygen vacancies. Reprinted with permission from ref. [75]. Copyright 2023 American Chemical Society. (**d**) Schematic of the fabrication of plasmonic Tin/semi-RGO nanohybrid solar absorbers. (**e**) Illustration of the light absorption and heat loss suppression in tin/semi-RGO absorbers. Reprinted with permission from ref. [39]. Copyright 2023 John Wiley and Sons. (**f**) Fabrication of Cu-BTC/RGO nanocomposite aerogels. Reprinted with permission from ref. [81]. Copyright 2023 Copyright Elsevier.

### 3.2. Enhancement of Photon Conversion Efficiency

To optimize the photon conversion efficiency of a system, it is essential to synergistically enhance photothermal–photochemical energy transfer pathways and regulate carrier spatial distribution. Common strategies include utilizing wide-bandgap semiconductor heterojunctions, coupling plasmonic metal nanostructures, and implementing defect engineering to suppress carrier recombination and improve full-spectrum photon utilization. Among these, constructing heterojunctions has proven effective for enhancing carrier separation efficiency.

For instance, Wang et al. [82] reported a $C_3N_4/NiIn_2S_4$ heterostructure synthesized via a facile solvothermal method, demonstrating optimized photon utilization efficiency in SDIE and photocatalytic $H_2O_2$ production (Figure 4a). The incorporation of $NiIn_2S_4$ extended light absorption and enhanced visible-light harvesting, while the heterojunction effectively suppressed photogenerated carrier recombination, enabling efficient charge separation and transport. Under one-sun illumination, the system achieved a solar steam evaporation rate of 3.25 $kg{\cdot}m^{-2}{\cdot}h^{-1}$ with 215.06% efficiency, along with a photocatalytic $H_2O_2$ production rate of 1080 $\mu mol{\cdot}L^{-1}{\cdot}h^{-1}$ under visible light. Gu et al. [83] developed a sunflower-inspired PVA-based hydrogel evaporator featuring vertically aligned microchannels and MXene@$TiO_2$@g-$C_3N_4$ heterojunctions to optimize photon utilization for efficient water management (Figure 4b). The vertical microchannels enhanced broadband absorption through multiple light reflections, while the synergistic photothermal effects of polypyrrole (PPy) and MXene enabled an evaporation rate of 3.0 $kg{\cdot}m^{-2}{\cdot}h^{-1}$ and 97.2% energy efficiency under one-sun irradiation. The heterojunction promoted carrier separation via band structure modulation and oxygen vacancy ($O_v$) generation, extended carrier lifetime, and reduced charge transfer resistance, achieving >94.2% degradation efficiency for high-concentration VOCs under visible light (Figure 4c). Mo et al. [84] designed a bioinspired SDIE system integrating photothermal and photocatalytic functions in $TiO_2/Ti_3C_2/C_3N_4$/PVA (TTCP) hydrogel (Figure 4d). The $Ti_3C_2/C_3N_4$ heterojunction broadened light absorption, while vertically aligned microchannels enhanced light trapping through multiple reflections. A Z-scheme heterojunction facilitated carrier separation, and the photothermal effect boosted reactive oxygen species generation (Figure 4e). This system achieved 1.54 $kg{\cdot}m^{-2}{\cdot}h^{-1}$ evaporation rate with 84.1% energy efficiency under one-sun irradiation, along with 69.4–100% removal efficiency for high-concentration VOCs.

### 3.3. Reducing the Enthalpy of Water Evaporation

The primary goal of enhancing photothermal conversion efficiency is to provide more energy to increase the solar evaporation rate. Similarly, reducing the energy required for water evaporation can achieve faster evaporation rates under the same energy input conditions. By decreasing the energy demand during the evaporation process, evaporation efficiency can be significantly improved without increasing energy input. This approach allows for the evaporation of more water molecules with limited energy input. Under normal conditions, due to the strong interactions between water molecules (i.e., hydrogen bonds), the enthalpy of water evaporation at 100 °C is approximately 2260 kJ $kg^{-1}$ [85,86]. Hydrophilic functional groups can effectively dissociate the hydrogen bond network of large water clusters, a mechanism closely related to the solvation effect. Studies have shown that surface polar groups preferentially adsorb adjacent water molecules by forming intermolecular hydrogen bonds, thereby weakening the original cooperative interactions between water molecules [87,88]. In addition to chemical modification strategies, capillary effects have also been shown to significantly reduce the energy required for water phase change When capillary nanochannels are constructed within materials, the resulting ultrathin water films induce dual effects: Firstly, they promote the formation of a three-phase contact line at the edges of nanostructures, and secondly, they generate microdroplets through surface tension, effectively dissociating water molecule aggregates [89]. For example, in graphene-based materials, there is bound water interacting directly with polymer chains, free water within the graphene lattice, and intermediate water between them. The strong interactions between hydrophilic functional groups and bound water weaken the hydrogen bonds of intermediate water molecules, further reducing the enthalpy of vaporization. As an effective strategy to accelerate water evaporation, the proportion of intermediate water can be increased by rationally designing graphene materials with appropriate surface functionalization and structural modulation [90].

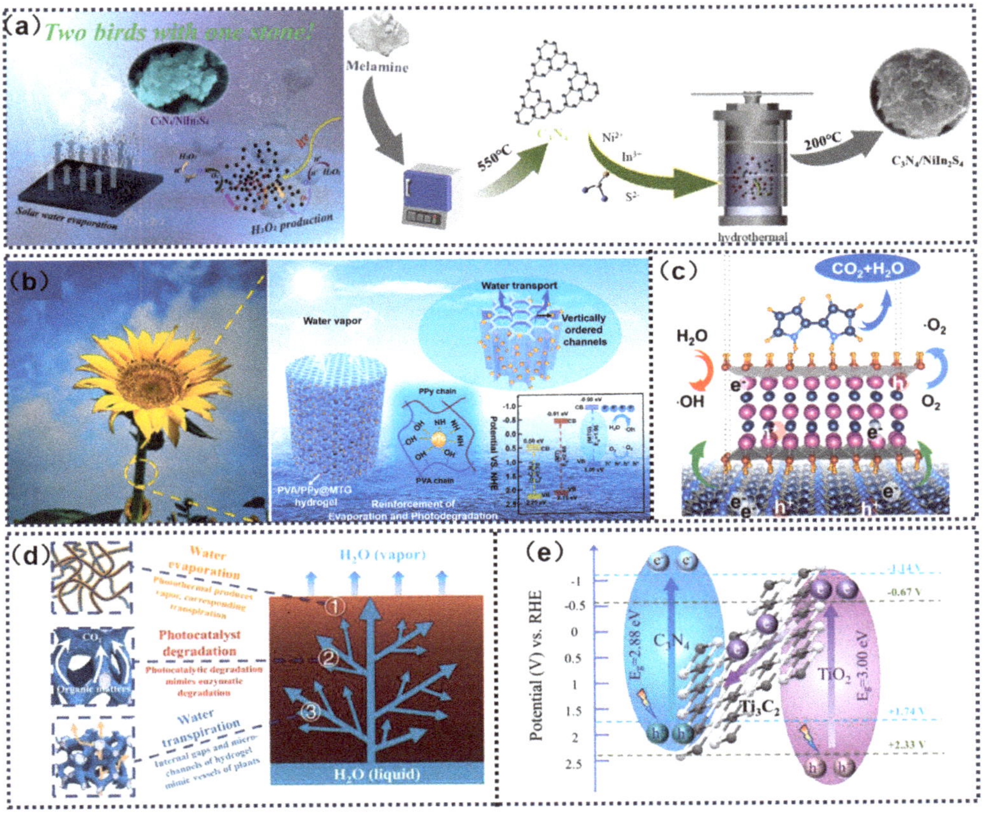


**Figure 4.** (**a**) Schematic diagram of $C_3N_4$/$NiIn_2S_4$ composite photothermal evaporation system. Reprinted with permission from ref. [82]. Copyright 2025 Elsevier. (**b**) The PVA/PPy@MTG hydrogel evaporator schematic diagram with vertically ordered channels for integrated solar-driven evaporation and photodegradation. (**c**) The catalytic degradation mechanism of the MTG-2 heterojunction for contaminant. Reprinted with permission from ref. [83]. Copyright 2025 Elsevier. (**d**) S Schematic diagram showing the similarities between a plant and the bionic system. (**e**) Schematic of the Z-scheme heterojunction of $C_3N_4$ and $TiO_2$/$Ti_3C_2$ hydrogels. Reprinted with permission from ref. [84]. Copyright 2025 Elsevier.

Li et al. [91] achieved efficient solar-driven water purification by constructing an MXene/graphene oxide/polyaniline (MGP) plastic composite with a patternable surface (Figure 5a,b). This material utilizes polyaniline to assist in assembling MXene and graphene oxide, forming a structure rich in surface functional groups. When water molecules come into contact with the functional groups on the surface of GO or MXene, the hydrogen bond network between them is rearranged. This new hydrogen bond network is more ordered than that in free water, reducing the number of hydrogen bonds between water molecules. As a result, the cohesive forces between water molecules are weakened, making it easier for them to transition into the gaseous state, thereby lowering the enthalpy of evaporation.

Additionally, the internal structure and channels of porous materials can provide more surface area for water molecules to interact. By optimizing the composition and structure of these materials, the proportion of intermediate-state water can be increased, further reducing energy consumption during the overall evaporation process. For example, through directional assembly methodologies, Xu et al. [92] reported a SDIE based on a reduced graphene oxide/sodium alginate (rGO/SA) hydrogel sponge (RSHS), constructed on a melamine sponge (MS) skeleton through a dipping–drying–crosslinking process (Figure 5c). This design significantly reduces the evaporation enthalpy of water and enhances evaporation efficiency. The porous structure plays a critical role: The interconnected porous network of the MS skeleton provides efficient channels for water transport and steam overflow, while hydrophilic groups in the rGO/SA hydrogel disrupt the equilibrium of water molecule clusters through hydrogen bonding, forming more intermediate water (IW) with lower evaporation enthalpy. LEE et al. [93] explored the impact of Å-scale oxidized functionalized graphene nanopores on water evaporation rates at atomically thin interfaces (Figure 5d). The study found that water molecules near these nanopores exhibited rapid rotational and translational motion, attributed to reduced hydrogen bonding and shorter bond lifetimes between water molecules. Furthermore, the presence of nanopores at the atomic thin-layer interface lowered the free energy barrier for water molecules transitioning from the liquid to the vapor phase, explaining why these nanopores can enhance evaporation flux (Figure 5e–g).

### 3.4. Optimization for Environmental Energy Applications

According to the second law of thermodynamics, heat naturally flows from regions of higher temperature to those of lower temperature [94,95]. However, during the evaporation process, the photothermal effect at the surface often results in a surface temperature that is higher than that of the underlying bulk water and surrounding air. Consequently, photothermal materials dissipate energy into the surrounding environment, preventing ideal solar thermal conversion applications. This necessitates the careful design and management of energy distribution and loss. When the evaporation system is passively activated by sunlight, three primary types of heat loss are considered: energy loss to the bulk water through thermal conduction, energy loss to the air through convection, and energy loss to the air through radiation [96,97].

Avoiding direct contact between photothermal materials and bulk water can effectively suppress conductive heat loss. Because water has a relatively high thermal conductivity (0.5–0.7 W $m^{-1}$ $K^{-1}$), research on heat conduction in the evaporation process mainly focuses on the energy transfer from photothermal materials to bulk water. Therefore, introducing a low thermal conductivity insulating layer to separate the photothermal materials from the bulk water can also prevent conductive heat loss. The heat flux of conduction to bulk water ($Q_{con}$) can be calculated using the following Equation (1):

$$Q_{con} = \frac{KA(T_m - T_w)}{l} \quad (1)$$

where $K$ represents the thermal conduction coefficient of the interfacial medium bridging the heated surface and cooler bulk water. The parameter $A$ denotes the effective contact area responsible for thermal energy transfer between the solid material and surrounding water body. $T_m$ indicates the temperature at the photothermally active interface of the material, while $T_w$ corresponds to the thermal condition at the water contact boundary. The variable $l$ quantifies the spatial separation between the heat-generating surface (material interface) and the cooling interface (water contact plane).

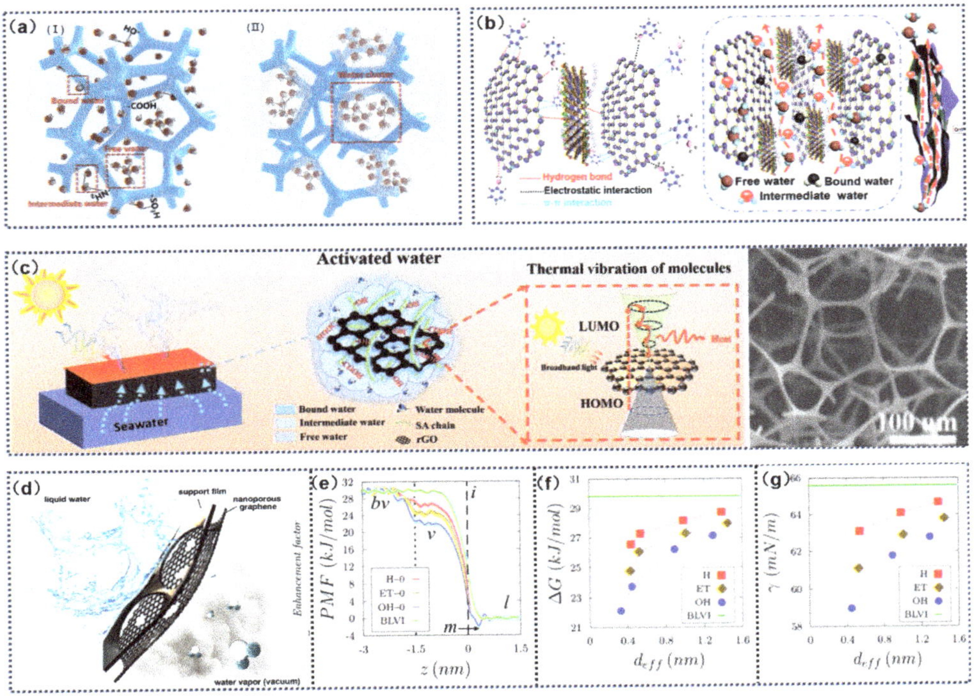


**Figure 5.** (**a**) Mechanism of (I) introducing hydrophilic functional groups and (II) constructing porous materials for enthalpy reduction. (**b**) Interaction between PANI, GO, and MXENE sheets and the interaction of oxygen-containing groups with water. Reprinted with permission from ref. [91]. Copyright 2021 John Wiley and Son. (**c**) Schematic diagram of RSHS evaporation system. Reprinted with permission from ref. [92]. Copyright 2023 Elsevier. (**d**) Schematic of Å-scale oxidized functionalized graphene nanopores. (**e**) Force curve of evaporating water molecules through nanopores with different regions. (**f**) Free energy barrier from bulk liquid water to vapor phase. (**g**) Surface tension at different water vapor interfaces. Reprinted with permission from ref. [93]. Copyright 2022 American Chemical Society.

Strategic engineering of absorber surface topography and dimensional hierarchy enables the precise regulation of thermal and molecular flux in solar-powered interfacial distillation systems. Gong et al. [98] fabricated an integrated carbon-based layered system via the controlled synthesis of orientation-engineered graphene architectures on graphite fibrous substrates, examining how multiscale graphene structural features (from micron-level patterning to nanoscale texturing) influence energy and mass transfer dynamics at absorber–liquid interfaces. The study revealed that, compared to smooth surfaces, nanostructured surfaces exhibited a higher density and smaller diameter of bubble distribution (Figure 6a–c). These characteristics significantly enhanced heat transfer efficiency through micro-convection and transient conduction mechanisms, thereby improving heat transfer and water evaporation efficiency. Additionally, constructing the macroscopic structure of the evaporator with insulating layers can reduce conductive heat loss. Liang et al. [99]

developed a solar steam generation system based on reduced rGO foam, utilizing an injection control technique (ICT) to regulate the capillary water state and optimize evaporation efficiency (Figure 6d). The rGO foam exhibits a full-spectrum high absorption rate and low thermal conductivity, effectively converting solar energy into thermal energy while minimizing heat dissipation. This design rapidly raises the surface temperature to 88.7 °C under illumination and maintains stability. By balancing the water supply rate and evaporation rate through ICT, the system prevents blockage of micrometer-sized pore channels, maximizes the evaporation area, and reduces heat absorption by excess water, thereby achieving thermal localization. Han et al. [100] demonstrated a graphene G@ZIF nano-hybrid structure using metal–organic framework (MOF) isolation. This material leveraged the synergistic effects of an ultrathin and insulating ZIF layer and the G@ZIF interfacial nano-cavities to enhance light absorption and concentrate heat. Under artificial solar illumination, G@ZIF composites reach a thermal peak of 120 °C under ambient conditions and exhibit a near-uniform (98%) photon-to-thermal energy conversion efficiency. The molecular screening porosity of the ZIF framework helps small molecules to selectively enter the thermal interface embedded in graphene, thus enabling guided heat flux propagation (Figure 6e–i).

In solar-driven water evaporation, the generated steam can potentially carry away some heat energy through thermal convection. Additionally, if the heated water does not evaporate promptly, it may transfer its energy to cooler water through convection. The core of SDIE is the separation of the heat source from the bulk water, but the heated water might still transfer heat to cooler water through free flow and diffusion. Regulating the direction of water transport and the water flux can help reduce the transfer of heat from high-temperature water to low-temperature water via convection (2).

$$Q_{cv} = hA(T_m - T_e) \tag{2}$$

In this thermodynamic model, $h$ is the thermal convection coefficient. Based on the coupling mechanism of multiple physical fields, the surface heat dissipation can be reduced by selecting functional materials with radiation suppression characteristics, and the directional regulation of heat-mass transfer can be achieved by optimizing the structure of the evaporation system driven by phase transformation to reduce heat loss. This cross-scale thermal control strategy achieves accurate control of the thermodynamic balance of the evaporation process by reducing the radiation/conduction loss coefficient at the material level and enhancing the convective heat transfer efficiency at the system level.

Finnerty et al. [101] constructed three-dimensional graphene oxide (3D GO) rods, which, due to their unique 3D structure, can more effectively compress the boundary layer thickness (Figure 6j). This reduces the distance that water molecules need to diffuse and enhances the evaporation driving force. Additionally, the longer average flow path length means that the boundary layer remains thin and continues to develop over a larger surface area of the 3D structure, making the 3D GO rods more responsive to increases in external convection speed. The 3D GO rods not only excel in light absorption and heat localization but also show great potential in promoting thermal convection, accelerating water evaporation, and optimizing the overall efficiency of the system (Figure 6k).

Since photothermal materials absorb light energy and convert it into heat, the temperature of the photothermal layer is generally higher than that of the surrounding environment. As a result, these materials release energy to their surroundings through radiation. The heat flux of radiation $Q_{rad}$ can be calculated using the Stefan–Boltzmann law (3) as follows:

$$Q_r = \varepsilon\sigma A\left(T_m^4 - T_e^4\right) \tag{3}$$

where ε represents the radiant efficiency of a photothermally active surface, and σ represents the universal radiant coefficient in the Stefan–Boltzmann law. $T_e$ reflects the ambient thermal distribution. The analysis shows that effective thermal radiation management can be achieved by optimizing the emissivity parameters of the energy harvesting surface, minimizing the radiation interface size and maintaining the thermal difference between the functional surface and the surrounding medium.

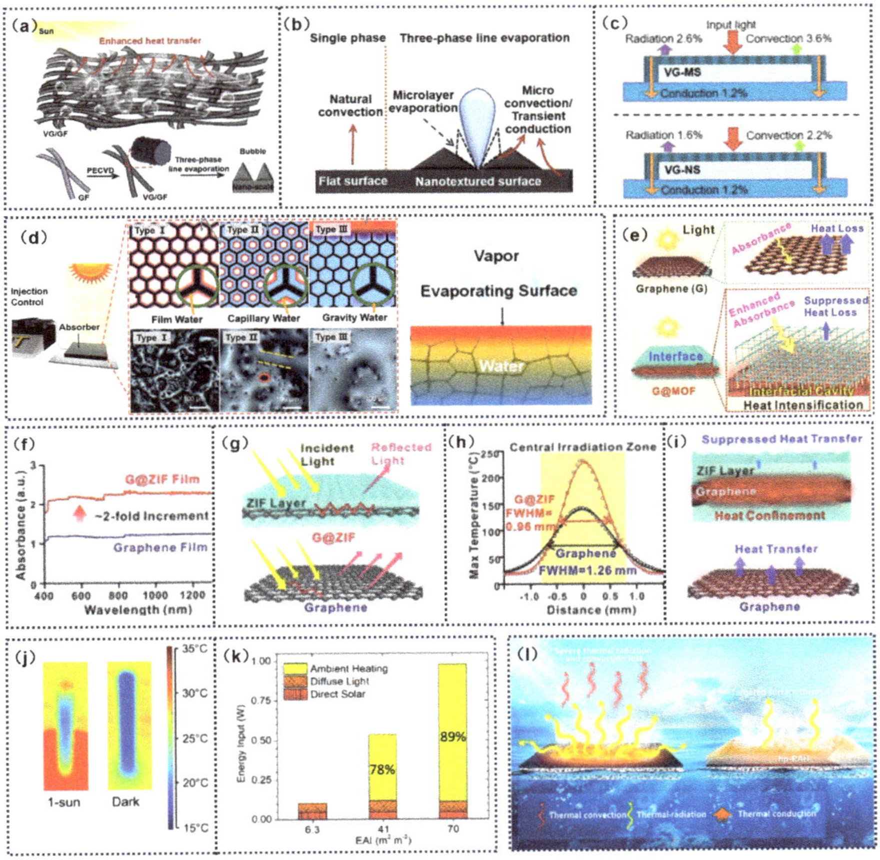


**Figure 6.** (**a**) Schematic illustration of bubble formation on a nanotextured surface during SDIE. (**b**) Mechanism of enhanced heat transfer through nanostructuring. (**c**) Heat transfer diagram of microstructured and nanostructured absorbers during evaporation. Reprinted with permission from ref. [98]. Copyright 2022 Elsevie; (**d**) Schematic of the formation process of a composite material made from hollow melamine foam and rGO. Reprinted with permission from ref. [99]. Copyright 2019 John Wiley and Sons. (**e**) Diagram of G@MOF, which enhances light absorption and suppresses heat loss, thereby concentrating thermal energy at the confined interface. (**f**) UV-Vis-NIR absorption of graphene and g@ZIF membranes. (**g**) Enhanced light absorption of g@ZIF. (**h**) Spatial temperature distribution in the central irradiated area under laser illumination at maximum equilibrium surface temperature. (**i**) Thermal localization and suppression of heat transfer in G@ZIF. Reprinted with permission from ref. [100]. Copyright 2021 John Wiley and Sons. (**j**) 3D thermal images of stems under light and dark conditions. (**k**) Contributions of direct solar energy, diffuse light, and ambient heating to energy input. Reprinted with permission from ref. [101]. Copyright 2021 American Chemical Society. (**l**) Schematic of hp RAH photothermal films. Reprinted with permission from ref. [102]. Copyright 2021 American Chemical Society.

Meng et al. [102] developed a hierarchical porous radiative absorption hydrogel membrane (hp-RAH), which significantly enhances solar evaporation efficiency by optimizing thermal radiation management in the photothermal layer (Figure 6l). This hydrogel membrane is coated on the evaporator surface, and the hp-RAH exhibits high absorption in the mid-infrared band. It captures the thermal radiation energy released from the photothermal layer and reuses it for water heating while blocking convective heat loss. Combined with its hierarchical porous structure, the system reduces the originally dissipated thermal radiation and convective losses from 6.6% to 0.39%, achieving effective thermal localization.

### 3.5. Water Transport Regulation

Adequate and continuous water supply is considered one of the key factors for achieving sustained evaporation in solar evaporators. However, the stacking or agglomeration of graphene nanosheets can lead to inefficient water transfer and vapor escape during the evaporation process. Therefore, ensuring good wettability, designing effective water transport channels, and regulating pumping mechanisms can not only reduce the accumulation of concentrated salts during evaporation but also adjust the heat distribution between the evaporator and the surrounding environment. This can create temperature gradients or cool evaporative surfaces, effectively enhancing the evaporation performance of the system [103,104].

A suitable pore size in micro/nanoporous structures facilitates rapid water transport within graphene driven by capillary action, ensuring efficient evaporation under solar irradiation [105]. Lee et al. [106] developed a nanoporous capillary membrane based on assemblies of highly porous graphene oxide (HGO) nanosheets, which features nanopores and nanochannels to ensure high capillary water flow through low-friction nanochannels. The study found that nanopores formed by highly oxidized sp3 regions and an increased proportion of graphitic regions favor the water transport process and enhance water stability. Additionally, the presence of large nanopores with a broad size distribution (1–40 nanometers) leads to faster evaporation rates and a larger vapor transport surface area (Figure 7a). Lim et al. [38] introduced a layered chitosan aerogel embedded with graphene nanoplatelets (GnP), featuring distinct density variations between its inner and outer sections. The differing densities in these regions result in specialized water transport properties: The low-density, large-pore area allows for swift movement and evaporation of water molecules, whereas the high-density, small-pore region boosts the material's structural integrity and minimizes surface energy, thus enhancing its ability to retain water (Figure 7b,c). This design effectively combines the advantages of both density variations, optimizing the aerogel's performance in water management and mechanical strength.

The continuous water-transport capability of graphene-based evaporators largely depends on their wettability. For photothermal interfaces, liquid water tends to spread and form a thin film on hydrophilic surfaces, while accumulating as thicker droplets on hydrophobic surfaces [107]. Previous studies have shown that due to the inherent non-polar carbon structure and rough surface of GO, GO composite aerogels typically exhibit some degree of hydrophobicity, with water contact angles ranging from 120° to 160°. Wettability can be adjusted by incorporating oxygen-containing functional groups or highly absorbent polymers to enhance water transport. Xu et al. [108] developed three-dimensional, hierarchically porous N-doped lignosulfonate/GO aerogels (NLGA) through a single-step solvothermal assembly protocol. By employing ethylenediamine-mediated nitrogen doping strategy and utilizing lignosulfonate's inherent hydrophilic functionalities as a multifunctional crosslinking agent bearing abundant hydrophilic moieties, this synthesis simultaneously achieved pore architecture engineering and surface chemistry optimization. The coordinated lignosulfonate–graphene interactions synergistically enhanced both surface wettability and hydraulic permeability through controlled capillary networks (Figure 7d).

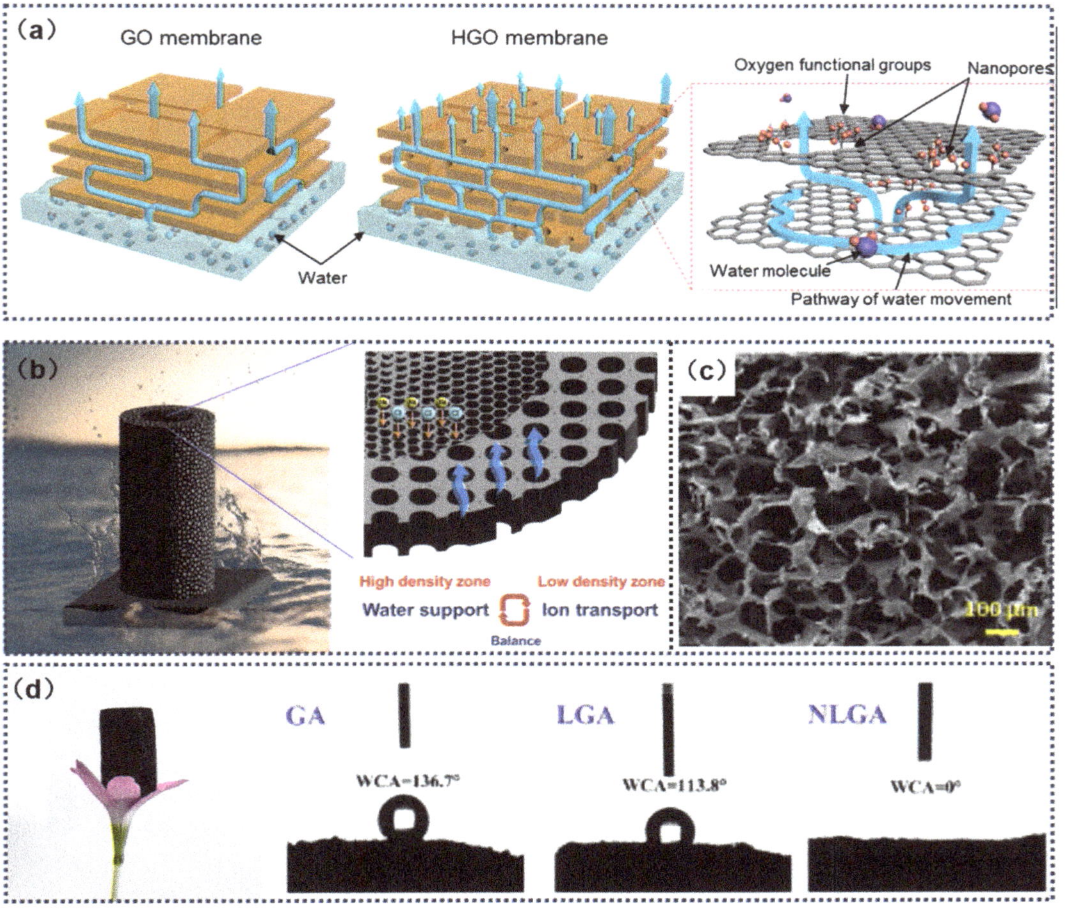


**Figure 7.** (**a**) High evaporation rate and accelerated water transport rate in HGO membranes. Reprinted with permission from ref. [106]. Copyright 2022 Elsevier. (**b**) Schematic of a 3D laminated chitosan/graphene nanoskeleton evaporator, which provides a balance between water transport and ion transfer. (**c**) Field emission scanning electron microscope (Fe-SEM) image of LCG. Reprinted with permission from ref. [38]. Copyright 2024 Elsevier. (**d**) Photograph of NLGA. Reprinted with permission from ref. [108]. Copyright 2023 Elsevier.

### 3.6. Desalination Strategies

During SDIE operation, convection mass transfer drives salt accumulation at the photothermal interface. When the thermodynamic supersaturation threshold is reached, the crystal precipitation induces a basting phenomenon, which reduces the optical collection efficiency and induces microfluidic channel obstruction, thus limiting the photothermal conversion and disrupting the liquid phase transport. Therefore, salt mitigation strategies are crucial for enhancing the operational lifespan and long-term stability of SDIE systems. While salt fouling can be addressed by rinsing and soaking the photothermal surface after evaporation cycles, this increases costs and may damage the photothermal surface [109].

To tackle this issue, researchers have developed various effective strategies to control salt fouling, which can be broadly categorized as follows: First, achieving salt resistance by regulating water/ion transfer within the photothermal evaporator; secondly, controlling salt precipitation at specific locations on the photothermal evaporator; and finally, implementing dynamic self-regulation of desalination through photothermal evaporators with specific morphologies.

Firstly, by optimizing the hydraulic transport and ion diffusion paths within the evaporator, the excessive accumulation of salt ions on the photothermal surface can be effectively suppressed. Zhang et al. [110] designed a salt-resistant $Fe^{3+}$@GOM heterogeneous evaporator, utilizing a Janus structure to achieve directional water transport through capillary force differences at the hydrophilic–hydrophobic interface (Figure 8a). Combined with a graphene oxide membrane, $Fe^{3+}$ cations further enhance the Donnan effect, effectively repelling multivalent salt ions and reducing advection-driven salt deposition from the source, thereby improving salt resistance. Secondly, controlled salt precipitation strategies guide salt crystallization to non-critical areas through structural design, thus preventing the photothermal surface from being covered by salt layers (Figure 8b). Zhou et al. [111] developed an innovative 3D Graded Graphene Helical Sponge (GGS Sponge) using an environmentally friendly fabrication process. This material demonstrates unique advantages in achieving zero liquid discharge (ZLD) desalination of high-concentration brine. The study shows that its helical structure significantly enhances energy capture efficiency during the photothermal conversion process, while the graded network system creates a synergistic effect by optimizing radial water transport paths and directional salt crystallization mechanisms. This structural coupling endows the material with multiple functional properties: It maintains a stable evaporation rate under solar radiation, efficiently collects salt, achieves water purification to ZLD standards, and exhibits excellent anti-fouling performance and structural durability. Additionally, the dynamic self-regulating morphology desalination strategy leverages the responsive characteristics of smart materials to actively disrupt salt crystal adhesion through interfacial deformation or changes in physical properties (Figure 8c). Liu et al. [112] introduced a water/light dual-responsive actuator composed of double-twisted hollow hydrogel fibers loaded with reduced graphene oxide (RGO@HHF). The double-twisted RGO@HHF can elongate to absorb water and contract to evaporate seawater under light stimulation due to differences in thermal expansion coefficients. During the actuation process, mechanical stress causes salt crystals to detach, and the accumulated salt dissolves back into the seawater (Figure 8d,e).

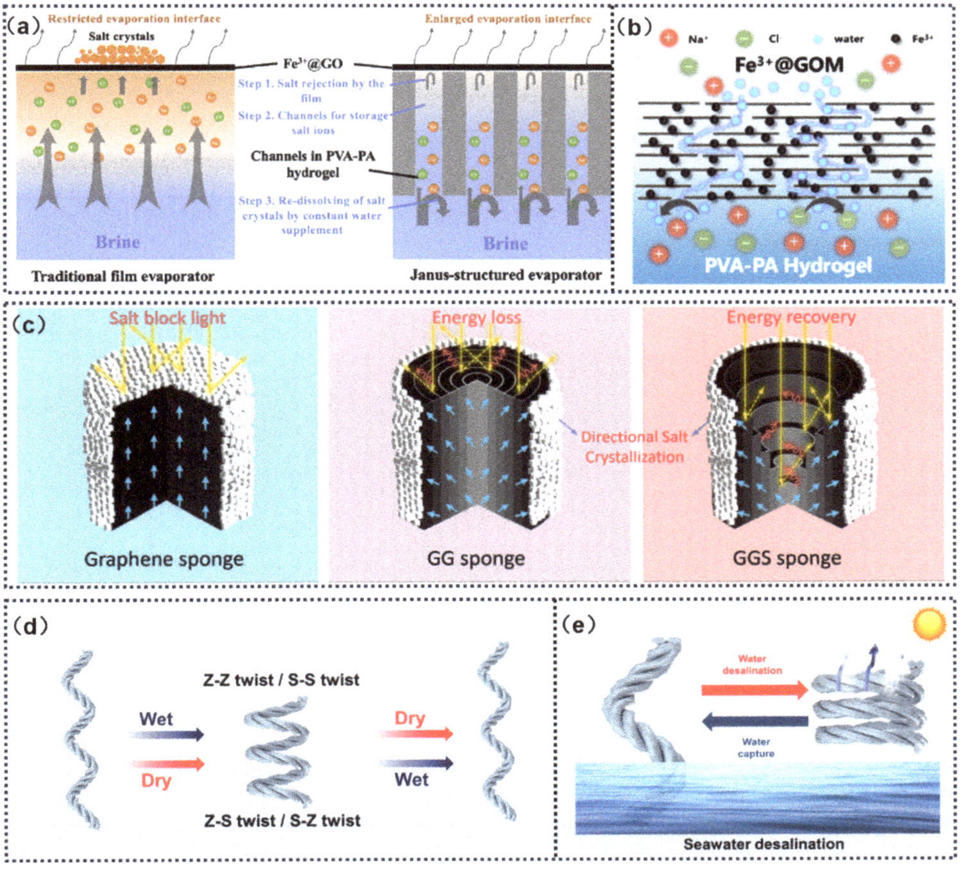


**Figure 8.** (**a**) Mechanism of effective salt resistance in FHJE. (**b**) Illustration of effective salt resistance mechanism via the Donnan effect with $Fe^{3+}$@GOM. Reprinted with permission from ref. [110]. Copyright 2023 American Chemical Society. (**c**) Salt crystallization on the surface of graphene sponge. Reprinted with permission from ref. [110,111]. Copyright 2024 American Chemical Society. (**d**) Schematic of the responsive behavior of RGO@HHF. (**e**) Application of RGO@HHF in seawater desalination. Reprinted with permission from ref. [110,112]. Copyright 2023 John Wiley and Sons.

## 4. Structural Design of Graphene-Based SDIE-Coupled Photocatalytic Systems

The concepts introduced above highlight the development of effective evaporation systems from the perspective of graphene material modification. In addition, researchers have also achieved significant improvements in evaporation efficiency by designing and enhancing the structure of evaporators [96,113]. These designs can be categorized into three types. (1) Single-Layer Structures: These are primarily membrane or thin sheet structures and represent the most basic design for SDIE. They typically consist of a single functional layer. (2) Double-Layer and Multi-Layer Structures: These designs separate the water molecule channel layer from the thermal insulation layer. Through functional stratification, they optimize light absorption, water transport, and thermal insulation properties. (3) 3D Structures: By employing three-dimensional designs, these structures increase the evaporation area, optimize light absorption paths and water transport channels, and effectively reduce heat loss from the device.

### 4.1. Monolayer Structure

The monolayer structure serves as the foundational configuration for graphene-based SDIE systems. Its core design principle is to achieve the synergistic coupling of photothermal conversion, water transport, and vapor release through a single functional layer, thereby simplifying system complexity and reducing fabrication costs [114,115]. Wo et al. [116] developed an integrated novel wood evaporator (DBW-GC) using ultrasound-assisted deposition. This evaporator is composed of defined camphor wood, reduced graphene oxide, and partially reduced $CuO/Cu_2O$ nanoparticles. The synergistic effect of the nanocomposite framework and heterojunction facilitates uniform distribution and sufficient interaction with water molecules, resulting in excellent structural and photothermal properties. Under one sun illumination, the energy conversion efficiency reached 83.39%, achieving an advanced pure water evaporation rate of 2.10 $kg{\cdot}m^{-2}{\cdot}h^{-1}$. Experimental and simulation tests further confirmed the strong hydration capability of the wood evaporator, which reduces the equivalent evaporation enthalpy (Figure 9a). Zang et al. [117] designed a nanofiber hydrogel-reduced graphene oxide (NHrG) composite membrane that integrates the porous hydrophilic network of electrospun hydrogel nanofibers with the efficient photothermal conversion properties of reduced rGO (Figure 9e). Specifically, the open-pore structure of the hydrogel accelerates water transport and vapor escape, while the broad-spectrum light absorption of rGO achieves a remarkable solar energy conversion efficiency of up to 95.4%.

### 4.2. Multilayer Structures

While single-layer structures offer significant advantages in terms of fabrication processes and water transport efficiency, their ability to localize heat is limited. Some heat is lost through conduction and convection, resulting in theoretical evaporation efficiencies typically below 80%. Additionally, prolonged operation can lead to pore blockage due to salt crystallization. In contrast, multilayer structures can enhance photothermal conversion efficiency by incorporating additional functional intermediate layers, such as insulation, light scattering, or reflective layers, which reduce radiative and convective heat losses. These structures are also better suited for high-salinity environments [118,119]. Sun et al. [120] designed a composite membrane made of graphene oxide (GO) and polytetrafluoroethylene (PTFE), consisting of a hydrophilic GO layer (153–210 nm) supported by a hydrophobic PTFE membrane (Figure 9g). The high thermal conductivity of the GO layer significantly reduces thermal boundary layer resistance, enhancing the temperature gradient. However, the Kelvin effect at the evaporation interface leads to a decrease in vapor pressure, partially offsetting the temperature gradient gain. Conversely, at the condensation interface, the Kelvin effect increases local vapor pressure, becoming the main factor for flux enhancement. Experimental and model validation showed that the flux enhancement at the evaporation interface with the GO layer was more significant (17.8–45.5% compared to 12.4–16.4% at the condensation interface). Wei et al. [121] proposed a hydrophilic composite graphene material combined with CuO, fabricated using a direct laser-induced graphene synthesis method on a polyimide film coated with $CuCl_2$. The enhanced hydrophilicity and layered structure provide rapid capillary action, allowing the assembled laser-induced lattice evaporator to achieve an evaporation rate of 2.54 kg $m^{-2}$ $h^{-1}$ under one sun illumination, with an efficiency of 91.1%. It also demonstrated excellent desalination performance at an evaporation rate of 1 kg $m^{-2}$ $h^{-1}$.

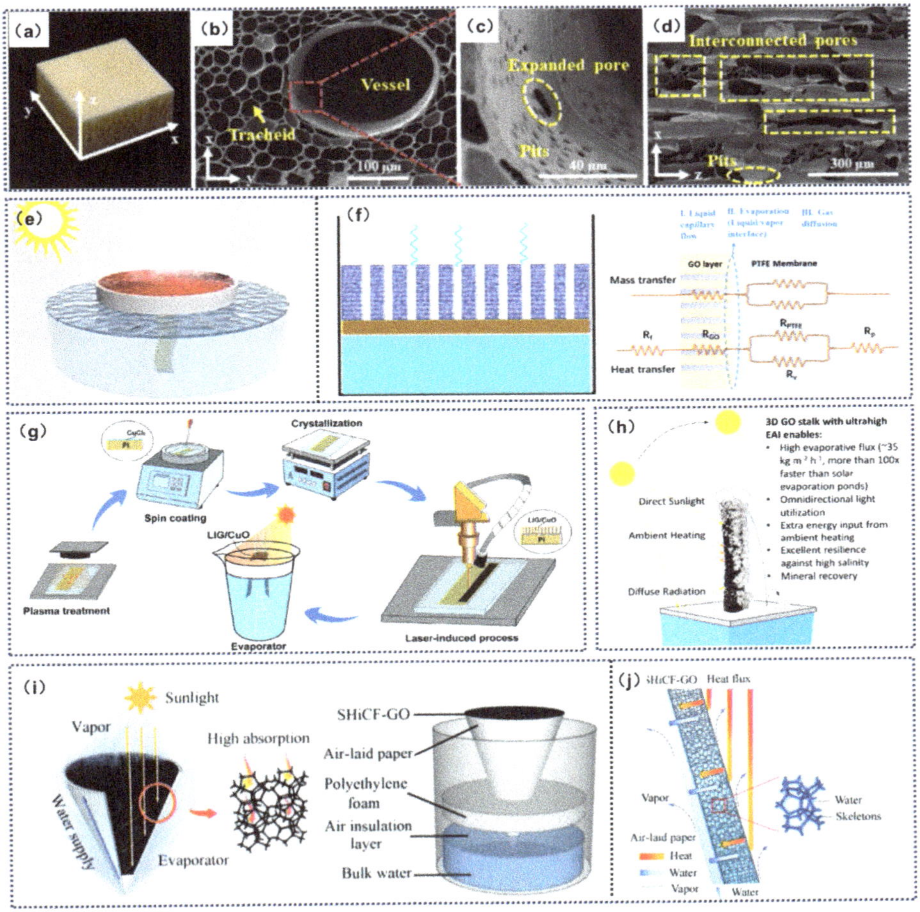


**Figure 9.** (**a–d**) DBW-GC micro-nano structures. Reprinted with permission from ref. [116]. Copyright 2025 Elsevier; (**e**) Schematic diagram for the NHrG membrane. Reprinted with permission from ref. [117]. Copyright 2021 Elsevier. (**f**) Schematic diagram of the evaporation test for the GO/PTFE membrane. Reprinted with permission from ref. [120]. Copyright 2022 Elsevier. (**g**) Process for fabricating laser-induced graphene/CuO composites using a continuous-wave CO laser on CuCl-coated PI films with different concentrations. Reprinted with permission from ref. [121]. Copyright 2024 John Wiley and Sons. (**h**) Performance comparison between the graphene oxide straw and other 3D evaporators. Reprinted with permission from ref. [101]. Copyright 2021 American Chemical Society. (**i**) Schematic diagram illustrating the solar evaporation and light absorption principles of the 3D inverted conical solar evaporator. (**j**) Schematic diagram of the synergistic regulation of thermal-mass water transport and vapor diffusion. Reprinted with permission from ref. [40]. Copyright 2023 John Wiley and Sons.

### 4.3. 3D Structures

In recent years, the study of 3D evaporators has garnered significant attention from researchers. Compared to 2D structures, 3D systems exhibit notable advantages in photothermal conversion efficiency, water transport dynamics, thermal localization control, and environmental adaptability. By optimizing spatial topology, 3D structures overcome the light absorption limitations of 2D planar structures. Their multi-level micro-nano morphologies (such as arrayed cones, gradient channels, or fractal networks) effectively suppress Fresnel reflection and enhance light capture capabilities. This is particularly beneficial under conditions of low solar incidence angles or diffuse light, where multiple internal reflections and scattering effects enable efficient broadband absorption (250–2500 nm) [122–124]. In contrast, 2D planar structures are limited by single-pass reflection absorption and surface flatness, making their photothermal conversion efficiency more susceptible to variations in incident angle and spectral distribution. Furthermore, 3D systems achieve spatial decoupling of water transport paths and evaporation interfaces by constructing vertically interconnected porous channels and lateral capillary force gradients. This design maintains rapid water replenishment while significantly reducing thermal convection losses caused by excessive heating of the bulk water. In comparison, 2D structures, due to the direct contact between the evaporation interface and the water body, are prone to thermal energy loss through heat conduction and natural convection to inactive areas during the evaporation process. This leads to the failure of thermal localization and a decrease in energy utilization efficiency.

Finnerty et al. [101] developed a three-dimensional graphene oxide (3D GO) stem, which confines the capillary core nearby and achieves an evaporation flux of 34.7 kg $m^{-2}$ $h^{-1}$ under one sun condition (1 kW/$m^2$). This represents nearly a 20-fold enhancement compared to a two-dimensional GO evaporator. The system also features omnidirectional sunlight utilization and maintains high evaporation flux in dark conditions by more effectively utilizing ambient heating. The evaporation rate is significantly increased by introducing wind, and the system demonstrates scalability and resistance to fouling in the evaporated liquid (Figure 9h). Lv et al. [40] developed a 3D hierarchical inverted cone solar evaporator, composed of a 3D copper foam skeleton cone decorated with micro/nanostructures and functionalized with graphene oxide (Figure 9i). This structure can be fabricated through a simple and scalable dip-coating process, followed by drying at room temperature. The proposed configuration enables efficient solar energy absorption, continuous liquid film diffusion and transport, enhanced localized SDIE, and rapid vapor diffusion through the pores. Notably, the 3D conical evaporator achieves thermal localization at the skeleton interface, allowing evaporation to occur along the entire structure with unobstructed rapid diffusion of liquid and vapor (Figure 9j).

## 5. Application of Graphene-Based SDIE-Coupled Photocatalytic Systems in Water Treatment

Graphene-based SDIE systems demonstrate significant advantages in water treatment. The core mechanism integrates the synergistic effects of photothermal–adsorption–catalysis and a multi-stage targeted removal mechanism for pollutants. The high specific surface area of graphene and its rich oxygen-containing functional groups endow it with excellent physical adsorption capabilities, enabling the efficient enrichment of heavy metal ions and organic pollutants. Simultaneously, its broad-spectrum absorption properties convert solar energy into heat, driving the generation of photogenerated carriers and activating the catalytic sites, which decompose pollutants via free radical oxidation or direct electron transfer pathways. Additionally, the vapor flow produced by evaporation can carry volatile pollutants to the condensation interface for selective separation. This adsorption–catalysis–evaporation synergy driven by solar energy provides a green solution for the efficient purification and resource recovery of complex wastewater.

### 5.1. Photocatalytic Degradation of Organic Pollutants

Graphene-based SDIE achieves the highly efficient utilization of light energy and promotes carrier separation in water treatment applications. The high conductivity and ultra-large specific surface area of graphene accelerate the transport of photogenerated electrons while suppressing electron–hole recombination, simultaneously enhancing interfacial adsorption capacity for pollutants. During the interfacial evaporation process, while solar energy is converted into thermal energy to drive water evaporation, the strongly oxidative reactive species generated on the photocatalyst surface can directionally attack organic pollutant molecules. Through reactions such as bond breaking and ring opening, these species gradually mineralize pollutants into $CO_2$ and $H_2O$ [125]. Furthermore, the localized thermal effect formed at the evaporation interface accelerates reaction kinetics, while the porous structure of graphene enriches pollutants through adsorption effects, further enhancing mass transfer efficiency. The conjugated $\pi$ structure of graphene can also adsorb and enrich pollutants through $\pi$–$\pi$ stacking, enhancing the efficiency of interfacial reactions. Its wide-spectrum response characteristic further expands the utilization rate of visible light.

Ma et al. [41] designed a novel reduced graphene oxide/FeOOH (RGO/FeOOH) asymmetric evaporator and demonstrated its excellent photocatalytic degradation ability for organic pollutants in water (Figure 10a). Yan et al. [70] developed a graphene oxide-polypyrrole (GR/PPY) composite aerogel that innovatively realized the dual functions of wastewater purification and freshwater regeneration. The nanocomposite material, prepared by a stepwise hydrothermal-reduction–freeze-drying method, effectively suppressed the stacking of graphene sheets through $\pi$–$\pi$ conjugation between components, constructing a cross-scale network structure with micron-scale fluid channels and nano active sites. This unique configuration significantly enhanced the adsorption–catalysis synergistic effect for organic pollutants by increasing the surface $\pi$–electron cloud density. The experimental data showed that the aerogel achieved degradation rates of 96.4%, 92.5%, and 82.6% for typical organic pollutants like Rhodamine B (RHB), phenol, and ciprofloxacin (CIP), respectively, confirming its multifunctional potential in water treatment applications (Figure 10b,c). Li et al. [126] developed a reduced graphene oxide–palladium (rGO–Pd) composite evaporator, which innovatively realized a synergy between SDIE and catalytic reactions. The system exhibited an enhanced apparent utilization rate of solar thermal energy by coupling the evaporation and catalytic energy conversion pathways. The experimental results indicated that thanks to the self-supply of reaction heat, this evaporator achieved a record-breaking water production rate of 12.7 $L \cdot m^{-2} \cdot h^{-1}$ under saline conditions, far superior to that of existing integrated SDIE systems. Its multi-porous rGO–Pd composite structure created efficient mass transfer channels, combining photothermal effects and rapid material transport, which improved the catalytic activity of Pd to 125.07 $min^{-1}$—an increase of 275% compared to dark-state conditions. This integrated SDIE system simultaneously enabled freshwater preparation, salt crystallization, and catalytic product recovery, achieving the full-resource utilization of aqueous media (Figure 10d). Huang et al. [127] successfully constructed a multifunctional nanocomposite system based on polyvinylidene fluoride (PVDF)/graphene–$BaTiO_3$ through $\beta$-phase self-assembly technology. This material integrated solar evaporation, piezoelectric power generation, pollutant degradation, and self-cleaning functions. Experiments confirmed that, due to the piezoelectric–photonic synergistic coupling effect, the material system effectively suppressed the recombination of photogenerated carriers. Under dynamic conditions, it achieved a 93% degradation efficiency for dyes while also enhancing membrane anti-fouling by 200%. This piezoelectric-catalysis and photocatalysis cross-scale energy synergy mechanism provides a new technological paradigm for the development of smart water treatment systems (Figure 10e).

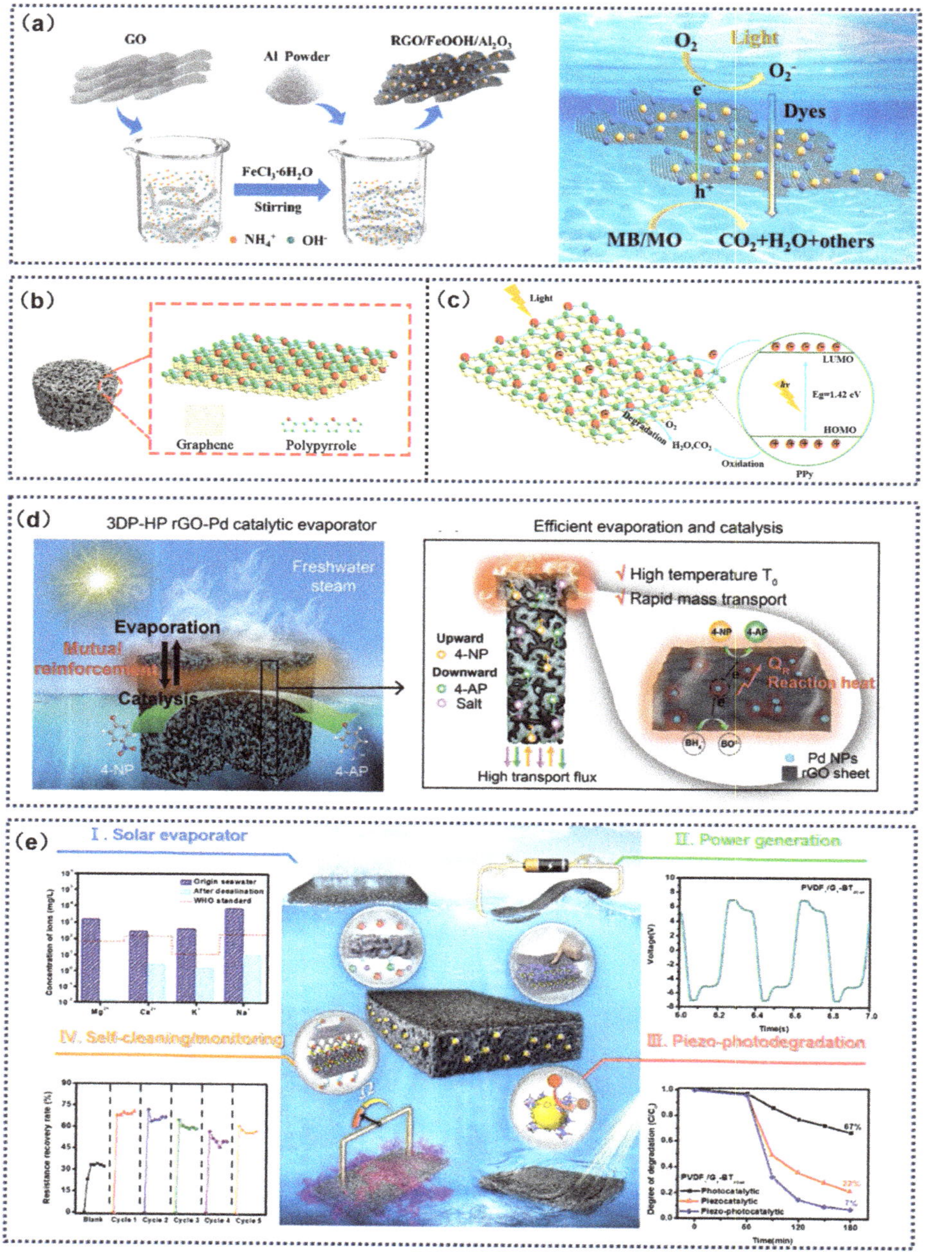


**Figure 10.** (**a**) Synthesis process and degradation mechanism of RGO/FeOOH. Reprinted with permission from ref. [41]. Copyright 2023 Elsevier. (**b**) Schematic diagram of GR/PPY aerogel. (**c**) Photocatalytic mechanism of GR/PPY. Reprinted with permission from ref. [70]. Copyright 2022 Elsevier. (**d**) Schematic diagram of interfacial evaporation and catalytic mechanism in the RGO-PD catalytic evaporator. Reprinted with permission from ref. [126]. Copyright 2023 John Wiley and Sons. (**e**) Four mechanisms of the PVDF/graphene-$BaTiO_3$ composite membrane. Reprinted with permission from ref. [127]. Copyright 2023 John Wiley and Sons.

### 5.2. Reduction in and Fixation of Heavy Metal Ions

Leveraging graphene's efficient photothermal conversion and synergistic reduction-adsorption mechanisms enables the reduction in and immobilization of heavy metal ions in water. With its high specific surface area and abundant surface functional groups, graphene-based SDIE effectively adsorbs and enriches heavy metal ions through electrostatic interactions, complexation, or ion exchange, significantly increasing their localized concentration. Under solar-driven conditions, the heterojunction structure formed between graphene and semiconductors facilitates the directional migration of photogenerated electrons to pollutant interfaces via Fermi-level alignment. This reduces higher-valent heavy metal ions to less toxic or easily precipitable lower-valent forms, while photogenerated holes participate in oxidation reactions to avoid electron–hole recombination. The localized high-temperature environment generated during interfacial evaporation not only accelerates reduction reaction kinetics but also induces the formation of stable coordination compounds or mineral phases between heavy metal ions and material surface functional groups, achieving long-term immobilization [128,129]. Additionally, evaporation-induced salt crystallization effects synergistically promote heavy metal ion co-precipitation, and graphene's hierarchical porous structure provides physical confinement spaces for immobilized products, preventing secondary release.

Fan et al. [43] innovatively converted waste polyethylene terephthalate (PET) into bismuth-based metal–organic framework (Bi-MOF) microcrystals and built a solar-driven system that simultaneously reduced hexavalent chromium and produced freshwater. The Bi-MOF/graphene composite evaporator demonstrated broad-spectrum light absorption (380–2500 nm), 92.3% photothermal conversion efficiency, and a 0.12° water contact angle. When coupled with a photo-Fenton catalytic system, the localized micro-effect generated during the interface evaporation process significantly enhanced hydrogen peroxide activation, achieving a hexavalent chromium reduction efficiency of 91.3%, with a reaction rate constant of 0.0548 $min^{-1}$—178% higher than that of traditional non-evaporation systems. This solar-driven microenvironment regulation mechanism provides a new technological coupling path for heavy metal pollution control and water resource regeneration (Figure 11a). Tian et al. [130] innovatively combined Ag-$MnO_2$ catalytic nanostructures with nitrogen-doped graphene aerogels, creating a three-dimensional cylindrical evaporator (SMNG) with both ultra-high evaporation performance and efficient water treatment capability. Under light irradiation, this device achieved an evaporation rate of 6.46 $kg \cdot m^{-2} \cdot h^{-1}$ with a 90.25% photothermal conversion efficiency. For seawater desalination applications, the SMNG exhibited an evaporation performance of 6.15 $kg \cdot m^{-2} \cdot h^{-1}$ and 99.9% ion removal efficiency. In outdoor tests, its evaporation performance was improved to 8.57 $kg \cdot m^{-2} \cdot h^{-1}$, confirming its practical feasibility. This design not only optimized evaporation efficiency and energy conversion performance but also significantly enhanced wastewater treatment ability through catalytic properties, achieving over 99.9% purification efficiency for organic dyes, heavy metal ions, formaldehyde, and *E. coli* (Figure 11b). Dong et al. [131] prepared a porous hydrophilic/hydrophobic cellulose graphene aerogel (CGA) for application in solar-driven water evaporation systems to achieve seawater desalination and heavy metal wastewater purification. By leveraging the synergistic effects of interfacial evaporation and adsorption, the CGA utilizes oxygen-containing functional groups to form hydrogen bonds and chelating bonds with heavy metal ions (e.g., $Ni^{2+}$, $Cu^{2+}$, $Pb^{2+}$, $Cd^{2+}$), effectively reducing the concentration of heavy metals in wastewater to trace levels and significantly decreasing the wastewater volume (Figure 11c,d).

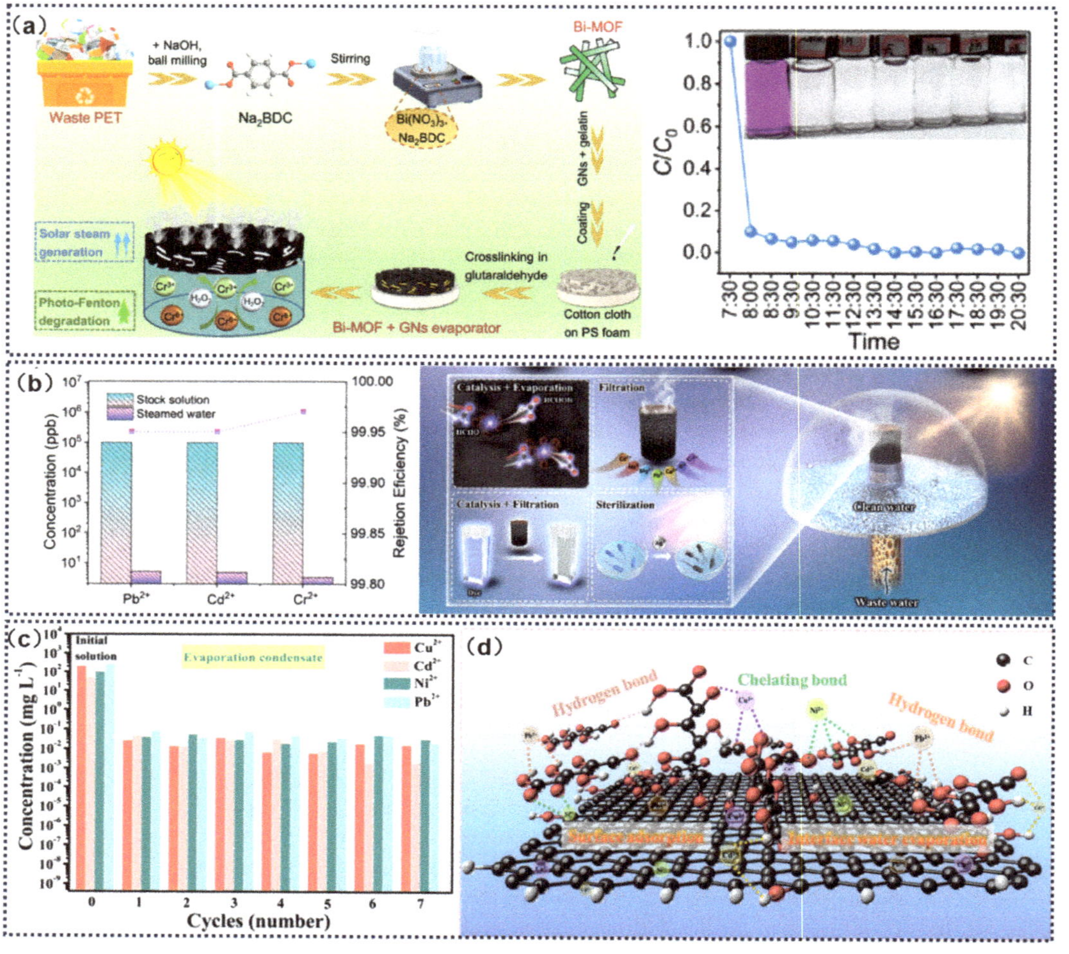


**Figure 11.** (**a**) Schematic diagram of BI-MOF/GNS evaporator preparation for ISS and photo-Fenton Cr(VI) reduction, and photo-Fenton reduction efficiency of Cr(VI). Reprinted with permission from ref. [43]. Copyright 2024 Elsevier. (**b**) Effect of SMNG system on heavy metal concentrations. Reprinted with permission from ref. [130]. Copyright 2023 Elsevier. (**c**) The influence of different cycles on the concentrations of $Ni^{2+}$, $Cu^{2+}$, $Pb^{2+}$ and $Cd^{2+}$ in CGA. (**d**) Mechanism diagram of CGA-8 purification of heavy metal wastewater. Reprinted with permission from ref. [131]. Copyright 2022 Elsevier.

### *5.3. Microbial Inactivation and Anti-Biofouling*

Graphene-based SDIE technology demonstrates significant application potential in water treatment, particularly showcasing unique advantages in microbial inactivation. Leveraging graphene's exceptional light absorption and thermal conduction properties, this technology achieves highly efficient water evaporation processes through solar energy conversion. During interfacial evaporation, graphene materials efficiently convert solar energy into thermal energy, creating localized high-temperature zones at the water-air interface that enable rapid inactivation of aquatic microorganisms. The high specific surface area and abundant surface functional groups of graphene facilitate the adsorption and enrichment of microorganisms in water, further enhancing inactivation efficiency. Moreover, graphene-based materials generate reactive oxygen species (ROS) such as hydroxyl radicals ($\cdot OH$) and superoxide anions ($O_2^-$) through photocatalytic activity [132,133]. These reactive substances can damage microbial cell membranes and DNA structures, substantially improving inactivation performance. By optimizing the structural design and surface chemistry of graphene-based materials, the highly efficient inactivation of diverse pathogenic microorganisms—including bacteria, viruses, and protozoa—can be achieved.

Lu et al. [134] developed a seawater desalination composite membrane material with enhanced antimicrobial properties. The study evenly dispersed GO nanosheets within a polymer matrix, utilizing magnetic field-controlled solvent evaporation dynamics to prepare vertically oriented GO nanocomposite films. This unique orientation significantly enhanced the interface effects at the edges of the nanosheets, fundamentally improving the material's antibacterial properties. This nanocomposite membrane is not only suitable for seawater desalination but also exhibits multifunctional potential in environmental engineering, such as serving as high-performance sensors, catalytic electrode materials, and efficient adsorption media (Figure 12a). Noureen et al. [42] developed a functionalized coating fabric based on an $Ag_3PO_4$-reduced graphene oxide ($Ag_3PO_4$-rGO) nanocomposite system. The material was loaded onto a cotton substrate to form a solar-driven floating water purification device. This fabric achieved an evaporation rate of 1.31 $kg \cdot m^{-2} \cdot h^{-1}$ by converting solar heat to enhance interface temperature, while integrating the photocatalytic degradation of organic dyes and microbial inactivation. Moreover, they proposed an innovative photothermal aerogel design by constructing a three-dimensional porous system with an $Ag_3VO_4$-rGO composite structure, which incorporated solar steam generation, pollutant photocatalysis, and seawater desalination functions, forming a highly efficient water treatment system (Figure 12b) [135]. Lin et al. [136] constructed a polyacrylamide-reduced graphene oxide composite hydrogel system based on radiation reduction technology, which simultaneously loaded silver nanoparticles to form a porous ArG–Ag composite hydrogel. This material integrated rapid water transport channels, localized thermal effects, and low evaporation enthalpy properties, achieving an evaporation rate of 3.089 $kg \cdot m^{-2} \cdot h^{-1}$ under standard illumination, with energy conversion efficiency exceeding 122.7%. Its structural design provided both mechanical strength and salt tolerance, effectively regulating ion migration for water purification. The system demonstrated a 99% inhibition rate against *E. coli* after 24 h and exhibited long-term antifungal properties, significantly enhancing anti-biofouling performance. This research innovatively integrated high-efficiency evaporation, salt regulation, broad-spectrum antibacterial, fungal inhibition, and catalytic functions into a single system, offering a new multifunctional solution for marine environmental applications (Figure 12c,d).

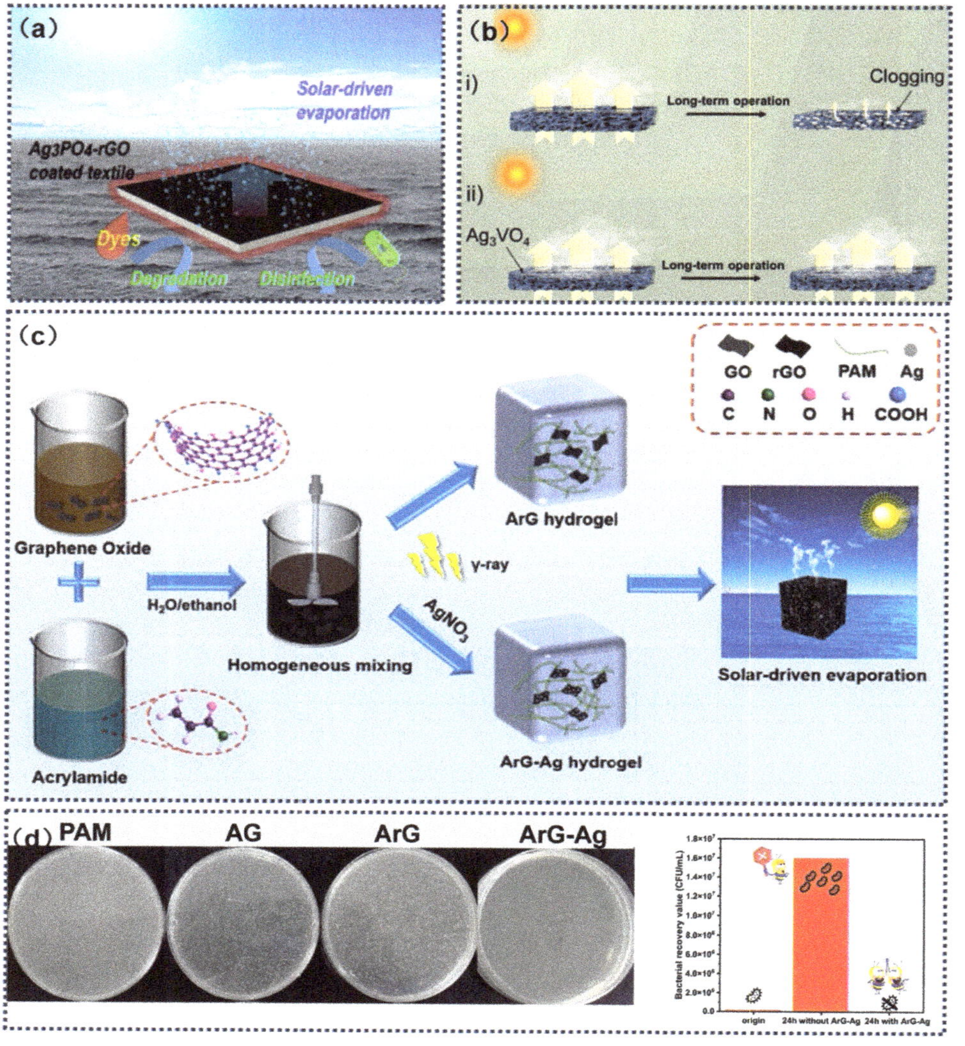


**Figure 12.** (**a**) Antibacterial performance of $Ag_3PO_4$-rGO against *E. coli*. Reprinted with permission from ref. [42]. Copyright 2020 American Chemical Society. (**b**) Photothermal membrane: (i) current issues (clogging/biofouling) and (ii) photothermal membrane with anti-clogging/anti-biofouling functionalities. Reprinted with permission from ref. [135]. Copyright 2024 Elsevier. (**c**) Synthesis process of ArG–Ag. (**d**) Antibacterial behavior of ArG–Ag toward *E. coli*. Reprinted with permission from ref. [136]. Copyright 2025 Elsevier.

## 6. Design and Scale-Up of Graphene-Based Evaporation–Catalytic Experimental Systems

In the previous section, we have systematically analyzed photothermal catalytic materials and conducted a comparative assessment of their performance and characteristics (Table 1). However, the ultimate objective of photothermal–photocatalytic systems is to achieve large-scale freshwater collection and water purification through efficient air–water interface evaporation. This necessitates the development of high-performance evaporators and well-designed purification–collection devices to provide an environmentally friendly and cost-effective sustainable solution for global freshwater scarcity. Laboratory-scale graphene-based photothermal evaporation–photocatalytic systems are typically composed of multiple functional layers: The photothermal conversion layer employs high-surface-area materials with nano-engineered structures to localize solar-to-thermal energy conversion; the water transport layer utilizes hydrophilic porous substrates to enable continuous water supply through capillary action while inhibiting salt crystallization; and the thermal insulation layer incorporates low-thermal-conductivity materials to minimize heat conduction losses and maintain elevated temperatures at the evaporation interface. Integrated photocatalytic components degrade organic pollutants through photogenerated carriers, achieving simultaneous evaporation and water purification. During operation, the photothermal layer generates localized thermal gradients under solar irradiation, driving the rapid vaporization of interfacial water molecules. The generated vapor is subsequently condensed and collected as freshwater, while photocatalytic reactions decompose contaminants either at the evaporation interface or in adjacent regions, forming a high-efficiency energy–water co-treatment system.

**Table 1.** Comparison of graphene-based solar driven interfacial evaporation coupled photocatalytic systems.

| Design Type | Evaporation Rate (kg m$^{-2}$ h$^{-1}$) | Energy Conversion Efficiency | Design Description | Durability | References |
|---|---|---|---|---|---|
| Reducing the Enthalpy of Water Evaporation | 3.94 | 135.6% | $Ti_3C_2T_x$/GO/polyaniline hybrids | 8 h | [91] |
| Reducing the Enthalpy of Water Evaporation | 3.39 | 104.1% | GO foam with numerous vertical channels | 10 days (10 h per day) | [92] |
| Optimization for Environmental Energy Applications | 1.42 | 94.2% | Vertically oriented graphene on graphite felt sheets | / | [98] |
| Optimization for Environmental Energy Applications | 1.48 | 92.9% | Bridge-arched and hollow melamine foam/rGO | Run the test 10 times | [99] |
| Optimization for Environmental Energy Applications | 34.70 | 95.3% | 3D GO stalk | 45 h | [101] |
| Water Transport Regulation | 5.98 | 73.0% | Laminated chitosan/graphene nanoplatelets | 30 h | [38] |
| Water Transport Regulation | 1.57 | 95.2% | N-doped lignosulfonate/GO | 6 cycles with 24 h | [108] |
| Monolayer Structure | 2.10 | 96.8% | CuO/$Cu_2O$ nanoparticles | 30 days | [116] |
| Monolayer Structure | 1.63 | 88.0% | Hydrogel-reduced GO composite membrane | 8 days (8 h per day) | [117] |
| Multilayer Structures | 2.54 | 91.1% | Graphene/CuO Composite | 8 h | [121] |
| Photocatalytic Degradation of Organic Pollutants | 1.94 | 84.0% | rGO/FeOOH-based asymmetric evaporator | Run the test 15 times | [41] |
| Photocatalytic Degradation of Organic Pollutants | 2.08 | 94.8% | GO/polypyrrole hybrid aerogel | 20 h | [126] |
| Photocatalytic Degradation of Organic Pollutants | 12.70 | 88.9% | rGO-Pd catalytic evaporator | Run the test 8 times | [127] |
| Reduction and Fixation of Heavy Metal Ions | 2.16 | 87.5% | Bi-MOF/graphene composite | Run the test 15 times | [43] |
| Reduction and Fixation of Heavy Metal Ions | 6.46 | 90.3% | Ag-$MnO_2$ nanostructure with N-doped graphene aerogel | 5-day cycle (8 h per day) | [130] |

**Table 1.** *Cont.*

| Design Type | Evaporation Rate (kg $m^{-2}$ $h^{-1}$) | Energy Conversion Efficiency | Design Description | Durability | References |
|---|---|---|---|---|---|
| Reduction and Fixation of Heavy Metal Ions | 1.80 | 70.0% | Cellulose graphene aerogel | Run the test 12 times | [131] |
| Microbial Inactivation and Anti-biofouling | 1.31 | 86.8% | $Ag_3PO_4$-rGO nanocomposite | Run the test 7 times | [42] |
| Microbial Inactivation and Anti-biofouling | 2.10 | 94.0% | $Ag_3VO_4$-rGO aerogel | Run the test 10 times | [135] |
| Microbial Inactivation and Anti-biofouling | 3.09 | 122.7% | Graphene/silver composite hydrogel | Run the test 8 times | [136] |

Although experimental systems demonstrate remarkable performance under controlled laboratory conditions, they face limitations in scale and material stability assessment. Moreover, laboratory environments may inadequately simulate real-world challenges such as complex water quality or fluctuating illumination conditions. In contrast, pilot-scale devices provide reliable transitional solutions for industrial applications through engineering optimizations. Li et al. [137] developed a three-dimensional interconnected porous carbon foam (3D IPCF) fabricated via gas-assisted expansion and perforation processes, achieving an exceptional evaporation rate of 10.9 kg $m^{-2}$ $h^{-1}$ under 1 sun irradiation with natural convection. The study emphasized that the material's interconnected pore architecture and millimeter-scale pore size significantly enhanced vapor diffusion efficiency, with theoretical modeling and experiments confirming the critical role of pore morphology. In scaled-up trials, a large-scale device comprising 20 independent 3D IPCF units demonstrated continuous outdoor operation for 13 h, accumulating an evaporation yield of 42.0 kg $m^{-2}$. The system maintained stable daytime and nighttime evaporation rates of 2.3–6.1 kg $m^{-2}$ $h^{-1}$ and 1.6–2.1 kg $m^{-2}$ $h^{-1}$ respectively, showcasing all-weather high-throughput capability. Furthermore, the material exhibited excellent performance in treating dye wastewater and heavy metal-ion contaminated water, achieving environmental discharge standards with good cyclic stability. Zhang et al. [138] engineered a vertically aligned graphene sheet membrane (VA-GSM) through anti-freezing-assisted freeze casting, where its through-channel structure dramatically improved water transport efficiency. The system attained evaporation rates of 1.62 kg $m^{-2}$ $h^{-1}$ and 6.25 kg $m^{-2}$ $h^{-1}$ under 1 sun and 4 sun irradiance, respectively, corresponding to solar-thermal conversion efficiencies of 86.5% and 94.2%. Scalability was confirmed through the successful fabrication of 16 cm × 16 cm large-area samples by adjusting mold dimensions (Figure 13d). When applied to Bohai seawater treatment under 4 sun conditions, the VA-GSM achieved continuous clean water production at 6.22 kg $m^{-2}$ $h^{-1}$ with ion rejection rates exceeding 99%, meeting international drinking water standards. Remarkably, the membrane maintained stable evaporation performance in strongly acidic (0.2–8 mol $mL^{-1}$ $H^+$) and alkaline (0.2–8 mol $mL^{-1}$ $OH^-$) environments, while reducing heavy metal-ion concentrations below 0.01 mg $kg^{-1}$, demonstrating broad-spectrum wastewater treatment capabilities for complex effluents.

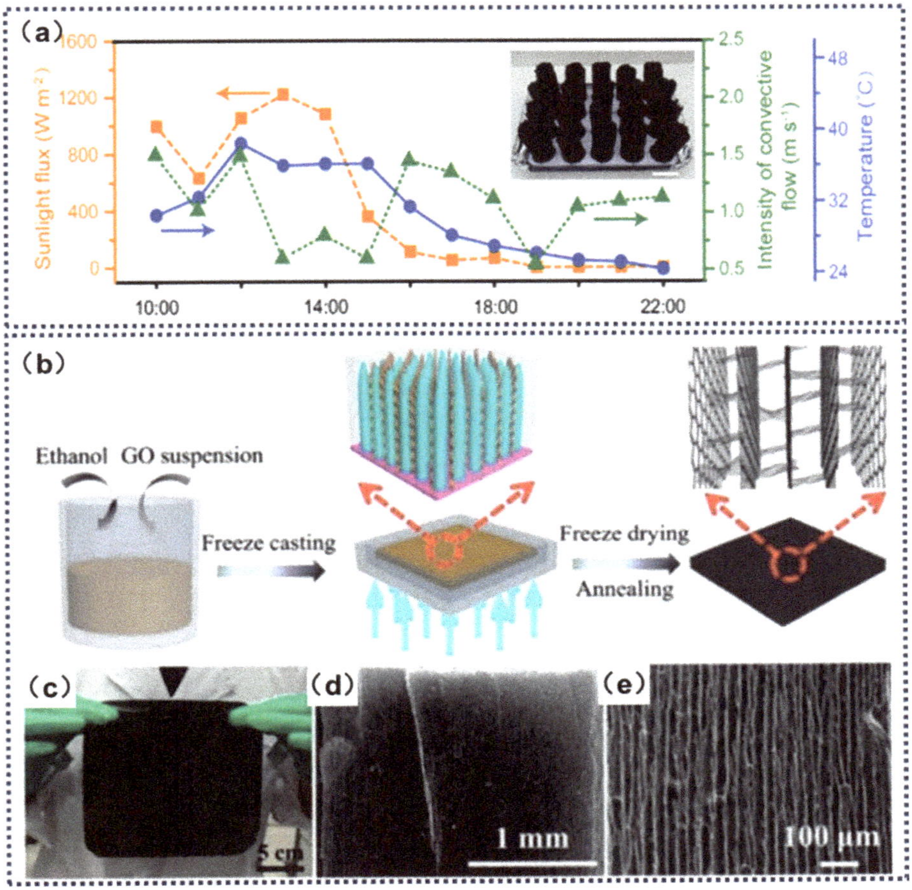


**Figure 13.** (**a**) Outdoor evaporation of large equipment settings. Outdoor conditions for evaporation: incident sunlight flux, intensity of natural convective flow, and ambient temperature over time. Reprinted with permission from ref. [137]. Copyright 2025 Elsevier. (**b**) Manufacturing process and characterization diagram of VA-GSM. (**c**) VA-GSM monocoque photo with a size of 16 cm × 16 cm. (**d**,**e**) SEM images of VA-GSM at different magnifications. Reprinted with permission from ref. [138] Copyright 2017 American Chemical Society.

## 7. Challenges and Future Perspectives

### *7.1. Low Steam Condensation Efficiency Affecting Subsequent Freshwater Collection*

Although graphene-based materials have demonstrated excellent photothermal conversion and steam generation performance in SDIE, the low efficiency of steam condensation remains a critical bottleneck limiting the actual yield of freshwater. Current research often focuses on thermodynamic optimization at the evaporation end, neglecting the dynamic regulation of latent heat release during the steam phase change and the enhancement of heat and mass transfer at the condensation end, which restricts the overall energy utilization efficiency of the system. On one hand, the spatial separation design between the evaporation system and the condensation module leads to significant heat dissipation during steam diffusion and transport due to heat exchange with the surrounding medium and turbulence disturbances, resulting in reduced latent heat recovery rates. On the other hand, traditional condenser surfaces lack the ability for directional adsorption of microscale steam molecules and rapid liquid drainage, causing the formation of insulating vapor films or liquid retention layers at the condensation interface, which severely hinders the improvement of phase change heat transfer coefficients. Additionally, the mismatch between the high-temperature steam (typically 60–90 °C) produced by graphene-based evaporation systems and the insufficient temperature difference for conventional condensation further reduces the condensation driving force. This issue is particularly pronounced in humid and hot climates or under high ambient temperature conditions, where freshwater collection efficiency may decrease by more than 40%.

### *7.2. Stability of Solar Evaporators in Long-Term Practical Operation*

The stability of graphene-based materials in solar-driven interfacial evaporators during long-term practical operation remains a core challenge for large-scale application. Firstly, material degradation under complex conditions is a prominent issue: Graphene nanosheets are prone to photo-oxidation or structural wrinkling during continuous photothermal cycling, leading to the attenuation of surface plasmon resonance effects and the disruption of thermal conduction networks. Hydrophilic polymer substrates (such as cellulose and polyacrylamide) are susceptible to hydrolysis or crosslinking failure under high temperature, high salinity, or UV radiation, causing the collapse of porous structures and blockage of water transport channels. Secondly, salt crystallization and pollutant deposition are significant problems, especially in desalination scenarios. Supersaturated salt precipitation near the evaporation interface can cover photothermal active sites, while the adsorption of impurities like oil and microorganisms alters surface wettability, resulting in a gradual decline in evaporation efficiency. Additionally, insufficient mechanical stability limits device lifespan; for example, structural stress from wave scouring or ice crystal expansion may compromise the integrity of the three-dimensional porous framework, and weak interfacial bonding between graphene and the substrate can lead to interlayer delamination under thermal stress.

### *7.3. Adaptive Limitations in Complex Water Bodies*

In terms of adaptability to complex water bodies, the graphene-based SDIE–photocatalytic coupling system still faces multiple technical bottlenecks: the salt crystallization behavior caused by ion enrichment in high-salinity water bodies easily leads to the blockage of the porous structure of the evaporation layer, significantly reducing the steam mass transfer efficiency and light penetration rate; low-concentration and highly toxic emerging pollutants such as perfluorinated compounds, antibiotic resistance genes, and nanoplastics are difficult to achieve efficient targeted degradation of due to the competition for interfacial adsorption sites and insufficient selective oxidation by free radicals; when multiple pollutants coexist, heavy metal ions and dissolved organic matter (DOM) may form antagonistic effects that inhibit photocatalytic kinetics through electron transfer competition, active site shielding, or free radical quenching effects; and in addition, the non-uniform deposition of suspended particles and colloidal substances at heterogeneous interfaces will aggravate light scattering loss and mass transfer resistance, leading to a significant deterioration in the long-term operational stability and anti-pollution ability of the system, which restricts its universal application in non-ideal water environments such as industrial wastewater and seawater desalination.

## 8. Conclusions

This paper systematically reviews the research progress of graphene-based materials in SDIE-coupled photocatalytic technology and their application potential in water treatment. Leveraging its broadband light absorption, ultrafast hot carrier dynamics, tunable electronic structure, and low evaporation enthalpy, graphene has emerged as a core material for photothermal–photochemical synergistic water treatment systems. Strategies such as surface morphology modulation, plasmonic material integration, and defect engineering have significantly enhanced light absorption efficiency. Concurrently, hydrophilic functional group modification, porous structure design, and thermal localization optimization effectively reduce water evaporation enthalpy while suppressing energy losses. Furthermore, a 3D structural design, salt-resistance strategies, and fluid channel engineering have improved system evaporation performance and environmental adaptability. Studies demonstrate that graphene-based SDIE systems efficiently degrade organic pollutants, immobilize heavy metal ions, and inactivate microorganisms, offering a green solution for the advanced purification and resource recovery of complex wastewater. However, challenges such as low steam condensation efficiency, insufficient long-term material stability, high costs in scalable manufacturing, and limited adaptability to complex water matrices remain bottlenecks for practical implementation. Future research should focus on enhancing heat-mass transfer in condensation systems, optimizing material environmental tolerance, developing low-cost fabrication processes, and exploring multi-pollutant synergistic removal mechanisms. These advancements will propel graphene-based SDIE technology from laboratory research to large-scale applications, ultimately contributing to global sustainable water resource management.

**Author Contributions:** Writing—original draft preparation, Y.Z.; writing—review and editing, H.W.; visualization, supervision, J.Z.; All authors have read and agreed to the published version of the manuscript.

**Funding:** This research received no external funding.

**Data Availability Statement:** No data was used for the research described in the article.

**Conflicts of Interest:** Author Jisheng Zhang was employed by the company Jiangsu Agrochem Laboratory Co., Ltd., Changzhou, China. The remaining authors declare that the research was conducted in the absence of any commercial or financial relationships that could be construed as a potential conflict of interest.

## References

1. Chen, A.; Liang, H.; Chen, T.; Yang, W.; Ding, C. Influence of long-term irrigation with treated papermaking wastewater on soil ecosystem of a full-scale managed reed wetland. *J. Soils Sediments* **2015**, *16*, 1352–1359. [CrossRef]
2. Magni, M.; Jones, E.R.; Bierkens, M.F.P.; van Vliet, M.T.H. Global energy consumption of water treatment technologies. *Water Res.* **2025**, *277*, 123245. [CrossRef] [PubMed]
3. Li, Z.; Qiu, J.; Xu, X.; Wan, R.; Yao, M.; Wang, H.; Zhou, Z.; Xu, J. Solar driven kaolin-based hydrogels for efficient interfacial evaporation and heavy metal ion adsorption from wastewater. *Sep. Purif. Technol.* **2025**, *354*, 129243. [CrossRef]
4. Jiang, Y.; Tang, Y.; Li, H. A review of trends in the use of sewage irrigation technology from the livestock and poultry breeding industries for farmlands. *Irrig. Sci.* **2022**, *40*, 297–308. [CrossRef]
5. Kama, R.; Liu, Y.; Aidara, M.; Kpalari, D.F.; Song, J.; Diatta, S.; Sulemana, H.; Li, H.; Li, Z. Plant-Soil Feedback Combined with Straw Incorporation Under Maize/Soybean Intercropping Increases Heavy Metals Migration in Soil-Plant System and Soil HMRG Abundance Under Livestock Wastewater Irrigation. *J. Soil Sci. Plant Nutr.* **2024**, *24*, 7090–7104. [CrossRef]
6. Yu, J.; Wang, C.; Liu, S.; Yang, S.; Du, X.; Liu, S.; Shao, C.; Kong, H.; Wang, B.; Wu, T.; et al. Solid waste-derived solar desalination devices: Enhanced efficiency in water vapor generation and diffusion. *Chem. Eng. J.* **2024**, *485*, 149870. [CrossRef]
7. Giwa, A.; Yusuf, A.; Dindi, A.; Balogun, H.A. Polygeneration in desalination by photovoltaic thermal systems: A comprehensive review. *Renew. Sustain. Energy Rev.* **2020**, *130*, 109946. [CrossRef]
8. Ali, A.; Tufa, R.A.; Macedonio, F.; Curcio, E.; Drioli, E. Membrane technology in renewable-energy-driven desalination. *Renew. Sustain. Energy Rev.* **2018**, *81*, 1–21. [CrossRef]
9. Richards, B.S.; Howard, I.A. Luminescent solar concentrators for building integrated photovoltaics: Opportunities and challenges. *Energy Environ. Sci.* **2023**, *16*, 3214–3239. [CrossRef]

10. Li, J.; Wang, J.; Zhang, J.; Han, T.; Hu, X.; Lee, M.M.S.; Wang, D.; Tang, B.Z. A Facile Strategy of Boosting Photothermal Conversion Efficiency through State Transformation for Cancer Therapy. *Adv. Mater.* **2021**, *33*, e2105999. [CrossRef]
11. Zhang, L.; Yang, Z.P.; Wu, X.; Wang, W.J.; Yang, C.; Xu, G.J.; Wu, C.A.; Bao, E.C. Open-Field Agrivoltaic System Impacts on Photothermal Environment and Light Environment Simulation Analysis in Eastern China. *Agronomy* **2023**, *13*, 1820. [CrossRef]
12. Bei, X.R.; Yu, X.J.; Li, D.Q.; Sun, Q.L.; Yu, Y.H.; Wang, Y.Q.; Okonkwo, C.E.; Zhou, C.S. Heat source replacement strategy using catalytic infrared: A future for energy saving drying of fruits and vegetables. *J. Food Sci.* **2023**, *88*, 4827–4839. [CrossRef] [PubMed]
13. Ren, J.; Xu, J.; Tian, S.; Shi, K.; Gu, T.; Zhao, J.; Li, X.; Zhou, Z.; Tijing, L.; Shon, H.K. Hydrodynamic solar-driven interfacial evaporation—Gone with the flow. *Water Res.* **2024**, *266*, 122432. [CrossRef]
14. Li, L.; Sun, X.S.; Miao, J.; Wang, H.N.; Song, Y.C.; Tang, D.W. Gradient engineering in interfacial evaporation for water, energy, and mineral harvesting. *Energy Environ. Sci.* **2025**, *18*, 1176–1190. [CrossRef]
15. Bi, X.Y.; Li, L.B.; Luo, L.J.; Liu, X.H.; Li, J.M.; You, T.Y. A ratiometric fluorescence aptasensor based on photoinduced electron transfer from CdTe QDs to $WS_2$ NTs for the sensitive detection of zearalenone in cereal crops. *Food Chem.* **2022**, *385*, 132657. [CrossRef]
16. Wang, J.; Cao, X.; Cui, X.; Wang, H.; Zhang, H.; Wang, K.; Li, X.; Li, Z.; Zhou, Y. Recent Advances of Green Electricity Generation: Potential in Solar Interfacial Evaporation System. *Adv. Mater.* **2024**, *36*, e2311151. [CrossRef]
17. Cai, M.; Li, C.; An, X.; Zhong, B.; Zhou, Y.; Feng, K.; Wang, S.; Zhang, C.; Xiao, M.; Wu, Z.; et al. Supra-Photothermal $CO_2$ Methanation over Greenhouse-Like Plasmonic Superstructures of Ultrasmall Cobalt Nanoparticles. *Adv. Mater.* **2024**, *36*, e2308859. [CrossRef] [PubMed]
18. Cai, R.; Xiang, H.; Yang, D.; Lin, K.T.; Wu, Y.; Zhou, R.; Gu, Z.; Yan, L.; Zhao, Y.; Tan, W. Plasmonic AuPt@CuS Heterostructure with Enhanced Synergistic Efficacy for Radiophotothermal Therapy. *J. Am. Chem. Soc.* **2021**, *143*, 16113–16127. [CrossRef]
19. Wan, X.; Li, Y.; Chen, Y.; Ma, J.; Liu, Y.A.; Zhao, E.D.; Gu, Y.; Zhao, Y.; Cui, Y.; Li, R.; et al. A nonmetallic plasmonic catalyst for photothermal $CO_2$ flow conversion with high activity, selectivity and durability. *Nat. Commun.* **2024**, *15*, 1273. [CrossRef]
20. He, S.S.; Jiang, Y.Y.; Li, J.C.; Pu, K.Y. Semiconducting Polycomplex Nanoparticles for Photothermal Ferrotherapy of Cancer. *Angew. Chem. Int. Ed.* **2020**, *59*, 10633–10638. [CrossRef]
21. Sun, L.; Li, Z.; Su, R.; Wang, Y.; Li, Z.; Du, B.; Sun, Y.; Guan, P.; Besenbacher, F.; Yu, M. Phase-Transition Induced Conversion into a Photothermal Material: Quasi-Metallic $WO_{2.9}$ Nanorods for Solar Water Evaporation and Anticancer Photothermal Therapy. *Angew. Chem. Int. Ed. Engl.* **2018**, *57*, 10666–10671. [CrossRef]
22. Yang, M.Q.; Tan, C.F.; Lu, W.; Zeng, K.; Ho, G.W. Spectrum Tailored Defective 2D Semiconductor Nanosheets Aerogel for Full-Spectrum-Driven Photothermal Water Evaporation and Photochemical Degradation. *Adv. Funct. Mater.* **2020**, *30*, 2004460. [CrossRef]
23. Hu, A.; Zhao, Y.; Hu, Q.; Chen, C.; Lu, X.; Cui, S.; Liu, B. Highly efficient solar steam evaporation via elastic polymer covalent organic frameworks monolith. *Nat. Commun.* **2024**, *15*, 9484. [CrossRef] [PubMed]
24. Huang, C.H.; Huang, J.X.; Chiao, Y.H.; Chang, C.M.; Hung, W.S.; Lue, S.J.; Wang, C.F.; Hu, C.C.; Lee, K.R.; Pan, H.H.; et al. Tailoring of a Piezo-Photo-Thermal Solar Evaporator for Simultaneous Steam and Power Generation. *Adv. Funct. Mater.* **2021**, *31*, 2010422. [CrossRef]
25. Qi, D.; Liu, Y.; Liu, Y.; Liu, Z.; Luo, Y.; Xu, H.; Zhou, X.; Zhang, J.; Yang, H.; Wang, W.; et al. Polymeric Membranes with Selective Solution-Diffusion for Intercepting Volatile Organic Compounds during Solar-Driven Water Remediation. *Adv. Mater.* **2020**, *32*, e2004401. [CrossRef] [PubMed]
26. Song, J.; Wang, F.; Yang, X.; Ning, B.; Harp, M.G.; Culp, S.H.; Hu, S.; Huang, P.; Nie, L.; Chen, J.; et al. Gold Nanoparticle Coated Carbon Nanotube Ring with Enhanced Raman Scattering and Photothermal Conversion Property for Theranostic Applications. *J. Am. Chem. Soc.* **2016**, *138*, 7005–7015. [CrossRef]
27. Yu, Y.; Liu, Y.; Zhang, X.; Lv, B.; Xu, Y.; Fan, X. Co-enhancing volatile organic compound degradation and steam generation in solar interfacial evaporation by integrating with electro-Fenton. *Water Res.* **2025**, *277*, 123348. [CrossRef]
28. Xia, Q.; Wang, C.; Xu, N.; Yang, J.; Gao, G.; Ding, J. A Floating Integrated Solar Micro-Evaporator for Self-Cleaning Desalination and Organic Degradation. *Adv. Funct. Mater.* **2023**, *33*, 2214769. [CrossRef]
29. Huang, L.; Liu, R.; Yang, J.; Shuai, Q.; Yuliarto, B.; Kaneti, Y.V.; Yamauchi, Y. Nanoarchitectured porous organic polymers and their environmental applications for removal of toxic metal ions. *Chem. Eng. J.* **2021**, *408*, 127991. [CrossRef]
30. Jing, Z.; Ding, J.; Zhang, T.; Yang, D.; Qiu, F.; Chen, Q.; Xu, J. Flexible, versatility and superhydrophobic biomass carbon aerogels derived from corn bracts for efficient oil/water separation. *Food Bioprod. Process.* **2019**, *115*, 134–142. [CrossRef]
31. Fang, S.; Huang, W.; Wu, J.; Han, J.; Wang, L.; Wang, Y. Separation and Purification of Recombinant beta-Glucosidase with Hydrophobicity and Thermally Responsive Property from Cell Lysis Solution by Foam Separation and Further Purification. *J. Agric. Food Chem.* **2023**, *71*, 3362–3372. [CrossRef]
32. Shoaib, M.; Li, H.; Khan, I.M.; Hassan, M.M.; Zareef, M.; Niazi, S.; Chen, Q. Emerging MXenes-based aptasensors: A paradigm shift in food safety detection. *Trends Food Sci. Technol.* **2024**, *151*, 104635. [CrossRef]

33. Li, W.T.; Zhang, X.A.; Shi, Y.Q.; Hu, X.T.; Wang, X.; Liang, N.N.; Shen, T.T.; Zou, X.B.; Shi, J.Y. A dual-modal biosensor coupling cooperative catalysis strategy for sensitive detection of AFB1 in agri-products. *Food Chem.* **2023**, *426*, 136553. [CrossRef] [PubMed]
34. Yin, M.; Wang, W.; Wei, J.; Chen, X.; Chen, Q.; Chen, X.; Oyama, M. Novel dual-emissive fluorescent immunoassay for synchronous monitoring of okadaic acid and saxitoxin in shellfish. *Food Chem.* **2022**, *368*, 130856. [CrossRef]
35. Guo, Z.; Chen, P.; Yin, L.; Zuo, M.; Chen, Q.; El-Seedi, H.R.; Zou, X. Determination of lead in food by surface-enhanced Raman spectroscopy with aptamer regulating gold nanoparticles reduction. *Food Control* **2022**, *132*, 108498. [CrossRef]
36. Kurapati, R.; Kostarelos, K.; Prato, M.; Bianco, A. Biomedical Uses for 2D Materials Beyond Graphene: Current Advances and Challenges Ahead. *Adv. Mater.* **2016**, *28*, 6052–6074. [CrossRef] [PubMed]
37. Li, Z.J.; Lei, H.; Kan, A.K.; Xie, H.Q.; Yu, W. Photothermal applications based on graphene and its derivatives: A state-of-the-art review. *Energy* **2021**, *216*, 119262. [CrossRef]
38. Lim, H.W.; Seung Lee, H.; Joon Lee, S. Laminated chitosan/graphene nanoplatelets aerogel for 3D interfacial solar desalination with harnessing wind energy. *Chem. Eng. J.* **2024**, *480*, 148197. [CrossRef]
39. Wang, Y.; Liu, X.; Zhang, Q.; Wang, C.; Huang, S.; Liu, Y.; Yu, T.; Yang, R.; Chen, G.; Chaker, M.; et al. Stable, Cost-Effective TiN-Based Plasmonic Nanocomposites with over 99% Solar Steam Generation Efficiency. *Adv. Funct. Mater.* **2023**, *33*, 2212301. [CrossRef]
40. Lv, F.; Miao, J.; Hu, J.; Orejon, D. 3D Solar Evaporation Enhancement by Superhydrophilic Copper Foam Inverted Cone and Graphene Oxide Functionalization Synergistic Cooperation. *Small* **2023**, *19*, e2208137. [CrossRef]
41. Ma, J.; Guo, Z.; Han, X.; Guo, K.; Li, H.; Fang, P.; Wang, X.; Xin, J. Reduced graphene oxide/FeOOH-based asymmetric evaporator for the simultaneous generation of clean water and electrical power. *Carbon* **2023**, *201*, 318–327. [CrossRef]
42. Fan, Z.; Liu, J.; Liu, H.; Liu, L.; She, Y.; Wen, X.; Wang, H.; Hu, G.; Niu, R.; Gong, J. Synergism of solar-driven interfacial evaporation and photo-Fenton Cr(VI) reduction by sustainable Bi-MOF-based evaporator from waste polyester. *J. Energy Chem.* **2024**, *94*, 527–540. [CrossRef]
43. Noureen, L.; Xie, Z.; Gao, Y.; Li, M.; Hussain, M.; Wang, K.; Zhang, L.; Zhu, J. Multifunctional $Ag_3PO_4$-rGO-Coated Textiles for Clean Water Production by Solar-Driven Evaporation, Photocatalysis, and Disinfection. *ACS Appl. Mater. Interfaces* **2020**, *12*, 6343–6350. [CrossRef]
44. Sharma, A.S.; Ali, S.; Sabarinathan, D.; Murugavelu, M.; Li, H.; Chen, Q. Recent progress on graphene quantum dots-based fluorescence sensors for food safety and quality assessment applications. *Compr. Rev. Food Sci. Food Saf.* **2021**, *20*, 5765–5801. [CrossRef]
45. Li, Y.; Li, Y.; Zhang, D.; Tan, W.; Shi, J.; Li, Z.; Liu, H.; Yu, Y.; Yang, L.; Wang, X.; et al. A fluorescence resonance energy transfer probe based on functionalized graphene oxide and upconversion nanoparticles for sensitive and rapid detection of zearalenone. *LWT* **2021**, *147*, 111541. [CrossRef]
46. Rong, J.; Zhou, Z.; Wang, Y.; Han, J.; Li, C.; Zhang, W.; Ni, L. Immobilization of Horseradish Peroxidase on Multi-Armed Magnetic Graphene Oxide Composite: Improvement of Loading Amount and Catalytic Activity. *Food Technol. Biotechnol.* **2019**, *57*, 260–271. [CrossRef] [PubMed]
47. Li, Y.; Luo, S.; Sun, L.; Kong, D.; Sheng, J.; Wang, K.; Dong, C. A Green, Simple, and Rapid Detection for Amaranth in Candy Samples Based on the Fluorescence Quenching of Nitrogen-Doped Graphene Quantum Dots. *Food Anal. Methods* **2019**, *12*, 1658–1665. [CrossRef]
48. Ren, H.Y.; Tang, M.; Guan, B.L.; Wang, K.X.; Yang, J.W.; Wang, F.F.; Wang, M.Z.; Shan, J.Y.; Chen, Z.L.; Wei, D.; et al. Hierarchical Graphene Foam for Efficient Omnidirectional Solar-Thermal Energy Conversion. *Adv. Mater.* **2017**, *29*, 1702590. [CrossRef]
49. Wei, N.; Peng, X.S.; Xu, Z.P. Understanding Water Permeation in Graphene Oxide Membranes. *Acs Appl. Mater. Interfaces* **2014**, *6*, 5877–5883. [CrossRef]
50. Chen, S.S.; Wu, Q.Z.; Mishra, C.; Kang, J.Y.; Zhang, H.J.; Cho, K.J.; Cai, W.W.; Balandin, A.A.; Ruoff, R.S. Thermal conductivity of isotopically modified graphene. *Nat. Mater.* **2012**, *11*, 203–207. [CrossRef]
51. Balandin, A.A.; Ghosh, S.; Bao, W.Z.; Calizo, I.; Teweldebrhan, D.; Miao, F.; Lau, C.N. Superior thermal conductivity of single-layer graphene. *Nano Lett.* **2008**, *8*, 902–907. [CrossRef] [PubMed]
52. Wu, S.; Duan, N.; Qiu, Y.; Li, J.; Wang, Z. Colorimetric aptasensor for the detection of *Salmonella enterica* serovar typhimurium using $ZnFe_2O_4$-reduced graphene oxide nanostructures as an effective peroxidase mimetics. *Int. J. Food Microbiol.* **2017**, *261*, 42–48. [CrossRef] [PubMed]
53. Li, Y.H.; Tang, Z.R.; Xu, Y.J. Multifunctional graphene-based composite photocatalysts oriented by multifaced roles of graphene in photocatalysis. *Chin. J. Catal.* **2022**, *43*, 708–730. [CrossRef]
54. Luo, L.; Liu, X.; Ma, S.; Li, L.; You, T. Quantification of zearalenone in mildewing cereal crops using an innovative photoelectrochemical aptamer sensing strategy based on ZnO-NGQDs composites. *Food Chem.* **2020**, *322*, 126778. [CrossRef] [PubMed]
55. Luo, L.; Ma, S.; Li, L.; Liu, X.; Zhang, J.; Li, X.; Liu, D.; You, T. Monitoring zearalenone in corn flour utilizing novel self-enhanced electrochemiluminescence aptasensor based on $NGQDs-NH_2-Ru@SiO_2$ luminophore. *Food Chem.* **2019**, *292*, 98–105. [CrossRef]

56. Zhang, W.; Xu, Y.; Zou, X. Rapid determination of cadmium in rice using an all-solid RGO-enhanced light addressable potentiometric sensor. *Food Chem.* **2018**, *261*, 1–7. [CrossRef]
57. Xu, Y.; Zhang, W.; Shi, J.; Zou, X.; Li, Y.; Haroon Elrasheid, T.; Huang, X.; Li, Z.; Zhai, X.; Hu, X. Electrodeposition of gold nanoparticles and reduced graphene oxide on an electrode for fast and sensitive determination of methylmercury in fish. *Food Chem.* **2017**, *237*, 423–430. [CrossRef]
58. Zhou, W.; Li, C.; Sun, C.; Yang, X. Simultaneously determination of trace $Cd_{2+}$ and $Pb_{2+}$ based on L-cysteine/graphene modified glassy carbon electrode. *Food Chem.* **2016**, *192*, 351–357. [CrossRef]
59. Zhang, J.; Lei, Y.; Lin, Z.; Xie, P.; Lu, H.; Xu, J. A novel approach to recovery of lithium element and production of holey graphene based on the lithiated graphite of spent lithium ion batteries. *Chem. Eng. J.* **2022**, *436*, 135011. [CrossRef]
60. Kim, S.H.; Kim, D.S.; Moradi, H.; Chang, Y.Y.; Yang, J.K. Highly porous biobased graphene-like carbon adsorbent for dye removal: Preparation, adsorption mechanisms and optimization. *J. Environ. Chem. Eng.* **2023**, *11*, 109278. [CrossRef]
61. Cao, L.Y.; Wang, C.W.; Huang, Y.X. Structure optimization of graphene aerogel-based composites and applications in batteries and supercapacitors. *Chem. Eng. J.* **2023**, *454*, 140094. [CrossRef]
62. Han, J.; Fang, S.; He, X.; Wang, L.; Li, C.; Wu, J.; Cai, Y.; Wang, Y. Combination of aqueous two-phase flotation and inverse transition cycling: Strategies for separation and purification of recombinant beta-glucosidase from cell lysis solution. *Food Chem.* **2022**, *373*, 131543. [CrossRef]
63. Duan, N.; Gong, W.; Wu, S.; Wang, Z. Selection and Application of ssDNA Aptamers against Clenbuterol Hydrochloride Based on ssDNA Library Immobilized SELEX. *J. Agric. Food Chem.* **2017**, *65*, 1771–1777. [CrossRef] [PubMed]
64. Zhao, Y.; Ma, Y.; Zhou, R.; He, Y.; Wu, Y.; Yi, Y.; Zhu, G. Highly sensitive electrochemical detection of paraoxon ethyl in water and fruit samples based on defect-engineered graphene nanoribbons modified electrode. *J. Food Meas. Charact.* **2022**, *16*, 2596–2603. [CrossRef]
65. Chao, Y.; Pang, J.; Bai, Y.; Wu, P.; Luo, J.; He, J.; Jin, Y.; Li, X.; Xiong, J.; Li, H.; et al. Graphene-like BN@$SiO_2$ nanocomposites as efficient sorbents for solid-phase extraction of Rhodamine B and Rhodamine 6G from food samples. *Food Chem.* **2020**, *320*, 126666. [CrossRef]
66. Zhu, Y.W.; Murali, S.; Cai, W.W.; Li, X.S.; Suk, J.W.; Potts, J.R.; Ruoff, R.S. Graphene and Graphene Oxide: Synthesis, Properties, and Applications. *Adv. Mater.* **2010**, *22*, 3906–3924. [CrossRef]
67. Han, B.; Zhang, Y.L.; Chen, Q.D.; Sun, H.B. Carbon-Based Photothermal Actuators. *Adv. Funct. Mater.* **2018**, *28*, 1802235. [CrossRef]
68. Xie, Z.; Duo, Y.; Lin, Z.; Fan, T.; Xing, C.; Yu, L.; Wang, R.; Qiu, M.; Zhang, Y.; Zhao, Y.; et al. The Rise of 2D Photothermal Materials beyond Graphene for Clean Water Production. *Adv. Sci.* **2020**, *7*, 1902236. [CrossRef]
69. Li, X.; Yu, J.; Wageh, S.; Al-Ghamdi, A.A.; Xie, J. Graphene in Photocatalysis: A Review. *Small* **2016**, *12*, 6640–6696. [CrossRef]
70. Yan, S.W.; Song, H.J.; Li, Y.; Yang, J.; Jia, X.H.; Wang, S.Z.; Yang, X.F. Integrated reduced graphene oxide/polypyrrole hybrid aerogels for simultaneous photocatalytic decontamination and water evaporation. *Appl. Catal. B-Environ. Energy* **2022**, *301*, 120820. [CrossRef]
71. Li, Q.; Lu, J.; Gupta, P.; Qiu, M. Engineering Optical Absorption in Graphene and Other 2D Materials: Advances and Applications. *Adv. Opt. Mater.* **2019**, *7*, 1900595. [CrossRef]
72. Zeng, L.; Deng, D.X.; Zhu, L.Y.; Zhang, Z.K.; Gu, X.; Wang, H.M.; Jiang, Y.J. Multi-scale CuS-rGO pyramidal photothermal structure for highly efficient solar-driven water evaporation and thermoelectric power generation. *Nano Energy* **2024**, *125*, 109531. [CrossRef]
73. Ma, H.D.; Yu, L.J.; Li, Z.Z.; Chen, J.L.; Meng, J.G.; Song, Q.W.; Liu, Y.M.; Wang, Y.Z.; Wu, Q.; Miao, M.H.; et al. A Lotus Seedpods-Inspired Interfacial Solar Steam Generator with Outstanding Salt Tolerance and Mechanical Properties for Efficient and Stable Seawater Desalination. *Small* **2023**, *19*, 2304877. [CrossRef]
74. Zhao, X.; Meng, X.; Zou, H.; Wang, Z.; Du, Y.; Shao, Y.; Qi, J.; Qiu, J. Topographic Manipulation of Graphene Oxide by Polyaniline Nanocone Arrays Enables High-Performance Solar-Driven Water Evaporation. *Adv. Funct. Mater.* **2022**, *33*, 2209207. [CrossRef]
75. Xiong, R.; Zhong, L.; Song, Y.; Xu, J.; Xiao, Y.; Cheng, B.; Lei, S. $CoCr_2O_4$ Nanoparticles with Abundant Oxygen Vacancies: A New Photothermal Platform for Efficient Solar Evaporation. *ACS Mater. Lett.* **2023**, *5*, 1992–2001. [CrossRef]
76. Li, G.Q.; Wang, Q.; Wang, J.; Ye, J.Y.; Zhou, W.W.; Xu, J.; Zhuo, S.; Chen, W.F.; Liu, Y. Carbon-supported nano tungsten bronze aerogels with synergistically enhanced photothermal conversion performance: Fabrication and application in solar evaporation. *Carbon* **2022**, *195*, 263–271. [CrossRef]
77. Meng, X.Y.; Yang, J.H.; Ramakrishna, S.; Sun, Y.M.; Dai, Y.Q. Gradient-aligned Au/graphene meshes with confined heat at multiple levels for solar evaporation and anti-gravity catalytic conversion. *J. Mater. Chem. A* **2020**, *8*, 16570–16581. [CrossRef]
78. Wang, X.L.; Lyu, Q.; Tong, T.Z.; Sun, K.; Lin, L.C.; Tang, C.Y.Y.; Yang, F.L.; Guiver, M.D.; Quan, X.; Dong, Y.C. Robust ultrathin nanoporous MOF membrane with intra-crystalline defects for fast water transport. *Nat. Commun.* **2022**, *13*, 266. [CrossRef]
79. Ren, L.T.; Yang, X.N.; Sun, X.; Yuan, Y.P. Synchronizing Efficient Purification of VOCs in Durable Solar Water Evaporation over a Highly Stable Cu/$W_{18}O_{49}$@Graphene Material. *Nano Lett.* **2023**, *24*, 715–723. [CrossRef]

80. Ji, M.Y.J.; Kim, J.H.; Jeon, H.Y.; Han, S.H.; Lee, D.H.; Lee, Y.I. High-performance interfacial water evaporation of black $TiO_2$-xwith high-concentration bulk oxygen vacancies. *Chem. Eng. J.* **2024**, *483*, 149435. [CrossRef]
81. Allahbakhsh, A.; Jarrahi, Z.; Farzi, G.; Shavandi, A. Solar-powered and antibacterial water purification via Cu-BTC-embedded reduced graphene oxide nanocomposite aerogels. *Chem. Eng. J.* **2023**, *467*, 143472. [CrossRef]
82. Wang, A.; Liang, H.; Chen, F.; Tian, X.; Yin, S.; Jing, S.; Tsiakaras, P. Facile synthesis of $C_3N_4$/$NiIn_2S_4$ heterostructure with novel solar steam evaporation efficiency and photocatalytic $H_2O_2$ production performance. *Appl. Catal. B Environ.* **2022**, *310*, 121336. [CrossRef]
83. Gu, J.; Zhang, Y.; Xiao, P.; Fan, Q.; Guan, J.; Liang, K.; Chen, T. Sunflower-inspired hydrogel evaporator with mutual reinforcement of evaporation and photodegradation for integrated water management. *Chem. Eng. J.* **2024**, *490*, 151550. [CrossRef]
84. Mo, H.; Wang, Y. A bionic solar-driven interfacial evaporation system with a photothermal-photocatalytic hydrogel for VOC removal during solar distillation. *Water Res.* **2022**, *226*, 119276. [CrossRef]
85. Zhang, C.Y.; Jiang, Y.; Zou, X.Q.; Xing, L.; Liu, W.X.; Huang, Z.H.; Feng, Y.W.; Wang, J.K. Biomass-Based Ferric Tannate Hydrogel with a Photothermal Conversion Function for Solar Water Evaporation. *Acs Appl. Polym. Mater.* **2023**, *5*, 9574–9584. [CrossRef]
86. Dao, V.-D.; Vu, N.H.; Thi Dang, H.-L.; Yun, S. Recent advances and challenges for water evaporation-induced electricity toward applications. *Nano Energy* **2021**, *85*, 105979. [CrossRef]
87. Lu, Y.; Fan, D.; Shen, Z.; Zhang, H.; Xu, H.; Yang, X. Design and performance boost of a MOF-functionalized-wood solar evaporator through tuning the hydrogen-bonding interactions. *Nano Energy* **2022**, *95*, 107016. [CrossRef]
88. Wu, X.; Lu, Y.; Ren, X.H.; Wu, P.; Chu, D.W.; Yang, X.F.; Xu, H.L. Interfacial Solar Evaporation: From Fundamental Research to Applications. *Adv. Mater.* **2024**, *36*, 2313090. [CrossRef]
89. Guo, Y.; de Vasconcelos, L.S.; Manohar, N.; Geng, J.; Johnston, K.P.; Yu, G. Highly Elastic Interconnected Porous Hydrogels through Self-Assembled Templating for Solar Water Purification. *Angew. Chem. Int. Ed. Engl.* **2022**, *61*, e202114074. [CrossRef]
90. Zhao, Q.; Yang, Y.; Zhu, B.; Sha, Z.; Zhu, H.; Wu, Z.; Nawaz, F.; Wei, Y.; Luo, L.; Que, W. Low vaporization enthalpy hydrogels for highly efficient solar-driven interfacial evaporation. *Desalination* **2023**, *568*, 116999. [CrossRef]
91. Li, X.P.; Li, X.; Li, H.; Zhao, Y.; Wu, J.; Yan, S.; Yu, Z.Z. Reshapable MXene/Graphene Oxide/Polyaniline Plastic Hybrids with Patternable Surfaces for Highly Efficient Solar-Driven Water Purification. *Adv. Funct. Mater.* **2021**, *32*, 2110636. [CrossRef]
92. Li, W.; Tian, X.; Li, X.; Han, S.; Li, C.; Zhai, X.-Z.; Kang, Y.; Yu, Z.-Z. Ultrahigh solar steam generation rate of a vertically aligned reduced graphene oxide foam realized by dynamic compression. *J. Mater. Chem. A* **2021**, *9*, 14859–14867. [CrossRef]
93. Lee, W.C.; Ronghe, A.; Villalobos, L.F.; Huang, S.; Dakhchoune, M.; Mensi, M.; Hsu, K.J.; Ayappa, K.G.; Agrawal, K.V. Enhanced Water Evaporation from A-Scale Graphene Nanopores. *ACS Nano* **2022**, *16*, 15382–15396. [CrossRef] [PubMed]
94. Chen, Q.; Burhan, M.; Shahzad, M.W.; Ybyraiymkul, D.; Akhtar, F.H.; Li, Y.; Ng, K.C. A zero liquid discharge system integrating multi-effect distillation and evaporative crystallization for desalination brine treatment. *Desalination* **2021**, *502*, 114928. [CrossRef]
95. Chen, Y.F.; Chen, H.; Dai, Y.T.; Xiao, J.L.; Qiu, F.X.; Zhang, T. Honeycomb porous regenerated cellulose aerogel films with enhanced thermal dissipation for agricultural mulch application. *Food Bioprod. Process.* **2024**, *147*, 418–427. [CrossRef]
96. Wang, J.; Kong, Y.; Liu, Z.; Wang, H. Solar-driven interfacial evaporation: Design and application progress of structural evaporators and functional distillers. *Nano Energy* **2023**, *108*, 108115. [CrossRef]
97. Ni, G.; Li, G.; Boriskina, S.V.; Li, H.X.; Yang, W.L.; Zhang, T.J.; Chen, G. Steam generation under one sun enabled by a floating structure with thermal concentration. *Nat. Energy* **2016**, *1*, 16126. [CrossRef]
98. Gong, B.; Yang, H.; Wu, S.; Tian, Y.; Yan, J.; Cen, K.; Bo, Z.; Ostrikov, K. Nanotexturing-enhanced heat transfer and interfacial evaporation for energy-efficient solar-thermal water desalination. *Int. J. Heat Mass Transf.* **2022**, *186*, 122462. [CrossRef]
99. Meng, S.; Zhao, X.; Tang, C.-Y.; Yu, P.; Bao, R.-Y.; Liu, Z.-Y.; Yang, M.-B.; Yang, W. A bridge-arched and layer-structured hollow melamine foam/reduced graphene oxide composite with an enlarged evaporation area and superior thermal insulation for high-performance solar steam generation. *J. Mater. Chem. A* **2020**, *8*, 2701–2711. [CrossRef]
100. Han, X.; Besteiro, L.V.; Koh, C.S.L.; Lee, H.K.; Phang, I.Y.; Phan-Quang, G.C.; Ng, J.Y.; Sim, H.Y.F.; Lay, C.L.; Govorov, A.; et al. Intensifying Heat Using MOF-Isolated Graphene for Solar-Driven Seawater Desalination at 98% Solar-to-Thermal Efficiency. *Adv. Funct. Mater.* **2021**, *31*, 2008904. [CrossRef]
101. Finnerty, C.T.K.; Menon, A.K.; Conway, K.M.; Lee, D.; Nelson, M.; Urban, J.J.; Sedlak, D.; Mi, B. Interfacial Solar Evaporation by a 3D Graphene Oxide Stalk for Highly Concentrated Brine Treatment. *Env. Sci. Technol.* **2021**, *55*, 15435–15445. [CrossRef]
102. Shi, L.; Wang, Y.; Zhang, L.; Wang, P. Rational design of a bi-layered reduced graphene oxide film on polystyrene foam for solar-driven interfacial water evaporation. *J. Mater. Chem. A* **2017**, *5*, 16212–16219. [CrossRef]
103. Song, H.M.; Liu, Y.H.; Liu, Z.J.; Singer, M.H.; Li, C.Y.; Cheney, A.R.; Ji, D.X.; Zhou, L.; Zhang, N.; Zeng, X.; et al. Cold Vapor Generation beyond the Input Solar Energy Limit. *Adv. Sci.* **2018**, *5*, 1800222. [CrossRef]
104. Wang, Y.D.; Wu, X.; Yang, X.F.; Owens, G.; Xu, H.L. Reversing heat conduction loss: Extracting energy from bulk water to enhance solar steam generation. *Nano Energy* **2020**, *78*, 105269. [CrossRef]
105. Kashyap, V.K.; Ghasemi, H.G. Solar heat localization: Concept and emerging applications. *J. Mater. Chem. A* **2020**, *8*, 7035–7065. [CrossRef]

106. Lee, K.H.; Kang, D.J.; Eom, W.; Lee, H.; Han, T.H. Holey graphene oxide membranes containing both nanopores and nanochannels for highly efficient harvesting of water evaporation energy. *Chem. Eng. J.* **2022**, *430*, 132759. [CrossRef]
107. Zhou, M.; Zhang, L.; Tao, S.; Li, R.; Wang, Y. Recent research advances in efficient solar-driven interfacial evaporation. *Chem. Eng. J.* **2024**, *489*, 151157. [CrossRef]
108. Xu, H.; Lu, Y.; Jiang, F.; Zhang, J.; Ge, Y.; Li, Z. 3D porous N-doped lignosulfonate/graphene oxide aerogel for efficient solar steam generation and desalination. *Int. J. Biol. Macromol.* **2023**, *233*, 123469. [CrossRef]
109. Liu, X.H.; Mishra, D.D.; Wang, X.B.; Peng, H.Y.; Hu, C.Q. Towards highly efficient solar-driven interfacial evaporation for desalination. *J. Mater. Chem. A* **2020**, *8*, 17907–17937. [CrossRef]
110. Zhang, H.; Du, Y.; Jing, D.; Yang, L.; Ji, J.; Li, X. Integrated Janus Evaporator with an Enhanced Donnan Effect and Thermal Localization for Salt-Tolerant Solar Desalination and Thermal-to-Electricity Generation. *ACS Appl. Mater. Interfaces* **2023**, *15*, 49892–49901. [CrossRef]
111. Zhao, D.; Ding, M.; Lin, T.; Duan, Z.; Wei, R.; Feng, P.; Yu, J.; Liu, C.Y.; Li, C. Gradient Graphene Spiral Sponges for Efficient Solar Evaporation and Zero Liquid Discharge Desalination with Directional Salt Crystallization. *Adv. Sci.* **2024**, *11*, e2400310. [CrossRef]
112. Liu, H.; Luo, H.; Huang, J.; Chen, Z.; Yu, Z.; Lai, Y. Programmable Water/Light Dual-Responsive Hollow Hydrogel Fiber Actuator for Efficient Desalination with Anti-Salt Accumulation. *Adv. Funct. Mater.* **2023**, *33*, 2302038. [CrossRef]
113. Sun, Y.; Tan, X.; Yuan, X.; Li, J. Solar-driven interfacial evaporation: Research advances in structural design. *Chem. Eng. J.* **2024**, *495*, 153316. [CrossRef]
114. Zhou, J.G.; Sun, Z.L.; Chen, M.Q.; Wang, J.T.; Qiao, W.M.; Long, D.H.; Ling, L.C. Macroscopic and Mechanically Robust Hollow Carbon Spheres with Superior Oil Adsorption and Light-to-Heat Evaporation Properties. *Adv. Funct. Mater.* **2016**, *26*, 5368–5375. [CrossRef]
115. Li, R.X.; Wu, M.R.; Ma, H.J.; Zhu, Y.Q.; Zhang, H.Y.; Chen, Q.M.; Zhang, C.H.; Wei, Y. A Single Component, Single Layer Flexile Foam Evaporator with the Higher Efficiency for Water Generation. *Adv. Mater.* **2024**, *36*, 2402016. [CrossRef]
116. Wo, Z.; Sun, X.; Sun, H.; Su, Y.; Xie, Y.; Yang, N.; Zhang, X. All-in-one design of wood evaporator with highly-efficient salt resistance for sustainable solar desalination and contaminated water purification. *Chem. Eng. J.* **2025**, *507*, 160715. [CrossRef]
117. Du, H.; Ge, C.; Xu, D.; Qian, Y.; Chen, Z.; Gao, C.; Song, B.; Shen, Z.; Chen, J.; Liu, K.; et al. Multifunctional woven fabric for integrated solar-driven water generation and personal thermal management. *Cellulose* **2023**, *30*, 9207–9220. [CrossRef]
118. Sun, Z.Z.; Han, C.A.L.; Gao, S.W.; Li, Z.X.; Jing, M.X.; Yu, H.P.; Wang, Z.A.K. Achieving efficient power generation by designing bioinspired and multi-layered interfacial evaporator. *Nat. Commun.* **2022**, *13*, 5077. [CrossRef]
119. Meng, F.L.; Gao, M.M.; Ding, T.P.; Yilmaz, G.; Ong, W.L.; Ho, G.W. Modular Deformable Steam Electricity Cogeneration System with Photothermal, Water, and Electrochemical Tunable Multilayers. *Adv. Funct. Mater.* **2020**, *30*, 2002867. [CrossRef]
120. Sun, N.; Li, J.; Ren, J.; Xu, Z.; Sun, H.; Du, Z.; Zhao, H.; Ettelatie, R.; Cheng, F. Insights into the enhanced flux of graphene oxide composite membrane in direct contact membrane distillation: The different role at evaporation and condensation interfaces. *Water Res.* **2022**, *212*, 118091. [CrossRef]
121. Wei, Y.; Li, W.; Zhang, S.; Yu, J.; Tang, Y.; Wu, J.; Yu, S. Laser-Induced Porous Graphene/CuO Composite for Efficient Interfacial Solar Steam Generation. *Adv. Funct. Mater.* **2024**, *34*, 2401149. [CrossRef]
122. Li, W.C.; Zheng, Z.H.; Qian, Z.Q.; Liu, H.; Wang, X.D. Phase Change Material Boosting Electricity Output and Freshwater Production through Hierarchical-Structured 3D Solar Evaporator. *Adv. Funct. Mater.* **2024**, *34*, 2316504. [CrossRef]
123. Zhang, Y.T.; Zhong, Q.Y.; Huang, Q.; Hu, M.; He, F.; Li, Y.X.; Wang, Z.X. Surface Engineering of 3D Solar Evaporator for Uncompromising Water Evaporation and Salt Production Toward High Concentration Brine. *Adv. Funct. Mater.* **2024**, *34*, 2408554. [CrossRef]
124. Eskafi, A.F.; De Finnda, C.; Garcia, C.A.; Mi, B.X. Mineral Scaling in 3D Interfacial Solar Evaporators-A Challenge for Brine Treatment and Lithium Recovery. *Environ. Sci. Technol.* **2025**, *59*, 892–901. [CrossRef] [PubMed]
125. Payanthoth, N.S.; Mut, N.N.N.; Samanta, P.; Li, G.L.; Jung, J.H. A review of biodegradation and formation of biodegradable microplastics in soil and freshwater environments. *Appl. Biol. Chem.* **2024**, *67*, 110. [CrossRef]
126. Li, D.; Liang, Z.; Yang, H.; Zhang, M.; Cao, K.; Zhao, B.; Wang, Y.; Peng, M.; Sun, Y.; Jiang, L. Mutual Reinforcement of Evaporation and Catalysis for Efficient Freshwater–Salt–Chemical Production. *Adv. Funct. Mater.* **2023**, *33*, 2300353. [CrossRef]
127. Huang, T.H.; Tian, X.Y.; Chen, Y.Y.; Widakdo, J.; Austria, H.F.M.; Setiawan, O.; Subrahmanya, T.M.; Hung, W.S.; Wang, D.M.; Chang, C.Y.; et al. Multifunctional Phra Phrom-like Graphene-Based Membrane for Environmental Remediation and Resources Regeneration. *Adv. Funct. Mater.* **2023**, *34*, 2308321. [CrossRef]
128. Han, Z.X.; Wang, N.; Zhang, H.L.; Yang, X.Y. Heavy metal contamination and risk assessment of human exposure near an e-waste processing site. *Acta Agric. Scand. Sect. B Soil Plant Sci.* **2017**, *67*, 119–125. [CrossRef]
129. Jiang, C.L.; Wang, X.H.; Hou, B.X.; Hao, C.; Li, X.; Wu, J.B. Construction of a Lignosulfonate-Lysine Hydrogel for the Adsorption of Heavy Metal Ions. *J. Agric. Food Chem.* **2020**, *68*, 3050–3060. [CrossRef]

130. Tian, Y.; Du, C.; Yong, S.; Zhou, X.; Zhou, C.; Yang, S. Catalysis-involved 3D N-doped graphene aerogel achieves a superior solar water purification rate and efficiency. *Chem. Eng. J.* **2023**, *453*, 139793. [CrossRef]
131. Dong, S.; Zhao, Y.; Yang, J.; Li, W.; Luo, W.; Li, S.; Liu, X.; Guo, H.; Yu, C.; Sun, J.; et al. Solar water recycling of carbonaceous aerogel in open and colsed systems for seawater desalination and wastewater purification. *Chem. Eng. J.* **2022**, *431*, 133824. [CrossRef]
132. Cui, H.Y.; Zhou, H.; Lin, L. The specific antibacterial effect of the Salvia oil nanoliposomes against *Staphylococcus aureus* biofiims on milk container. *Food Control* **2016**, *61*, 92–98. [CrossRef]
133. Hu, W.; Li, C.Z.; Dai, J.M.; Cui, H.Y.; Lin, L. Antibacterial activity and mechanism of *Litsea cubeba* essential oil against methicillin-resistant *Staphylococcus aureus* (*MRSA*). *Ind. Crops Prod.* **2019**, *130*, 34–41. [CrossRef]
134. Lu, X.; Feng, X.; Zhang, X.; Chukwu, M.N.; Osuji, C.O.; Elimelech, M. Fabrication of a Desalination Membrane with Enhanced Microbial Resistance through Vertical Alignment of Graphene Oxide. *Environ. Sci. Technol. Lett.* **2018**, *5*, 614–620. [CrossRef]
135. Noureen, L.; Wang, Q.; Ismail, P.M.; Alomar, M.; Arshad, N.; Irshad, M.S.; Xu, Q.; Wang, X. Multifunctional aerogel with antibiofouling properties for efficient solar steam generation and seawater desalination. *Nano Today* **2024**, *54*, 102130. [CrossRef]
136. Lin, L.; Wang, W.; Li, D.; Xu, S.; Sun, Y.; Li, L.; Fan, K.; Xing, C.; Zhang, L.; Li, J. Multifunctional graphene/Ag hydrogel with antimicrobial and catalytic properties for efficient solar-driven desalination and wastewater purification. *Chem. Eng. J.* **2023**, *478*, 147249. [CrossRef]
137. Li, J.; Wang, X.; Lin, Z.; Xu, N.; Li, X.; Liang, J.; Zhao, W.; Lin, R.; Zhu, B.; Liu, G.; et al. Over 10 kg $m^{-2}$ $h^{-1}$ Evaporation Rate Enabled by a 3D Interconnected Porous Carbon Foam. *Joule* **2020**, *4*, 928–937. [CrossRef]
138. Zhang, P.; Li, J.; Lv, L.; Zhao, Y.; Qu, L. Vertically Aligned Graphene Sheets Membrane for Highly Efficient Solar Thermal Generation of Clean Water. *ACS Nano* **2017**, *11*, 5087–5093. [CrossRef]

catalysts

MDPI

*Article*

# Synthesis of Black g-$C_3N_4$ and Exploration of the Mechanism Underlying the Enhancement of Photocatalytic $CO_2$ Reduction

**Shaokun Lv, Jun Zhang, Xiaoke Chen, Yue Zou, Qiuli Chen, Yongsheng Yan * and Pengxin Li ***

School of Chemistry and Chemical Engineering, Jiangsu University, Zhenjiang 212013, China; 18154177730@163.com (S.L.); 18896661191@163.com (J.Z.); 15515038037@163.com (X.C.); 19707205273@163.com (Y.Z.); 15539676362@163.com (Q.C.)

* Correspondence: yys@ujs.edu.cn (Y.Y.); pengxinli121@163.com (P.L.)

**Abstract:** The use of solar energy to convert $CO_2$ into value-added chemicals is a promising sustainable development strategy. In this study, a black graphitic carbon nitride (CN-B) photocatalyst was fabricated through a single-step calcination process, employing phloxine B and urea as the precursor materials. The catalysts were characterized using TEM, XRD, FTIR, XPS and so on. The amount of prepolymer phloxine B was 25 mg, 35 mg and 45 mg, respectively, and the obtained samples were CN-B-0.025, CN-B-0.035 and CN-B-0.045. All samples were used for visible-catalyzed $CO_2$ reduction. The experimental findings indicate that the CO evolution rate of the optimal photocatalyst CN-B-0.035 reaches 27.56 μmol $g_{cat.}^{-1}$ $h^{-1}$. This value is nine-fold higher than that of pure CN, which has a CO evolution rate of 3.22 μmol $g_{cat.}^{-1}$ $h^{-1}$. The excellent photocatalytic reduction performance is due to the following factors: Firstly, the exceedingly thin nanosheet structure of the catalyst enhances the velocity of the charge transfer, and transmission electron microscopy (TEM) analysis shows that the nanosheet thickness of the catalyst CN-B is significantly thinner. Secondly, the light absorption capacity of the catalyst is enhanced. The absorbance of CN-B increases significantly in the ultraviolet region and extends to the near-infrared region, as shown with UV diffuse reflection spectroscopy. Finally, the photothermal effect of CN-B causes the catalyst temperature to rise rapidly from 20 °C to 131 °C within 120 s, which further promotes photogenerated carrier separation. This research offers a novel approach to the development of photocatalysts aimed at the photothermal-assisted photocatalytic conversion of $CO_2$.

**Keywords:** carbon dioxide reduction; product selectivity; black g-$C_3N_4$; photocatalysis

## 1. Introduction

Rapid population growth and industrialization have led to a dramatic increase in global energy demand. Simultaneously, the combustion of fossil fuels releases significant quantities of carbon dioxide ($CO_2$), a greenhouse gas that profoundly affects the global ecosystem. The effective utilization of $CO_2$ has become a hot topic in current research [1,2]. Due to the abundance of solar energy resources, photocatalysis presents a promising application prospect [3]. Therefore, photocatalytic $CO_2$ reduction holds significant potential for future applications. Through the photocatalytic reduction of carbon dioxide, the concentration of carbon dioxide in the atmosphere can be reduced, which helps to slow down the trend of global warming and has positive significance for environmental protection and coping with climate change.

Graphitic carbon nitride (g-$C_3N_4$) is a common chip semiconductor photocatalyst [4–9]. Due to its compatible energy gap (Eg = 2.7 eV), excellent physical and chemical stability, low cost, non-toxic properties and easy-to-adjust synthesis methods, it has attracted great attention in photocatalytic $CO_2$ reduction applications. However, challenges faced by g-$C_3N_4$ include poor visible light response, the continuous recombination of excitons and short specific surface area. All these hinder g-$C_3N_4$ to obtain the best photocatalytic performance [10]. Despite extensive research on photocatalytic $CO_2$ reduction, achieving high selectivity and activity in CN photocatalysts often requires the use of sacrificial agents or cocatalysts. This is mainly due to their high electron-hole recombination rate, narrow light absorption range and low $CO_2$ adsorption capacity, which limit their performance [11–13]. To tackle these challenges, multiple approaches have been employed to improve the photocatalytic performance of pristine g-$C_3N_4$, including morphology regulation, elemental doping and the creation of heterostructures [14–16]. Within these approaches, the separation of photoinduced electron-hole pairs is regarded as one of the most efficient ways to enhance photocatalytic performance [17]. For instance, Hu and his team fabricated a BaCN-$C_3N_4$ photocatalyst. This photocatalyst facilitated partial ring opening by acting on the heptazine ring. The presence of $Ba^{2+}$ not only enhanced photocatalytic activity but also improved visible light absorption [18]. Nevertheless, because of the low temperature in the surrounding environment and the restricted utilization of photons, the efficiency of photocatalytic reduction achieved by the photocatalyst remains insufficient.

As luck would have it, the development of a photocatalytic system synergized with photothermal function can increase the temperature within the reaction space. This increase in temperature accelerates the chemical reaction kinetics on the photocatalyst's surface, promotes the transfer of photogenerated charges and ultimately improves the photocatalytic degradation efficiency. A less explored aspect in developing $CO_2$ photo reduction systems involves photo-to-thermal conversion, a process in which materials absorb light and convert it into heat. Enhanced photo-to-thermal conversion could improve photocatalytic efficiency by supplying the energy needed to overcome activation barriers and enable reaction pathways [19]. For example, Zhang et al. developed an enhanced NdO/TiO photothermal catalyst using the sol–gel method. Due to the Z-type heterojunction formed at the interface between NdO and TiO and the photothermal effect that promoted charge separation, it showed excellent $CO_2$ reduction performance under light [20]. However, much of the research on photothermal-assisted photocatalysis involves the integration of photothermal materials, which often face challenges such as inconsistent preparation stability, complex processes and limited $CO_2$ adsorption capacity [21,22]. Therefore, the design of an ultra-thin g-$C_3N_4$ nanosheet photocatalytic material with its own photothermal effect is of great significance for the photothermal-assisted photocatalytic reduction of $CO_2$.

In this study, the red dye phloxine B was added during the preparation of carbon nitride, and the black carbon nitride (CN-B) photocatalyst was synthesized through a single-step calcination process. Our results show that the constructed CN-B ultra-thin nanosheet structures shorten the path length of charge transport. Moreover, the addition of phloxine B significantly enhanced the absorbance of the catalyst. As a result, the black substance showed a potent photothermal property. The temperature of the reaction surroundings increased, which in turn advanced the efficiency of photocatalytic reactions.

## 2. Results and Discussion

The morphology of the photocatalyst was tested. Transmission electron microscopy (TEM) images showed that there was little difference in the morphology of CN and CN-0.035 photocatalysts, both of which exhibited coiled, layered and stacked nanostructures (Figure 1a,b). TEM was employed to characterize both the microstructure and elemental composition of CN-B-0.035. Through TEM analysis, the morphological features and elemental makeup of CN-B-0.035 were determined. TEM was utilized to conduct an in-depth analysis of CN-B-0.035's microstructure and elemental constituents. As seen in Figure 1c,d, a large sheet grid structure is evident. Upon enlarging the image, as shown in Figure 1e,f, the large sheet structure remains visible. From the above analysis, it is clear that the thickness of CN-B-0.035 is much smaller than that of CN, indicating that the addition of phloxine B not only maintains the original layer structure of CN but also reduces the charge transfer distance due to the ultra-thin thickness of nanoparticles and their sheet structure. This reduction in the charge transfer distance facilitates the transport of photogenerated charge carriers, thereby improving photocatalytic activity. Furthermore, from the AFM test image in Figure S1 (Supplementary Materials), it can be clearly seen that after adding phloxine B, the sample thickness decreases from 1.762 nm to 1.423 nm. Additionally, as shown in Figure 1g–j, the element mapping images confirm that the C, N and O elements are distributed on the surface of CN-B nanosheets.

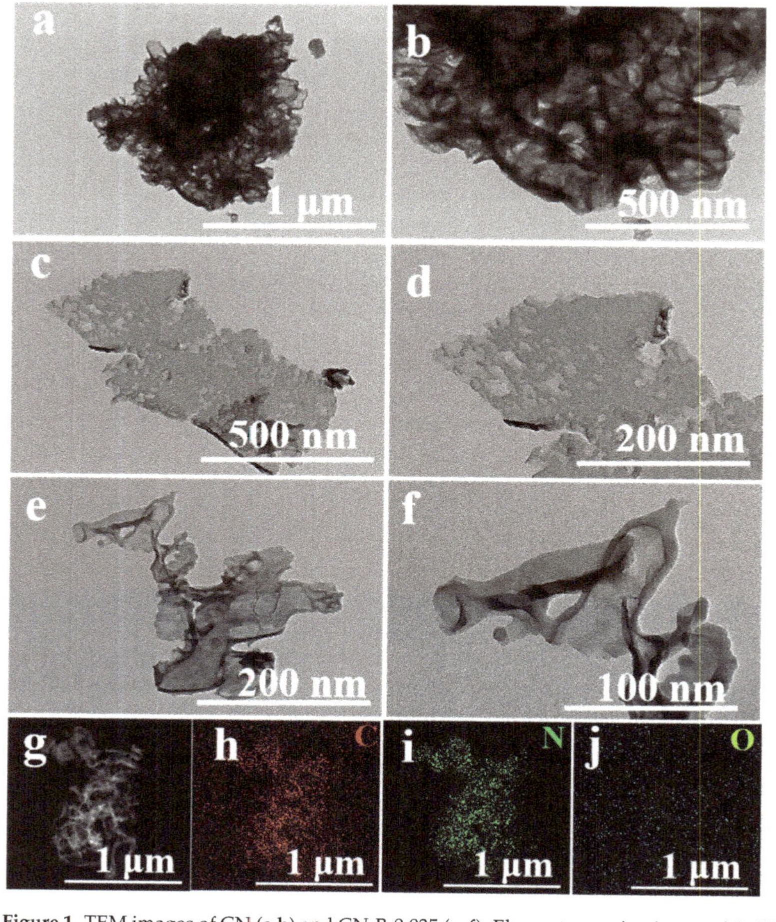


**Figure 1.** TEM images of CN (**a**,**b**) and CN-B-0.035 (**c**–**f**). Element mapping images (**g**): (**h**) C, (**i**) N and (**j**) O.

The crystal structure of the sample was analyzed with XRD, as shown in Figure 2a. The diffraction peaks at 27.2° can be assigned to the (002) planes of the sample CN, respectively [23]. The initial characteristic peak uniquely corresponds to the organization of the triazine units in the CN network [24,25]. Conversely, the next characteristic peak indicates the ordered aggregation of carbon nitride nanosheets in the C-axis [26,27]. Moreover, when the XRD patterns of CN and CN-B photocatalysts are contrasted, it is evident that as the amount of phloxine B increases, the intensity of the (002) characteristic peak of the CN-B photocatalyst steadily declines. This decline suggests that the periodicity of the catalyst's planar structure is disrupted [28]. Furthermore, due to the enhanced interfacial interaction between CN-B nanosheets and the reduced layer spacing, the (001) crystal plane diffraction peak of photocatalyst CN-B shows a slight blue shift [29,30]. The functional groups of the synthesized samples were further analyzed using Fourier transform infrared spectroscopy (FTIR). The peaks located between the 810 $cm^{-1}$ and 900–1800 $cm^{-1}$ regions correspond to the out-of-plane breathing vibrations of the triazine unit and the stretching vibrations of N–C=N/N–$C_3$, respectively. This indicates that the characteristic structure of CN is retained in both samples CN-B and CN [31]. These findings demonstrated that following the introduction of the precursor phloxine B, there was no substantial alteration in the functional-group configuration of CN [32,33].

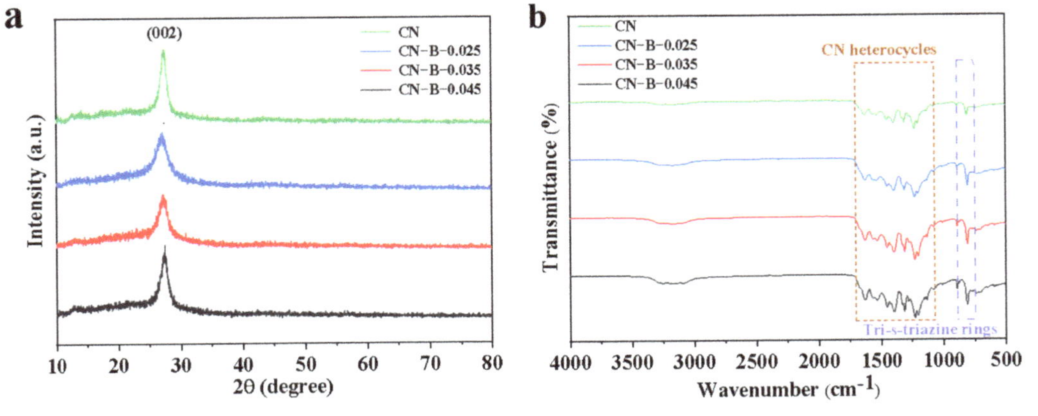


**Figure 2.** XRD patterns of (**a**) CN and CN-B photocatalysts and (**b**) FTIR spectra CN and CN-B photocatalysts.

To examine the chemical and elemental states of the photocatalyst samples, X-ray photoelectron spectroscopy (XPS) was utilized to characterize the electronic structure of both CN and CN-B. In Figure 3a, the peaks of C 1s, N 1s and O 1s can be clearly observed, which is consistent with the results of the element distribution mapping graphs. Moreover, the presence of a trace amount of element O stems from water adsorbed on the surface of the catalyst. When contrasted with CN, CN-B-0.35 exhibits a greater C atom ratio. Also, because phloxine B is added during preparation, its C/N atom ratio is elevated. Figure 3b illustrates that the C 1s spectrum for the CN-B-0.35 photocatalyst displays three distinct peaks. At 288.2 eV, this peak corresponds to N–C=N bonds, the 285.9 eV peak aligns with C–N–H bonds and the 284.8 eV peak is associated with C–C bonds [34,35]. In Figure 3c, regarding the N 1s spectrum of CN-B-0.035, three peaks stand out. The peak at 400.7 eV is assignable to the N–H bond. The 399.7 eV peak correlates with the triazine ring featuring C–N=C and $N_2C$ structures. Meanwhile, the 398.89 eV peak corresponds to nitrogen N-$(C)_3$ groups [36]. In the O 1s spectrum of CN-B-0.035, two prominent characteristic peaks are seen at 532.1 and 533.5 eV, as shown in Figure 3d. The 532.1 eV peak corresponds to C=O bonds, with the 533.5 eV peak assignable to C–O–H bonds [37]. In summary, compared with CN, the positions of the peaks in the C 1s, N 1s and O 1s spectra of the CN-B-0.035 photocatalyst were slightly shifted, indicating that the addition of phloxine B may alter the internal structural environment of CN [38,39]. This alteration can affect the electron transfer within the catalyst network to a certain extent.

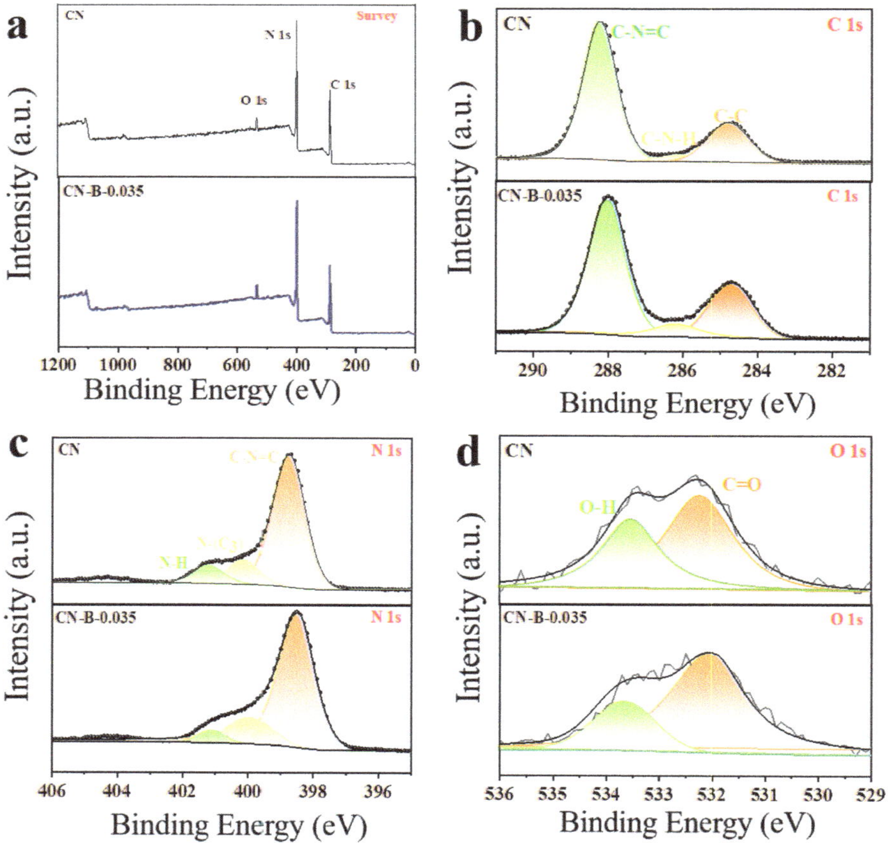


**Figure 3.** XPS spectra and magnetization curves of CN and CN-B-0.035 composite samples. (**a**) Survey spectrum of CN and CN-B-0.035. (**b**) C 1s spectrum of CN and CN-B-0.035. (**c**) N 1s spectrum of CN and CN-B-0.035. (**d**) O 1s spectrum of CN and CN-B-0.035.

The light absorption capacity of the photocatalyst was studied using UV diffuse reflection (DRS) spectroscopy. As shown in Figure 4a, CN and CN-B exhibit different light absorption ranges and capacities. CN-B demonstrates a strong light absorption capacity from 450 nm to 800 nm, showing better light absorption in the visible light and near-infrared regions, thereby enhancing the utilization of light [40,41]. Additionally, based on the Tauc function $(\alpha h)^2 = A(h\nu\text{-}Eg)$ [42,43], the band gaps (Eg) of CN and CN-B are determined to be 2.75 eV and 2.62 eV, respectively (Figure 4b). The results indicate that the band gap value of CN-B (2.62 eV) is narrower than that of CN (2.75 eV), suggesting that more electrons can be generated under visible light. According to the results of XPS-valence band spectroscopy, the addition of prepolymer phloxine B significantly changes the valence band edge position of the sample. Figure 4c shows that the CN and CN-B-0.035 photocatalysts possess valence band (VB) values of 1.73 eV and 1.55 eV in sequence. This shift to a lower binding energy enhances the reducing power of the semiconductor. Using the position of the flat band potential and the equation $E_{VB} = E_{CB} + Eg$, the conduction band (CB) positions of CN and

CN-B-0.035 are calculated to be −1.02 eV and −1.07 eV, respectively. This indicates that CN-B has a more negative conduction band position [44].

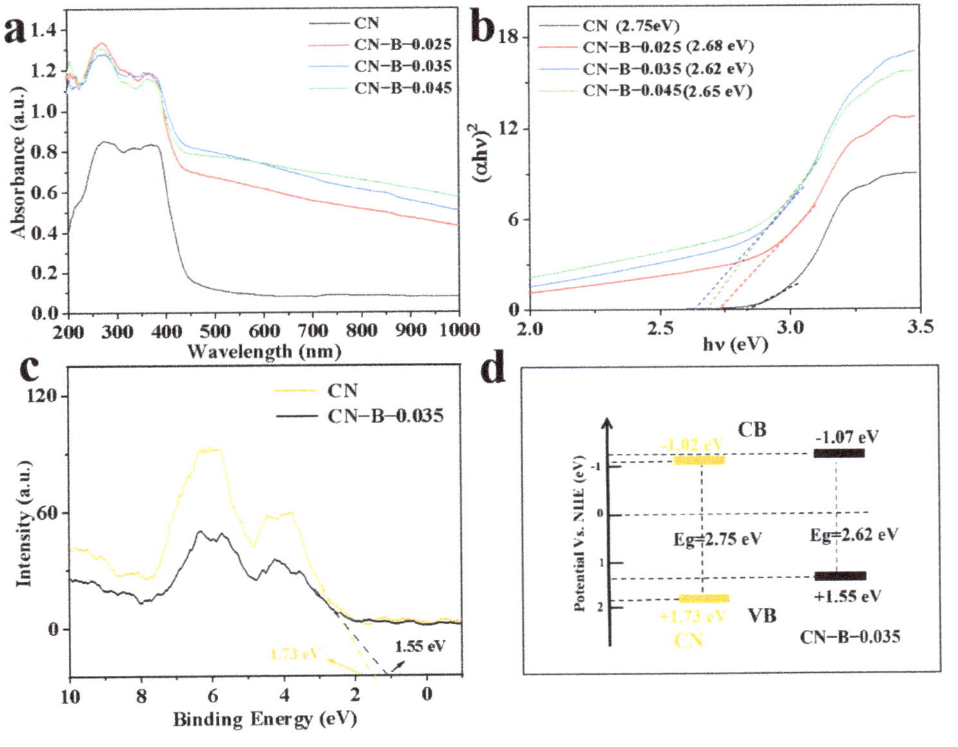


**Figure 4.** (**a**) UV–Vis diffuse reflectance spectra of as-synthesized photocatalysts. (**b**) Band gap values of as-synthesized photocatalysts. (**c**) XPS valence band spectra of CN-B-0.035 and CN-B. (**d**) Electronic band structures of CN-B-0.035 and CN.

To assess the photocatalytic $CO_2$ reduction efficiency of the synthesized photocatalyst, a 300 W xenon lamp was used to test the photocatalytic $CO_2$ reduction of the sample without adding sacrificial agents and photosensitizers. As shown in Figure 5a, CN-B samples were prepared at different temperatures of 480 °C, 500 °C, 520 °C, 540 °C and 560 °C, and their carbon dioxide reduction properties were also tested. It is evident that the samples calcined at 540 °C exhibit the best photocatalytic $CO_2$ reduction performance. As shown in Figure 5b, the CO formation rate of CN within 4 h is only 3.12 μmol $g_{cat.}^{-1}$ $h^{-1}$, while the CO formation rates of CN-B-0.025, CN-B-0.035 and CN-B-0.045 are 13.78, 27.56 and 15.63 μmol $g_{cat.}^{-1}$ $h^{-1}$, respectively. The CO formation rate of the photocatalyst CN-B-0.035 is 8.8 times that of CN. In conclusion, an appropriate amount of phloxine B can enhance the migration of photogenerated electrons and boost the photocatalytic degradation performance of CN. However, an excessive amount of phloxine B may result in the formation of new recombination centers for photogenerated carriers, thereby decreasing the generation of active species. Additionally, photocatalyst stability is a key factor in practical applications. Thus, we conducted six cycles of photocatalytic carbon dioxide reduction experiments. Notably, after six cycles of experiments, the CN-B-0.035 photocatalyst still maintained a high CO yield (Figure 5c), indicating that the CN-B-0.035 photocatalyst has superior stability.

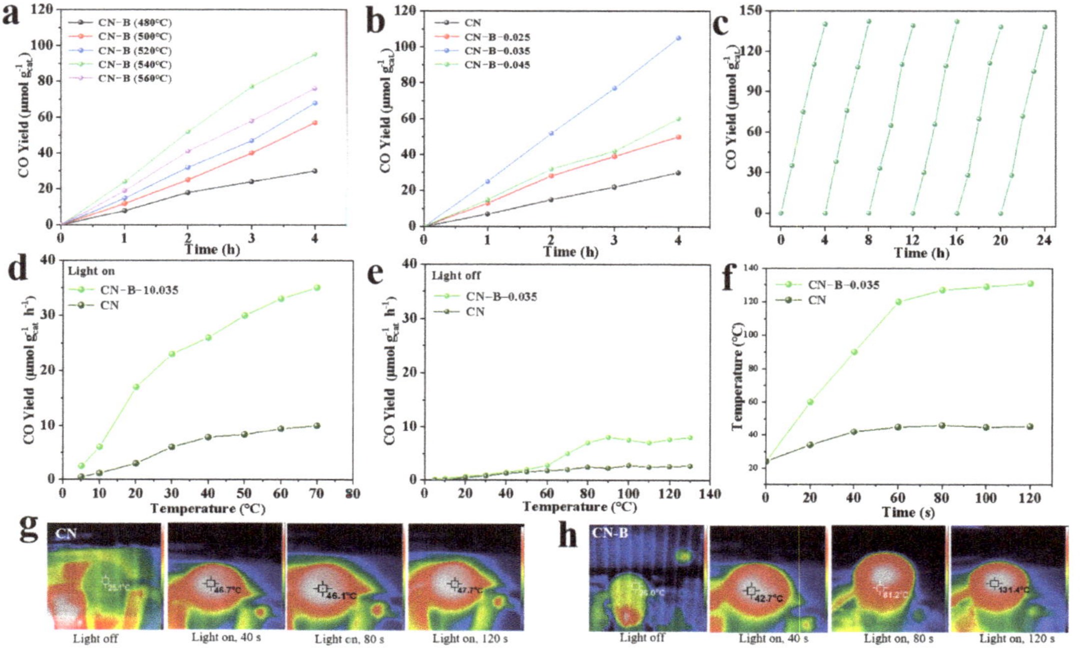


**Figure 5.** (**a**) The photocatalytic $CO_2$ reduction performance of the catalyst prepared with the same prepolymer at 35 mg at different temperatures. (**b**) Photocatalytic $CO_2$ reduction performance of catalysts with different prepolymer contents at the same temperature 540 °C. (**c**) Six cycle experiments of photocatalytic carbon dioxide reduction on CN-B-0.035 photocatalyst. (**d**,**e**) The comparison of gas yields at different temperatures without and with white light using CN and CN-B-0.035. (**f**) The corresponding temperature curves of CN and CN-B-0.035. (**g**,**h**) Photothermal infrared images of CN and CN-B.

To reveal the contribution of phloxine B precursor addition to enhanced photocatalytic $CO_2$ reduction, CN and CN-B-0.035 photocatalyst powders were irradiated with a 300 mW $cm^{-2}$ Xenon lamp for 120 s, and the surface temperatures at 40 s intervals were recorded with infrared thermography, as shown in Figure 5g,h. Figure 5f illustrates the temperature variation of the CN and CN-B-0.035 photocatalysts. The temperature of CN-B-0.035 jumps swiftly from 26.0 °C to 131.4 °C and then stays steady. This final temperature significantly exceeds that of pure CN, which reaches only 47.7 °C. The temperature rise is attributed to the addition of the phloxine B precursor. This addition darkens the color of CN, enhancing the photothermal performance of CN-B. As a consequence, there is an elevation in temperature. Based on the above analysis, the $CO_2$ reduction properties of pure CN and CN-B-0.035 photocatalysts were tested at different temperatures by controlling the reaction temperature in a circulating condensing unit system and an oil bath. As seen in Figure 5d, both original CN and CN-B-0.035 exhibit some degree of temperature responsiveness, with an upward trend in temperature, and there is a corresponding increase in the CO production rate, indicating that heat can improve their performance. Additionally, as shown in Figure 5d,e, the CO generation rate of the CN-B-0.035 photocatalyst with a light source is 18.3 μmol $g_{cat.}^{-1}$ $h^{-1}$, compared to 1.3 μmol $g_{cat.}^{-1}$ $h^{-1}$ without a light source, demonstrating that light can significantly enhance its photocatalytic reduction performance. These findings suggest that by introducing the phloxine B precursor, the photocatalytic $CO_2$

reduction capabilities of CN are strengthened. This occurs via the combined action of the photothermal effect, enabling CN to more effectively convert $CO_2$ during photocatalysis.

Accordingly, the photoelectrochemical characterization of different samples was carried out to study the effect of the photothermal effect on the catalyst carrier kinetics. As shown in Figure 6a, CN-B-0.035 exhibits a lower PL signal than pure CN due to the introduction of a cyanogen defect that increases the photon-generated carrier separation rate of CN-B-0.035. In Figure 6b, CN-B-0.035 shows the highest photocurrent response intensity, which demonstrates that the modified photocatalyst has a strong electron transfer ability [45,46]. Furthermore, in the electrochemical impedance spectroscopy (EIS) diagram shown in Figure 6c, the CN-B-0.035 sample has the smallest arc radius, indicating lower charge transfer resistance [47].

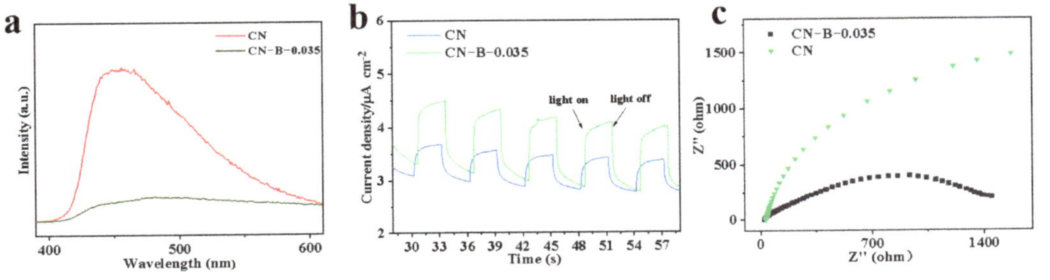


**Figure 6.** (**a**) PL chart. (**b**) Momentary photocurrent response graphs. (**c**) EIS of CN-B-0.035 and CN.

In order to understand the conversion mechanism of $CO_2$ molecules on CN-B-0.035, FT-IR spectroscopy was used to study the intermediate products in the photoreduction process. As shown in Figure 7a, the $CO_2$ stretching vibration (2350–2250 $cm^{-1}$), H–O–H bending vibration band (1700–1600 $cm^{-1}$) and H–O–H stretching vibration band (3700–3200 $cm^{-1}$) are observed. This indicates the reduction of $CO_2$ and the dissociation of $H_2O$. Figure 7b shows that the $CO_2$ reduction signal of CN-B is enhanced. Evidently, introducing phloxine B boosts its reduction effectiveness. It can be observed that the two vibration absorption peaks of $H_2O$ have different intensities. This difference may be due to the O–H bond stretching vibration in water, which involves a large relative displacement of oxygen and hydrogen atoms, resulting in a significant change in charge distribution, strong infrared absorption and high absorption peak intensity. Conversely, the dipole moment of the flexural vibration molecule changes little. Similar to the bending vibration of water molecules, the displacement of hydrogen atoms relative to oxygen atoms is small, the change in charge distribution is minimal, the infrared absorption is weak and the absorption peak intensity is low [48–50]. It is generally believed that *COOH is a key intermediate in the conversion of $CO_2$ to CO [50–53]. Based on the above analysis, the preliminary reaction mechanism of the CN-B photocatalytic reduction of $CO_2$ is proposed: First, $CO_2$ is adsorbed on the catalyst surface ($CO_2 \rightarrow {}^*CO_2$), followed by the reaction of adsorbed activated $^*CO_2$ with $H^+$ and $e^-$ to produce the important intermediate COOH*. Then, *COOH is converted into $H_2O$ and CO* through the proto-electron transfer reduction process. Finally, CO* desorbs to release CO (CO* → CO). Figure 7c illustrates the mechanism of the $CO_2$ reduction process.

$$\text{CN-B} \overset{h\nu}{\rightarrow} e^- + h^+ \quad (1)$$

$$CO_2 \rightarrow CO_2* \quad (2)$$

$$H_2O + h^+ \rightarrow OH* + H^+ \quad (3)$$

$$OH* + h^+ \rightarrow 1/2O_2(g) + H^+ \quad (4)$$

$$CO_2 * + e^- + h^+ \rightarrow COOH* \quad (5)$$

$$COOH * + e^- + H^+ \rightarrow CO * (g) + H_2O \quad (6)$$

$$CO* \rightarrow CO \quad (7)$$

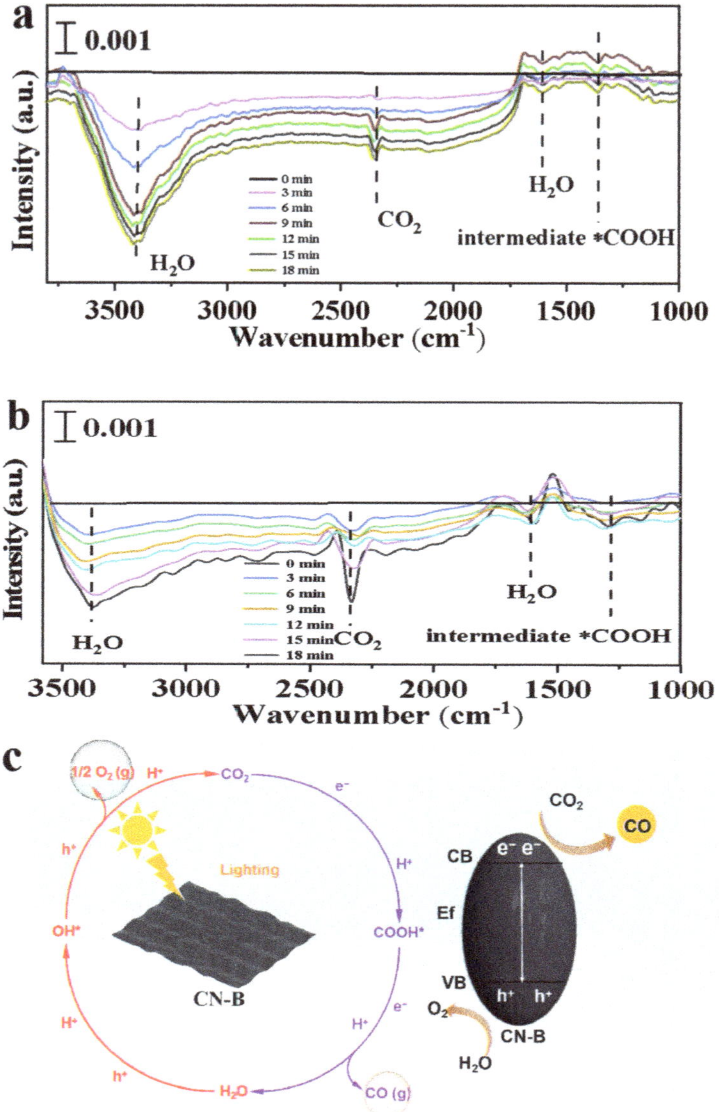


**Figure 7.** (**a**) In situ FT-IR spectrum of CN. (**b**) In situ FT-IR spectrum of CN-B-0.035. (**c**) Schematic illustration for photocatalytic mechanism of CN-B-0.035.

# 3. Experimental

## 3.1. Chemicals and Reagents

For specific details on the chemicals and reagents, please refer to Section S1.1 of the Supporting Information.

## 3.2. Synthesis and Characterizations

For the synthesis procedures of black g-$C_3N_4$ (CN-B) and pure CN, as well as detailed characterization, please refer to Section S1.2 of the Supplementary Materials.

## 3.3. $CO_2$ Photoreduction Experiments

$CO_2$ reduction experiments were conducted in a 200 mL quartz reaction vessel using a 350 W xenon lamp (1000 mW·cm$^{-2}$) as a white light source. The catalyst, amounting to 0.01 g, was made to disperse in two milliliter unit volumes of water free of ions. The mixture was deposited on a 1 cm × 2 cm quartz glass and dried under an infrared lamp at 60 °C. The dehydrated catalyst was positioned at the base of the quartz reaction apparatus. $H_2O$ steam was introduced into the reactor for 20 min along with the $CO_2$ stream to completely purify the air. The light intensity of 1000 mW·cm$^{-2}$ on the photocatalyst was measured with an automatic optical power meter (CEL-NP200010A, Zolix Instruments, Beijing, China). Following a 60 min irradiation period (using a 300 W xenon lamp at 1000 mW· cm$^{-2}$), the gaseous products were analyzed using a gas chromatograph (GC-7920, Nanjing Kejie Analytical Instrument Co., Ltd., Nanjing, China) fitted with a hydrogen flame ionization detector (FID).

## 3.4. Photo-Electrochemical Evaluations

For detailed procedures of the Photoelectrochemical Measurements, please refer to Section S1.3 of the Supplementary Materials.

# 4. Conclusions

Overall, a highly efficient photothermal-aided photocatalytic system for $CO_2$reduction was engineered. CN-B came into being through an on-pot calcination approach. Phloxine B along with urea were deployed as precursors in the synthetic process. The experimental results show that the CO yield of the optimal sample, CN-B-0.035, is 27.56 μmol $g_{cat.}^{-1}$ h$^{-1}$ within 4 h under the irradiation of a 500 mW cm$^{-2}$ Xenon lamp. The excellent photocatalytic reduction performance is mainly attributed to the following three reasons: (1) the ultra-thin nanosheets of black CN-B-0.035 reduce the charge transfer distance; (2) the addition of prepolymers promotes the separation and migration of photogenerated carriers; (3) due to the high-efficiency photothermal conversion effect of the catalyst, the temperature of the reaction environment is increased, so the photocatalytic degradation performance is further improved. This investigation presents a viable method and systematic approach for the development of photothermal-assisted photocatalytic $CO_2$ reduction photocatalysts based on CN.

**Supplementary Materials:** The following supporting information can be downloaded at: https://www.mdpi.com/article/10.3390/catal15040349/s1, Figure S1: (a–d) AFM of CN, CN-B-0.025, CN-B-0.035 and CN-B-0.045.

**Author Contributions:** S.L. and X.C.: investigation, data curation and writing of the original draft. Y.Z. and Q.C.: resources. J.Z. writing—review. P.L. and Y.Y.: conceptualization, supervision, writing—review and editing and funding acquisition. All authors have read and agreed to the published version of the manuscript.

**Funding:** This research was funded by the Senior Talent Research Foundation of Jiangsu University (No. 22JD017, 23JDG030). This work was also supported by the China Postdoctoral Science Foundation (2022M711373) and the Jiangsu Funding Program for Excellent Postdoctoral Talent (2022ZB637).

**Data Availability Statement:** The authors confirm that the data supporting the findings of this study are available within the article and its Supporting Information. Raw data that support the findings of this study are available from the corresponding author, upon reasonable request.

**Conflicts of Interest:** The authors declare no conflicts of interest.

## References

1. Chen, Y.; Guan, B.; Wu, X.; Guo, J.; Ma, Z.; Zhang, J.; Huang, Z. Research status, challenges and future prospects of renewable synthetic fuel catalysts for $CO_2$ photocatalytic reduction conversion. *Environ. Sci. Pollut. Res.* **2023**, *30*, 11246–11271. [CrossRef] [PubMed]
2. Ye, W.; Zhang, Y.; Chen, L.; Wu, F.; Yao, Y.; Wang, W.; Wang, G. A strongly coupled metal/hydroxide heterostructure cascades carbon dioxide and nitrate reduction reactions toward efficient urea electrosynthesis. *Angew. Chem. Int. Ed.* **2024**, *63*, e202410105.
3. Khan, K.Y.; Tang, Y.; Cheng, P.; Song, Y.; Li, X.; Lou, J.; Du, D. Effects of degradable and non-degradable microplastics and oxytetracycline co-exposure on soil $N_2O$ and $CO_2$ emissions. *Appl. Soil Ecol.* **2024**, *197*, 105331.
4. Han, C.; Zhang, R.; Ye, Y.; Wang, L.; Ma, Z.; Su, F.; Ye, L. Chainmail co-catalyst of NiO shell-encapsulated Ni for improving photocatalytic $CO_2$ reduction over g-$C_3N_4$. *J. Mater. Chem. A* **2019**, *7*, 9726–9735.
5. Fu, J.; Zhu, B.; Jiang, C.; Cheng, B.; You, W.; Yu, J. Hierarchical porous O-doped g-$C_3N_4$ with enhanced photocatalytic $CO_2$ reduction activity. *Small* **2017**, *13*, 1603938.
6. Wang, K.; Fu, J.; Zheng, Y. Insights into photocatalytic $CO_2$ reduction on $C_3N_4$: Strategy of simultaneous B, K co-doping and enhancement by N vacancies. *Appl. Catal. B Environ.* **2019**, *254*, 270–282.
7. Liu, Z.; Wang, L.; Liu, P.; Zhao, K.; Ye, S.; Liang, G. Rapid ultrasensitive and non-enzyme electrochemiluminescence detection of hydrogen peroxide in food based on the ssDNA/g-$C_3N_4$ nanosheets hybrid. *Food Chem.* **2021**, *357*, 129753.
8. Truc NT, T.; Bach, L.G.; Hanh, N.T.; Pham, T.D.; Le Chi NT, P.; Tran, D.T.; Nguyen, M.V. The superior photocatalytic activity of Nb doped $TiO_2$/g-$C_3N_4$ direct Z-scheme system for efficient conversion of $CO_2$ into valuable fuels. *J. Colloid Interface Sci.* **2019**, *540*, 1–8.
9. Zhang, X.; Wang, L.; Du, Q.; Wang, Z.; Ma, S.; Yu, M. Photocatalytic $CO_2$ reduction over $B_4C/C_3N_4$ with internal electric field under visible light irradiation. *J. Colloid Interface Sci.* **2016**, *464*, 89–95.
10. Gu, W.; Lu, D.; Kondamareddy, K.K.; Li, J.; Cheng, P.; Ho, W.; Wang, Z. Efficient photocatalytic decomposition of NO and mechanism insight enabled by NaBH4-reduced N (ligancy-3)-vacancy-rich-graphitic carbon nitride. *Mater. Today Phys.* **2024**, *46*, 101487.
11. Talapaneni, S.N.; Singh, G.; Kim, I.Y.; AlBahily, K.; Al-Muhtaseb AA, H.; Karakoti, A.S.; Vinu, A. Nanostructured carbon nitrides for $CO_2$ capture and conversion. *Adv. Mater.* **2020**, *32*, 1904635. [CrossRef] [PubMed]
12. Kong, T.; Jiang, Y.; Xiong, Y. Photocatalytic $CO_2$ conversion: What can we learn from conventional COx hydrogenation. *Chem. Soc. Rev.* **2020**, *49*, 6579–6591. [CrossRef]
13. Wang, Q.; Fang, Z.; Zhang, W.; Zhang, D. High-efficiency g-$C_3N_4$ based photocatalysts for $CO_2$ reduction: Modification methods. *Adv. Fiber Mater.* **2022**, *4*, 342–360. [CrossRef]
14. Li, Y.; Fang, Y.; Cao, Z.; Li, N.; Chen, D.; Xu, Q.; Lu, J. Construction of g-$C_3N_4$/PDI@ MOF heterojunctions for the highly efficient visible light-driven degradation of pharmaceutical and phenolic micropollutants. *Appl. Catal. B Environ.* **2019**, *250*, 50–162. [CrossRef]
15. Wu, M.; He, X.; Jing, B.; Wang, T.; Wang, C.; Qin, Y.; An, T. Novel carbon and defects co-modified g-$C_3N_4$ for highly efficient photocatalytic degradation of bisphenol A under visible light. *J. Hazard. Mater.* **2020**, *384*, 121323. [CrossRef] [PubMed]
16. Wu, H.; Han, D.; Hu, B.; Liang, J.; Tang, X.; Zhu, Z.; Huo, P. Insights into enhanced photocatalytic $CO_2$ reduction on carbon nitride: A strategy of simultaneous O, S co-do. *J. Environ. Chem. Eng.* **2025**, *13*, 116121. [CrossRef]
17. Liu, G.; Dong, G.; Zeng, Y.; Wang, C. The photocatalytic performance and active sites of g-$C_3N_4$ effected by the coordination doping of Fe (III). *Chin. J. Catal.* **2020**, *41*, 1564–1572. [CrossRef]
18. Hu, X.; Lu, P.; Pan, R.; Li, Y.; Bai, J.; He, Y.; Zhang, C.; Fu, M. Metal-ion-assisted construction of cyano group defects in g-$C_3N_4$ to simultaneously degrade wastewater and produce hydrogen. *Chem. Eng. J.* **2021**, *423*, 130278. [CrossRef]
19. Chen, Y.; Lin, X.; Li, W.; Sun, H.; Jia, S.; Zhou, Y.; Yan, Y. Harnessing photo-to-thermal conversion in sulfur-vulcanized Mxene for high-efficiency solar-to-carbon-fuel synthesis. *Adv. Funct. Mater.* **2024**, *34*, 2400121. [CrossRef]
20. Zhang, S.M.; He, Z.H.; Shi, J.J.; Wang, S.W.; Liu, J.; Wang, K.; Liu, Z.T. Photothermal-driven $CO_2$ reduction over $Nd_2O_3/TiO_2$ heterojunction catalysts in aqueous medium. *Mol. Catal.* **2024**, *559*, 114078. [CrossRef]

21. Zhang, X.; Ma, Y.; Zhang, X.; Pang, X.; Yang, Z. Bio-inspired self-assembled bacteriochlorin nanoparticles for superior visualization and photothermal ablation of tumors. *Biomed. Pharmacother.* **2023**, *165*, 115014.
22. Hou, B.; Shi, Z.; Kong, D.; Chen, Z.; Yang, K.; Ming, X.; Wang, X. Scalable porous Al foil/reduced graphene oxide/$Mn_3O_4$ composites for efficient fresh water generation. *Mater. Today Energy* **2020**, *15*, 100371.
23. Sun, Z.; Fischer JM, T.A.; Li, Q.; Hu, J.; Tang, Q.; Wang, H.; Wang, L. Enhanced $CO_2$ photocatalytic reduction on alkali-decorated graphitic carbon nitride. *Appl. Catal. B Environ.* **2017**, *216*, 146–155.
24. Wang, Y.; Yang, W.; Chen, X.; Wang, J.; Zhu, Y. Photocatalytic activity enhancement of core-shell structure g-$C_3N_4$@$TiO_2$ via controlled ultrathin g-C3N4 layer. *Appl. Catal. B Environ.* **2018**, *220*, 337–347.
25. Guo, F.; Sun, H.; Huang, X.; Shi, W.; Yan, C. Fabrication of $TiO_2$/high-crystalline g-$C_3N_4$ composite with enhanced visible-light photocatalytic performance for tetracycline degradation. *J. Chem. Technol. Biotechnol.* **2020**, *95*, 2684–2693. [CrossRef]
26. Tan, X.; Jiang, K.; Zhai, S.; Zhou, J.; Wang, J.; Cadien, K.; Li, Z. X ray spectromicroscopy investigation of heterogeneous sodiation in hard carbon nanosheets with vertically oriented (002) Planes. *Small* **2021**, *17*, 2102109.
27. Guo, F.; Wang, L.; Sun, H.; Li, M.; Shi, W. High-efficiency photocatalytic water splitting by a N-doped porous g-$C_3N_4$ nanosheet polymer photocatalyst derived from urea and N, N-dimethylformamide. *Inorg. Chem. Front.* **2020**, *7*, 1770–1779.
28. Ji, Y.; Larsen, D.H.; Huang, Y.; Heuvelink, E.; Marcelis, L.F. Gradually increasing light intensity during the growth period increases dry weight production compared to constant or gradually decreasing light intensity in lettuce. *Sci. Hortic.* **2023**, *311*, 111807.
29. Shi, Y.; Li, L.; Sun, H.; Xu, Z.; Cai, Y.; Shi, W.; Du, X. Engineering ultrathin oxygen-doped g-$C_3N_4$ nanosheet for boosted photoredox catalytic activity based on a facile thermal gas-shocking exfoliation effect. *Sep. Purif. Technol.* **2022**, *292*, 121038.
30. Shi, W.; Shu, K.; Sun, H.; Ren, H.; Li, M.; Chen, F.; Guo, F. Dual enhancement of capturing photogenerated electrons by loading CoP nanoparticles on N-deficient graphitic carbon nitride for efficient photocatalytic degradation of tetracycline under visible light. *Sep. Purif. Technol.* **2020**, *246*, 116930.
31. Lin, J.; Tian, W.; Zhang, H.; Duan, X.; Sun, H.; Wang, S. Graphitic carbon nitride-based Z-scheme structure for photocatalytic $CO_2$ reduction. *Energy Fuels* **2020**, *35*, 7–24. [CrossRef]
32. Zeng, D.; Ong, W.J.; Chen, Y.; Tee, S.Y.; Chua, C.S.; Peng, D.L.; Han, M.Y. $Co_2P$ nanorods as an efficient cocatalyst decorated porous g-$C_3N_4$ nanosheets for photocatalytic hydrogen production under visible light irradiation. *Part. Part. Syst. Charact.* **2018**, *35*, 1700251. [CrossRef]
33. Shi, W.; Yang, S.; Sun, H.; Wang, J.; Lin, X.; Guo, F.; Shi, J. Carbon dots anchored high-crystalline g-$C_3N_4$ as a metal-free composite photocatalyst for boosted photocatalytic degradation of tetracycline under visible light. *J. Mater. Sci.* **2021**, *56*, 2226–2240. [CrossRef]
34. Yu, M.; Liang, H.; Zhan, R.; Xu, L.; Niu, J. Sm-doped g-$C_3N_4$/$Ti_3C_2$ MXene heterojunction for visible-light photocatalytic degradation of ciprofloxacin. *Chin. Chem. Lett.* **2021**, *32*, 2155–2158. [CrossRef]
35. Zhao, X.; Liu, Q.; Li, X.; Ji, H.; Shen, Z. Two-dimensional g-$C_3N_4$ nanosheets-based photo-catalysts for typical sustainable processes. *Chin. Chem. Lett.* **2023**, *34*, 108306. [CrossRef]
36. Qin, J.; Chen, J.; Salisbury, J.B. Photon transferred TL signals from potassium feldspars and their effects on post-IR IRSL measurements. *J. Lumin.* **2015**, *160*, 1–8. [CrossRef]
37. Liu, G.; Yan, S.; Shi, L.; Yao, L. The improvement of photocatalysis $H_2$ evolution over g-$C_3N_4$ with Na and cyano-group co-modification. *Front. Chem.* **2019**, *7*, 639. [CrossRef] [PubMed]
38. Ge, J.; Zhang, L.; Xu, J.; Liu, Y.; Jiang, D.; Du, P. Nitrogen photofixation on holey g-$C_3N_4$ nanosheets with carbon vacancies under visible-light irradiation. *Chin. Chem. Lett.* **2020**, *31*, 792–796. [CrossRef]
39. Shi, Y.; Li, L.; Xu, Z.; Sun, H.; Guo, F.; Shi, W. One-step simple green method to prepare carbon-doped graphitic carbon nitride nanosheets for boosting visible-light photocatalytic degradation of tetracycline. *J. Chem. Technol. Biotechnol.* **2021**, *96*, 3122–3133. [CrossRef]
40. Wang, Y.; Shang, X.; Shen, J.; Zhang, Z.; Wang, D.; Lin, J.; Li, C. Direct and indirect Z-scheme heterostructure-coupled photosystem enabling cooperation of $CO_2$ reduction and $H_2O$ oxidation. *Nat. Commun.* **2020**, *11*, 3043. [CrossRef]
41. Wang, Y.; He, F.; Chen, L.; Shang, J.; Wang, J.; Wang, S.; Sun, H. Acidification and bubble template derived porous g-$C_3N_4$ for efficient photodegradation and hydrogen evolution. *Chin. Chem. Lett.* **2020**, *31*, 2668–2672.
42. Shen, W.; Qi, Q.; Hu, B.; Zhu, Z.; Huo, P.; Jiang, J.; Tang, X. Studying bimetals copper-indium for enhancing PCN photocatalytic $CO_2$ reduction activity and selectivity mechanism. *J. Ind. Eng. Chem.* **2024**, *145*, 384–394.
43. Sun, X.; Shi, Y.; Lu, J.; Shi, W.; Guo, F. Template-free self-assembly of three-dimensional porous graphitic carbon nitride nanovesicles with size-dependent photocatalytic activity for hydrogen evolution. *Appl. Surf. Sci.* **2022**, *606*, 154841. [CrossRef]
44. Shi, W.; Cao, L.; Shi, Y.; Chen, Z.; Cai, Y.; Guo, F.; Du, X. Environmentally friendly supermolecule self-assembly preparation of S-doped hollow porous tubular g-$C_3N_4$ for boosted photocatalytic $H_2$ production. *Ceram. Int.* **2023**, *49*, 11989–11998.
45. Zhang, Z.; Gao, Y.; Li, P.; Qu, B.; Mu, Z.; Liu, Y.; Jing, L. Highly sensitive fluorescence detection of chloride ion in aqueous solution with Ag-modified porous g-$C_3N_4$ nanosheets. *Chin. Chem. Lett.* **2020**, *31*, 2725–2729.

46. Guo, F.; Li, L.; Shi, Y.; Shi, W.; Yang, X. Synthesis of N-deficient g-$C_3N_4$/epoxy composite coating for enhanced photocatalytic corrosion resistance and water purification. *J. Mater. Sci.* **2023**, *58*, 4223–4239.
47. Yu, X.; Tang, X.; Luo, H.; Mao, Y. Construction of biochar assisted S-scheme of $CeO_2$/g-$C_3N_4$ with enhanced photoreduction $CO_2$ to CO activity and selectivity. *Mater. Res. Bull.* **2025**, *181*, 113085.
48. Zhu, Z.; Ye, J.; Tang, X.; Chen, Z.; Yang, J.; Huo, P.; Crittenden, J. Vacancy-rich CoSx@LDH@Co-NC catalytic membrane for antibiotic degradation with mechanistic insights. *Environ. Sci. Technol.* **2023**, *57*, 16131–16140.
49. Zhu, Z.; Xing, X.; Qi, Q.; Shen, W.; Wu, H.; Li, D.; Huo, P. Fabrication of graphene modified $CeO_2$/g-$C_3N_4$ heterostructures for photocatalytic degradation of organic pollutants. *Chin. J. Struct. Chem.* **2023**, *42*, 100194. [CrossRef]
50. Zhu, Z.; Xing, X.; Qi, Q.; Li, H.; Han, D.; Song, X.; Huo, P. Regulation CN reduction of $CO_2$ products selectivity by adjusting the number of V sites and mechanism exploration. *Fuel* **2025**, *388*, 134509.
51. Xing, Y.; Wang, X.; Hao, S.; Zhang, X.; Wang, X.; Ma, W.; Xu, X. Recent advances in the improvement of g-$C_3N_4$ based photocatalytic materials. *Chin. Chem. Lett.* **2021**, *32*, 13–20.
52. Zhu, X.; Yang, J.; Zhu, X.; Yuan, J.; Zhou, M.; She, X.; Xu, H. Exploring deep effects of atomic vacancies on activating $CO_2$ photoreduction via rationally designing indium oxide photocatalysts. *Chem. Eng. J.* **2021**, *422*, 129888.
53. Qi, Q.; Shen, W.; Cai, M.; Cai, J.; Hu, B.; Han, D.; Zhu, Z.; Huo, P. Construction of Cu-Modified g-$C_3N_4$ nanosheets for photoinduced $CO_2$ reduction to CO and selectivity mechanism insight. *ACS Appl. Nano Mater.* **2024**, *7*, 24788–24797. [CrossRef]

*catalysts*

MDPI

*Article*

# Interfacial Engineering of S-Scheme $WO_3/In_2S_3$ Heterojunction for Efficient Solar-Driven $CO_2$ Photoreduction

**Yameng Wang [1], Ao Xu [1], Jihui Lang [1], Bin Zuo [1], Zihan Yu [1], Keyu Cui [2], Xuefei Li [1], Kewei Zhang [3], Xin Li [1,*], Maobin Wei [1,*] and Jian Cao [1,*]**

[1] Key Laboratory of Functional Materials Physics and Chemistry of Ministry of Education, Jilin Normal University, Changchun 130103, China; wym10247@163.com (Y.W.); xuaotg@126.com (A.X.); langjihui80@126.com (J.L.); zuobin1027@163.com (B.Z.); yzh181226977@163.com (Z.Y.); xuefeili@jlnu.edu.cn (X.L.)

[2] Key Laboratory of Preparation and Applications of Environmental Friendly Material of the Ministry of Education, College of Chemistry, Jilin Normal University, Changchun 130103, China; hellocuikeyu@163.com

[3] State Key Laboratory of High Pressure and Superhard Materials, College of Physics, Jilin University, Changchun 130012, China; kwzhang@jlu.edu.cn

* Correspondence: xlwl@jlnu.edu.cn (X.L.); jlsdzccw@126.com (M.W.); caojian_928@163.com (J.C.)

**Abstract:** $CO_2$ photoreduction technology offers significant potential for addressing energy and environmental challenges, though its practical application is hindered by insufficient photo-absorption and rapid carrier recombination. Herein, we constructed the $WO_3/In_2S_3$ S-scheme heterojunction through hydrothermal assembly of two-dimensional $WO_3$ nanosheets and scale-like $In_2S_3$ nanoflakes. Systematic characterization via XRD, XPS, SEM, and TEM verified the successful preparation of hierarchical nanostructures with optimized interfacial contact in the $WO_3/In_2S_3$ composites. UV-Vis DRS analysis showed that the photo-absorption range of the catalyst was significantly widened. Photoelectrochemical investigations (EIS, TPR, PL, and LSV) revealed enhanced carrier separation efficiency and reduced recombination kinetics in the heterojunction system. The optimized $WO_3/In_2S_3$ (WI-60) catalyst had a CO evolution efficiency of 55.14 $\mu mol \cdot g^{-1}$ under the UV-Vis light, representing a 3.9-fold enhancement over the pure $In_2S_3$ (14.08 $\mu mol \cdot g^{-1}$). Mechanistic studies through the XPS and band-structure analysis confirmed the establishment of an S-scheme carrier' transfer pathway, which simultaneously preserved strong redox potentials and promoted the separation process of carriers. This research provides a validated strategy for developing efficient S-scheme photocatalytic systems for solar fuel generation.

**Keywords:** $CO_2$ photoreduction; S-scheme; $WO_3$; $In_2S_3$

## 1. Introduction

The excessive combustion of fossil fuels has triggered an alarming surge in atmospheric $CO_2$ concentrations, precipitating severe environmental crises, including ocean acidification, glacier retreat, and ecosystem collapse [1–3]. Photocatalytic $CO_2$ reduction, harnessing inexhaustible solar energy to convert $CO_2$ into value-added hydrocarbons (CO, $CH_4$, and $CH_3OH$), emerges as a promising carbon-neutral strategy. Nevertheless, this approach faces three fundamental challenges: (1) The thermodynamic stability of linear $CO_2$ molecules (C=O bond energy: 750 kJ/mol) impedes effective adsorption and activation on catalyst surfaces [4]; (2) multi-electron transfer pathways ($2e^- \rightarrow 8e^-$) yield unpredictable product selectivity [5]; and (3) conventional wide-bandgap semiconductors ($TiO_2$, ZnO, $SnO_2$) suffer from limited visible-light utilization (<5%) and rapid carrier recombination [6–8], resulting in unsatisfactory quantum yields. Thus, it is very important to design and construct new catalytic materials for the $CO_2$ photoreduction process.

In recent years, researchers have carried out a lot of research on the design of new photocatalytic materials. Yu et al. proposed a new concept of S-scheme heterojunction in view of the problems and challenges in thermodynamics and kinetics of the traditional Z-type photocatalytic mechanism, which provided a new vision for the preparation of a new generation of photocatalyst [9–11]. Unlike conventional type-II heterojunctions that sacrifice redox potential, S-scheme configurations strategically combine oxidation photocatalyst (OP) and reduction photocatalyst (RP) semiconductors through Fermi-level ($E_f$) equilibration. This process generates an internal electric field (IEF) and band bending, enabling the selective recombination of low-energy electrons (OP) with low-energy holes (RP) while preserving high-potential carriers [12]. The resultant synergy of enhanced charge separation and retained strong redox capacity makes S-scheme systems particularly suited for $CO_2$ photoreduction.

Indium sulfide ($In_2S_3$, $Eg \approx 2.07$ eV), a prominent RP candidate, exhibits exceptional $CO_2$ activation ability due to its highly negative conduction band (−0.86 V vs. NHE) and sulfur-rich surface sites that facilitate *COOH intermediate formation [13], attracting extensive attention in the $CO_2$ photoreduction field. For example, we previously used $In_2S_3$ as the main catalyst and used Au and rGO modification to construct a multi-channel electron transport mode ternary composite material photocatalyst for efficient $CO_2$ reduction [14]. Its photocatalytic performance remains constrained by severe bulk charge recombination. The construction of the S-scheme heterojunction is an effective means to enhance its catalytic activity, so the selection of OP materials is particularly important. Tungsten trioxide ($WO_3$, $Eg \approx 2.63$ V), with its positive valence band (+2.93 V) and exceptional hole mobility, serves as an ideal OP partner to construct S-scheme systems. In the previous research, we used CdS as the RP and WO as the OP to construct the S-scheme, which effectively promoted the separation of carriers and obtained good photocatalytic activity [15]. The complementary band structures of $WO_3$ and $In_2S_3$ suggest the potential for creating an IEF-driven heterojunction that simultaneously enhances visible-light harvesting, suppresses carrier recombination, and optimizes surface redox dynamics.

In the present study, we selected the $In_2S_3$ semiconductor, known for its excellent visible light response, to construct a $WO_3/In_2S_3$ S-scheme heterojunction for efficient $CO_2$ photoreduction through a controlled hydrothermal reaction. A series of characterization methods were carried out to probe the structural and photoelectrochemical properties of the materials. By adjusting the loading amount of $In_2S_3$, we aimed to identify the optimal ratio for $CO_2$ photoreduction. The photocatalytic performance of the materials was tested without the use of any molecular cocatalysts or sacrificial agents. The remarkable performance was attributed to the S-scheme carrier transfer mechanism. This work provides a new perspective for designing S-scheme heterojunction photocatalysts, promoting effective $CO_2$ activation and enhancing the conversion of $CO_2$ to CO.

## 2. Results and Discussion

The crystalline structure and phase composition of prepared materials were systematically characterized using X-ray diffraction (XRD). The diffraction patterns were recorded in the 2θ range of 10–80°, with a scanning rate of 5° $min^{-1}$. Figure 1a presents the XRD pattern of pristine $WO_3$ NSs, where the distinct diffraction peaks at 23.1°, 23.7°, 24.1°, 34.1°, and 49.3° can be unambiguously indexed to the (001), (020), (200), (220), and (400) crystallographic planes of monoclinic $WO_3$ (JCPDS No. 20-1324), respectively [16]. For pure $In_2S_3$, three prominent diffraction peaks at 27.3°, 33.1°, and 47.7° were observed, corresponding to the (311), (400), and (440) planes of cubic $\beta$-$In_2S_3$ (JCPDS No. 65-0459) [17]. The absence of secondary phases confirmed the phase purity of the synthesized indium sulfide. The XRD pattern of the WI-60 composite (Figure 1a) clearly exhibited characteristic diffraction peaks from both the $WO_3$ and $In_2S_3$ phases. The slight XRD peak displacement of $In_2S_3$ in the composite can be attributed to the interfacial interaction between the W atoms in $WO_3$ and the In or S atoms in $In_2S_3$. This interaction modified the local atomic arrangement of $In_2S_3$, thereby influencing the lattice spacing. Furthermore, during the formation of the heterojunction, the mutual diffusion of atoms, chemical bonding, and stress effects at the interface collectively induced lattice distortion in $In_2S_3$, ultimately causing the observed peak shift. This dual-phase coexistence indicates that the crystalline structures of both components remained intact during the modified hydrothermal synthesis reaction. In Figure 1b, the intensity evolution of $In_2S_3$ diffraction peaks (particularly the (311) plane at 27.3°) in all $WO_3/In_2S_3$ binary composites exhibited a positive correlation with increasing $In_2S_3$ content in the composite series, confirming the controllable composition modulation through our synthesis protocol.

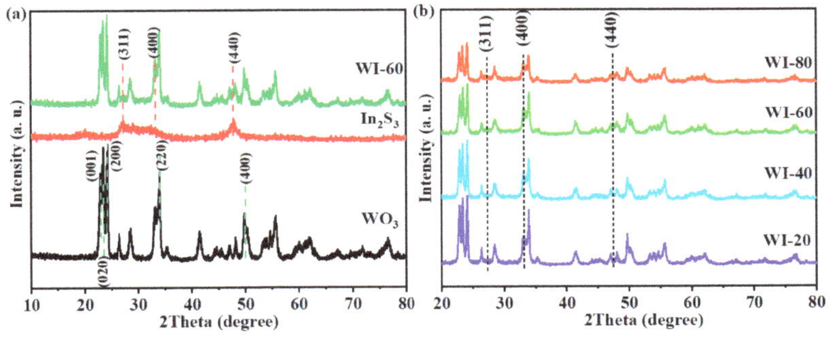


**Figure 1.** XRD patterns of prepared $WO_3$, $In_2S_3$, WI-60 (**a**), and the binary composites with different amounts of $In_2S_3$ (**b**).

The surface morphology and nano-structural feature of the materials were investigated through scanning electronic microscopy (SEM), transmission electron microscopy (TEM), and high-resolution transmission electron microscopy (HR-TEM). As shown in Figure 2a, pristine $WO_3$ exhibited uniform square-shaped nanosheets (NSs) with smooth surfaces, with a thickness of about 50 nm and lateral dimensions of about 300 nm. In contrast, pure $In_2S_3$ displayed a well-defined micro-flower spherical structure and an obvious flaky structure on its surface (Figure 2b), where the hierarchical structure provided abundant active sites for surface reactions [18,19]. For $WO_3/In_2S_3$ composites (Figure 2c–f), progressive decoration of $WO_3$ NSs with $In_2S_3$ nanoflakes was observed with increasing $In_2S_3$ content. At optimal loading (WI-60), the $WO_3$ NSs were uniformly coated with $In_2S_3$ nanoflakes, forming an intimate heterointerface (Figure 2e).

Figure 2g exhibits the TEM and HR-TEM images of pure $WO_3$. The $WO_3$ had regular square-shaped nanosheets, just like the results in the SEM. The HR-TEM result revealed distinct lattice fringes with interplanar spacings of 0.365 nm, corresponding to the (200) plane of the monoclinic $WO_3$ [20]. Additionally, the TEM result of WI-60 is displayed in Figure 2h, and we can ascertain that the loading of $In_2S_3$ did not change the nanosheets structure of $WO_3$ NSs. Moreover, a large number of flocculent structures appeared on the surface of the $WO_3$ NSs, which corresponded to the flaky structure of $In_2S_3$. The HR-TEM image of WI-60 is shown in Figure 2h1, and it reveals that the crystal plane spacing of the lattice was about 0.322 nm, corresponding to the (311) plane of cubic $In_2S_3$ [21]. The SAED result from the HR-TEM of $WO_3/In_2S_3$ is shown in Figure 2h2. It can be seen that the main body of the SAED data was a regular lattice structure, which corresponded to the single-crystal structure of $WO_3$. In addition, we can also observe relatively regular but low-brightness concentric diffraction rings, which corresponded to the (311) and (440) crystal planes of $In_2S_3$ with a small amount of loading in WI-60. The simultaneous existence of a uniform single-crystal diffraction lattice and concentric circular diffraction rings further indicated the good crystal symmetry of the $WO_3/In_2S_3$ composite. The above results showed that the $WO_3/In_2S_3$ composites were successfully prepared.

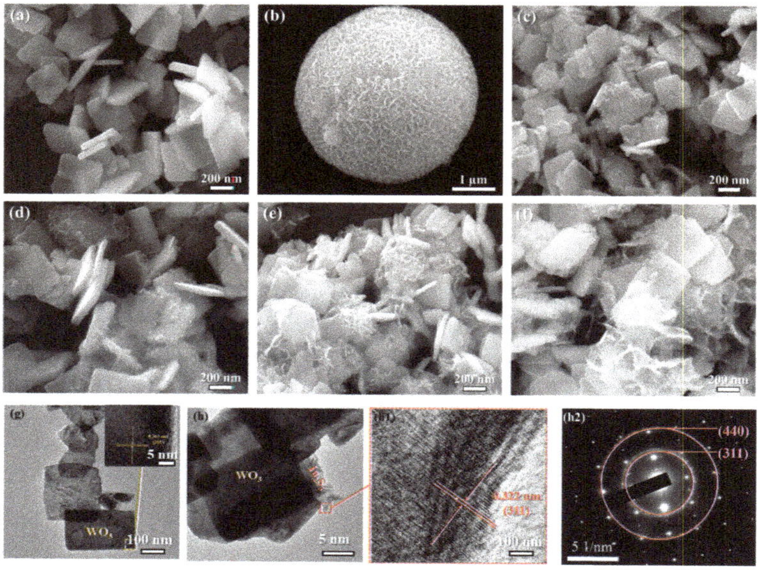


**Figure 2.** SEM images prepared of $WO_3$ (**a**), $In_2S_3$ (**b**), and the binary composites with different amounts of $In_2S_3$ (**c–f**); TEM and HR-TEM images of $WO_3$ (**g**) and $In_2S_3$ (**h**,**h1**); SAED of WI-60 (**h2**).

To research the surface elemental composition, chemical states, and interfacial electron transfer between $WO_3$ and $In_2S_3$, X-ray photoelectron spectroscopy (XPS) was tested, as shown in Figure 3. The survey spectrum of $WO_3/In_2S_3$ in Figure 3a proves the co-existence of the C, O, W, In, and S elements. The high-resolution C 1s spectra of all samples (Figure 3b) exhibited characteristic peaks at about 284.8, 286.4, and 289.0 eV, corresponding to the typical C-C, C-O, and C=O bonds in the previous research, respectively [22]. In the In 3d spectrum of pure $In_2S_3$ (Figure 3c), the binding energies at 444.9 eV and 452.3 eV were caused by the In $3d_{5/2}$ and In $3d_{3/2}$ splitter peaks, which verified the +3 state of indium. The S 2p spectrum of pure $In_2S_3$ in Figure 3d displayed doublet peaks at 161.3 eV (S $2p_{3/2}$) and 162.6 eV (S $2p_{1/2}$), characteristic of the $S^{2-}$ species [23]. Notably, the W 4f and O 1s spectra of pure $WO_3$ (Figure 3e,f) all revealed two distinct peaks. The peaks at about 35.4 eV (W $4f_{7/2}$) and 37.50 eV (W $4f_{5/2}$) were caused by the presence of the $W^{6+}$ species. The components at 530.1 eV and 532.2 eV can be attributed to the lattice $O^{2-}$ and adsorbed oxygen species [24]. In addition, it can be seen that after the combined process, the In 3d XPS peaks moved towards the direction of low-binding energy, while W XPS peaks shifted towards the direction of high-binding energy, which indicates that the In element lost electrons and the W element gained electrons after coupling. These binding energy shifts suggest electron transfer from $In_2S_3$ to $WO_3$ during hybridization, generating an IEF from $In_2S_3$ to $WO_3$ at the heterojunction interface. The formed IEF effectively promoted the spatial separation of photogenerated carriers, accounting for the enhanced $CO_2$ photoreduction activity [25].

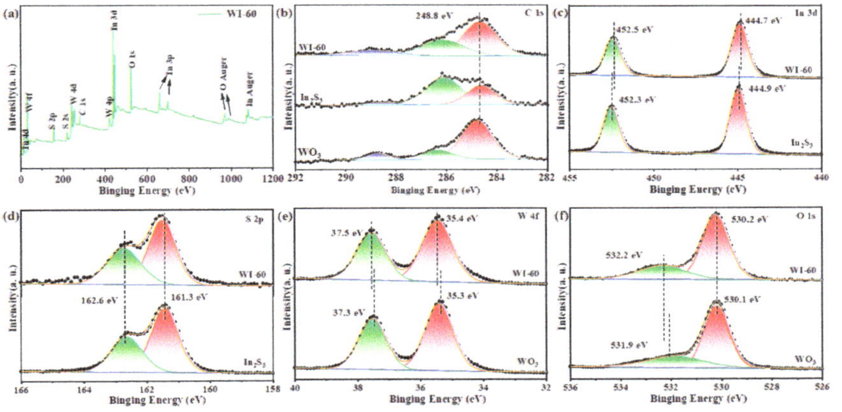

**Figure 3.** XPS total spectrum of WI-60 (**a**); high-resolution XPS spectra of C 1s (**b**), In 3d (**c**), S 2d (**d**), W 4f (**e**), and (**f**) O 1s.

Brunauer–Emmett–Teller (BET) analysis was carried out to research the surface characterization of samples, and the corresponding curves are shown in Figure 4. $N_2$ adsorption/desorption results in Figure 4a showed that the composite had a type-IV isotherm with an H3 hysterescence ring, a feature of the mesoporous material that may be due to the flaky structure of $In_2S_3$ [26]. The specific surface area was augmented from 32.1 $m^2/g$ ($WO_3$) to 68.7 $m^2/g$ (WI-60). $CO_2$ adsorption isotherms in Figure 4b demonstrated WI-60's superior $CO_2$ uptake compared to $WO_3$, attributed to synergistic effects between hierarchical porosity and surface structure. The great surface characteristics were conducive to the $CO_2$ photoreduction process [27].

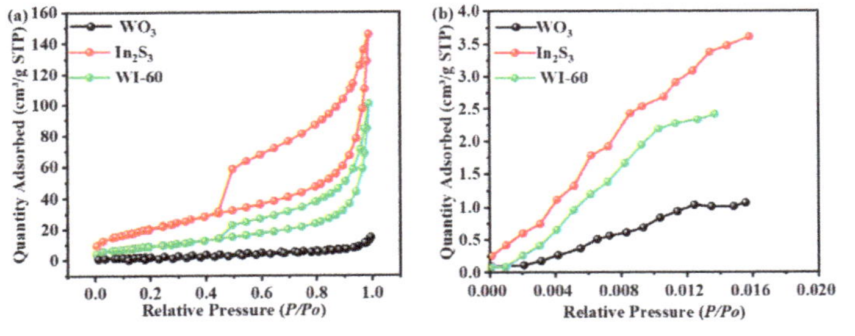

**Figure 4.** $N_2$ adsorption–desorption isotherm (**a**) and $CO_2$ adsorption curve (**b**) of pure $In_2S_3$, $WO_3$, and WI-60.

To explore the $CO_2$ photoreduction ability of the prepared catalyst, the $CO_2$ photocatalytic reduction experiment over the prepared catalysts was carried out under the irradiation of a 300 W Xe lamp (purchased from Beijing Zhongjiao Jinyuan Technology Co., Ltd., Beijing, China) within the UV-Vis light range. The photocatalytic $CO_2$ reduction performances of materials are presented in Figure 5. The $CO_2$ photoreduction activities of diverse materials were investigated under UV-Vis light irradiation for 4 h, and the main gas product was CO. Figure 5a compares the CO generation rates of $In_2S_3$ and $WO_3/In_2S_3$ composites with varying ratios. The CO yield of $In_2S_3$ as the catalyst was approximately 14.08 $\mu mol \cdot g^{-1}$. The CO production rates of $WO_3/In_2S_3$ composite photocatalysts significantly surpassed those of pure-phase $In_2S_3$. However, the catalytic activity of composite materials first increased and then decreased with the loading amount of $In_2S_3$, with WI-60 demonstrating the highest CO yield (55.14 $\mu mol \cdot g^{-1}$). Further increasing the $In_2S_3$ content reduced the CO yield, potentially due to excessive $In_2S_3$ affecting the photogenerated carrier transfer processes between $WO_3$ and $In_2S_3$ in the composite. These results confirmed the critical importance of intimate interfacial contact and strong interactions in the composite for efficient $CO_2$ photoreduction. During the photocatalytic $CO_2$ reduction process, the utilization efficiency of photogenerated electrons ($R$(electron)) played a decisive role in reaction kinetics. Therefore, the $R$(electron) value was systematically investigated (Figure 5b). The calculation formula was defined as: $R(\text{electron}) = 2r(CO) + 8r(CH_4)$ [28]. Based on this equation, the electron utilization efficiencies of different catalysts were calculated. The $CO_2$ photoreduction process clearly showed that WI-60 achieved the highest electron utilization efficiency (27.57 $\mu mol\cdot g^{-1}\cdot h^{-1}$). These results further demonstrated that WI-60 possessed superior $CO_2$ photoreduction performance, compared to the pure $In_2S_3$ and other composite catalysts. To further investigate whether CO came from the $CO_2$ photoreduction process, the photoreduction performances of WI-60 were examined under different conditions. As shown in Figure 5c, no CO production can be found in the absence of either the catalyst or $CO_2$, confirming that the detected reduction products originated from the photocatalytic $CO_2$ reduction over WI-60 [29]. Meanwhile, catalyst stability represents a crucial parameter for practical photocatalytic $CO_2$ reduction applications. The stability of WI-60 was evaluated through four consecutive photocatalytic $CO_2$ reduction cycles. The results in Figure 5d indicate that the composite maintained excellent catalytic stability throughout the cycling tests [30].

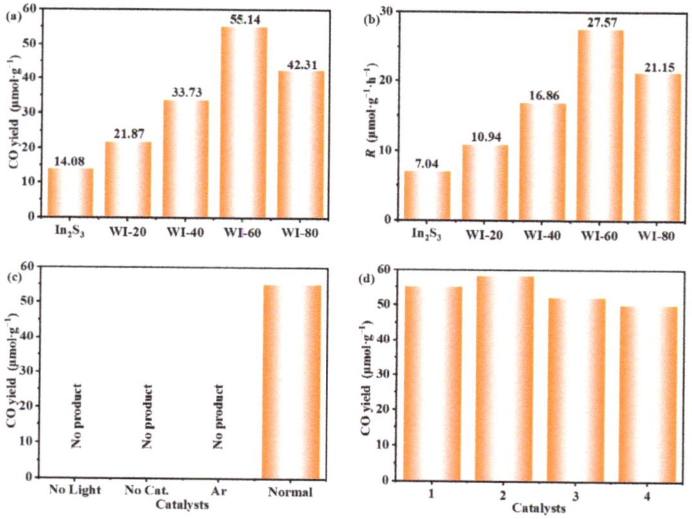


**Figure 5.** Performance of $CO_2$ conversion to CO after 4 h via all samples (**a**), electron utilization rate of each sample (**b**), control experiment under different experimental conditions (**c**), and performance of the photoreduction of $CO_2$ to CO performed 4 times, with WI-60 as catalyst (**d**).

To further compare the catalytic performance of the WI-60 sample, we compared its photocatalytic $CO_2$ reduction performance with that of other sulfide-based photocatalytic materials reported recently. The specific results are shown in Table 1. It can be clearly seen from the data in the table that the catalyst we prepared performed outstandingly in several key indicators. Compared with the other sulfide-based photocatalysts, WI-60 exhibited a significant advantage in terms of $CO_2$ conversion efficiency. Its CO generation rate reached 55.14 $\mu mol\cdot g^{-1}$, which was higher than that of most of the comparative samples.

**Table 1.** A comparative analysis of the performance of our catalyst vs. reported catalysts production from $CO_2$ photoreduction.

| Sample | Light Source | Sacrificial Agent | Evolution Rate μmol/g (4 h) | Ref. |
|---|---|---|---|---|
| $MoS_2/In_2S_3$ | 300 W Xe-lamp (Beijing Zhongjiao Jinyuan CEL-HXF300, Beijing, China) | Without | CO: 40.36 | [31] |
| Atomic strain $In_2S_3$ | 300 W Xe-lamp (Chengdu Lihang PLS-SXE300D), AM 1.5 G filter, Chengdu, China | Without | CO: 20.64 | [32] |
| $Cu_7S_4@Cu_2O$ | 300 W Xe Arc lamp (Beijing Bofeilai, Beijing, China) | Without | CO: 24.76 | [33] |
| $ZnIn_2S_4$/MOF-808 | OFR Xe-lamp (Beijing Education Au-light, Beijing, China) | Without | CO: 32.84 | [34] |
| CDs/CdS | 300 W Xe-lamp, AM 1.5 G filter. (Xi'an TopTION Instrument Co., Ltd., Xi'an, China) | Triethanolamine | CO: 48.16 | [35] |
| $MgCr_2O_4/MgIn_2S_4$ | 300 W Xe-lamp (Chengdu Lihang PLS-SXE 300+, Chengdu, China) | Without | CO: 32.12 | [36] |
| $ZnIn_2S_4/Cu_2S$ | 300 W Xe-lamp (Beijing China Education Au-light, Beijing, China) | Without | CO: 13.35/$CH_4$: 34.81 | [37] |
| $NiTiO_3$/CdS | 300 W Xe-lamp (Beijing China Education Au-light, Beijing, China) | Triethanolamine | CO: 20.8 | [38] |
| $CuInSnS_4$ | 300 W Xe-lamp, 420 nm cut-off wavelength filter (Beijing China Education Au-light, Beijing, China) | Without | $CH_4$: 23.32 | [39] |
| $WO_3/In_2S_3$ | 300W Xe-lamp (Beijing China Education Au-light, Beijing, China) | Without | CO: 55.2 | This work |

The photo-absorption properties of the synthesized catalysts constituted a critical factor influencing their photocatalytic property. The photo-absorption performance directly determines the photon utilization efficiency of photocatalysts [40]. Thus, we employed UV-Vis diffuse reflectance spectroscopy (UV-vis DRS) to analyze the photo-absorption characteristics and bandgap width of the obtained samples, with detailed results presented in Figure 6. As shown in Figure 6a, pure $WO_3$ NSs and $In_2S_3$ exhibited absorption edges at about 462 nm and 594 nm, respectively. The WI-60 composite demonstrated combined absorption features from both constituent materials. Notably, in contrast to pure $WO_3$, the WI-60 composite exhibited a discernible red shift in its absorption edge. This phenomenon served as a clear indication of its augmented visible-light utilization capacity, which empowered a greater number of photons to partake in the $CO_2$ photoreduction process. The calculated bandgap widths of $WO_3$ and $In_2S_3$ are shown in Figure 6b, yielding values of approximately 2.89 eV and 2.30 eV, respectively. Similarly, the composite material exhibited hybrid band characteristics associated with both components, providing further evidence for the successful fabrication of the heterostructure. This synergistic combination of optical properties in the composite aligns with the observed enhancement in photocatalytic performance.

The photoelectrochemical (PEC) properties of catalysts are the key factors affecting their catalytic activity. Therefore, the PEC performances of $WO_3$, $In_2S_3$, and WI-60 were systematically investigated through electrochemical impedance spectroscopy (EIS), transient photocurrent response (TPR), photoluminescence (PL) spectroscopy, and linear sweep voltammetry (LSV). As shown in Figure 7a, the WI-60 composite demonstrated a smaller Nyquist arc radius, compared to pristine $WO_3$ and $In_2S_3$ in EIS curves. This suggested a decrease in the resistance to carrier transfer at the heterojunction interface. (The equivalent circuit is shown as the inset in Figure 2). This structural configuration facilitated the separation and transfer of photogenerated electron–hole pairs [41]. TPR analysis in Figure 7b revealed an enhanced transient photocurrent response for WI-60 under the UV-Vis light illumination, suggesting superior charge carrier generation and separation efficiency, compared to the pure $WO_3$ and $In_2S_3$. The rapid current stabilization further confirmed efficient charge transfer kinetics in the composite system [42]. In Figure 7c, the PL spectra of prepared samples exhibited characteristic emission peaks at 468 nm ($WO_3$) and 529 nm ($In_2S_3$), with a significantly quenched intensity observed in the composite material. This fluorescence suppression directly evidenced the effective separation of photogenerated carriers through interfacial charge transfer processes [43]. LSV measurement results of pure $WO_3$, $In_2S_3$, and WI-60 are shown in Figure 7d. Compared with pure-phase $WO_3$, WI-60 had lower overpotential at the same current density, which contributed to the occurrence of the $CO_2$ reduction reaction. The overpotential of WI-60 was lower under light, which proved that the $CO_2$ reduction reaction was easier to carry out under the same conditions. The enhanced catalytic activity originated from the synergistic effects of improved charge separation and optimized surface reaction kinetics [44].

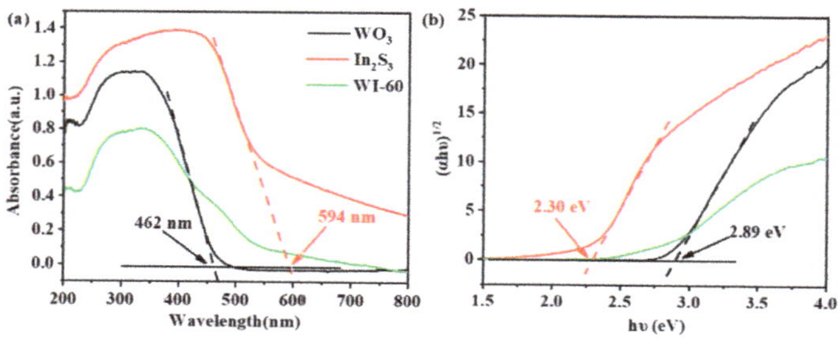


**Figure 6.** UV-Vis DRS spectra of $WO_3$ NSs, $In_2S_3$, and $WO_3/In_2S_3$ (**a**); $WO_3$ and $In_2S_3$ bandgap widths were calculated by Tauc curves (**b**).

To elucidate the photogenerated carrier transfer mechanism at the $WO_3/In_2S_3$ heterojunction interface, the Mott–Schottky test was carried out to determine the flat-band potentials. As shown in Figure 8, the flat-band potentials of $WO_3$ and $In_2S_3$ were measured to be +0.58 V and −1.51 V vs. Ag/AgCl in Figure 8a,b, respectively. Thus, the conduction band minima edge positions ($E_{CB}$) of $WO_3$ and $In_2S_3$ can be determined to be +0.78 V and −1.31 V (vs. NHE) by the conversion process ($E_{NHE} = E_{Ag/AgCl} + 0.197$ V) [45]. The valence band maxima edge positions ($E_{VB}$) of $WO_3$ and $In_2S_3$ can be calculated as +3.67 V and +0.99 V (vs. NHE) using the formula: $E_{VB} = E_{CB} + Eg$ [46]. Additionally, the corresponding band structures are depicted in Figure 8c. As the typical N-type semiconductors, the Fermi level ($E_f$) positions of $WO_3$ and $In_2S_3$ were slightly lower than the $E_{CB}$. Since the $E_{CB}$ of $In_2S_3$ was much higher than that of $WO_3$, we can infer that their $E_f$ were in the same trend, that is, the $E_f$ of $In_2S_3$ was much higher than that of $WO_3$. Owing to the higher $E_f$ of $In_2S_3$ compared to $WO_3$, electrons spontaneously transferred from the semiconductor with a higher $E_f$ to that with a lower $E_f$ [47–49]. Consequently, upon dark contact, electron migration occurred from $WO_3$ to $In_2S_3$ until $E_f$ equilibration was achieved. This process established an electron accumulation layer at the $WO_3$ interface and an electron depletion layer at the $In_2S_3$ interface, generating an IEF and band bending (Figure 8d). Under light irradiation, the IEF-driven band bending facilitated the recombination of electrons photogenerated in the CB of $WO_3$, with holes photogenerated in the VB of $In_2S_3$ [50–52]. Meanwhile, the retained photogenerated electrons in the CB of $In_2S_3$ and holes in the VB of $WO_3$ exhibited enhanced redox capabilities, actively participating in the photocatalytic $CO_2$ reduction to CO, as illustrated in Figure 8e. These experimental and theoretical discoveries collectively demonstrated the S-scheme electron transfer mechanism in the $WO_3/In_2S_3$ heterojunction. The S-scheme band alignment in the $WO_3/In_2S_3$ photocatalyst promoted spatial separation and the directional transfer of charge carriers, thereby prolonging charge lifetime and significantly improving the $CO_2$-to-CO conversion efficiency (Figure 8f).

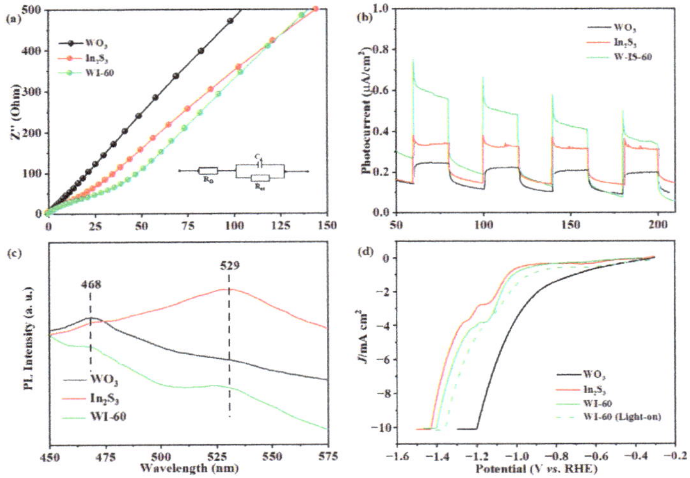


**Figure 7.** EIS (**a**), TPR (**b**), PL (**c**), and LSV (**d**) results of the prepared sample.

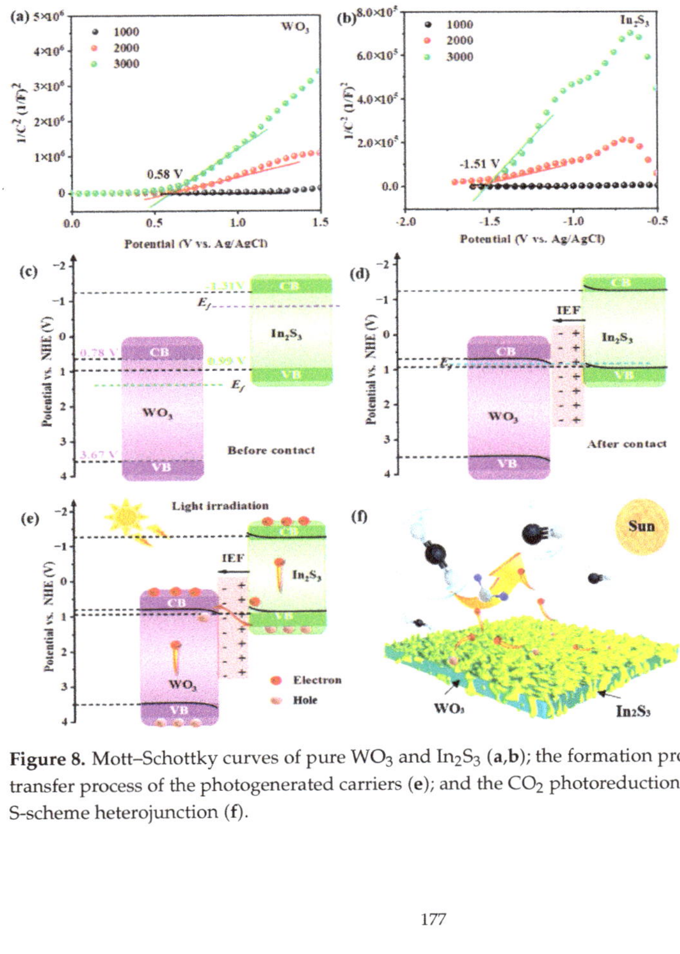


**Figure 8.** Mott–Schottky curves of pure $WO_3$ and $In_2S_3$ (**a**,**b**); the formation process of IEF (**c**,**d**); the transfer process of the photogenerated carriers (**e**); and the $CO_2$ photoreduction process of the WI-60 S-scheme heterojunction (**f**).

## 3. Materials and Methods

### *3.1. Materials*

The ammonium paratungstate hydrate ($(NH_4)_{10}W_{12}O41{\cdot}5H_2O$, APT, 99.9%, Aladdin, Nashville, TN, USA), hydrochloric acid (HCl, Sinopharm, Beijing, China), oxalic acid dihydrate (OA, $C_2H_2O_4{\cdot}2H_2O$, 99.5%, Macklin, Shanghai, China), indium chloride tetrahydrate ($InCl_3{\cdot}4H_2O$, 99.99%, Alfa Aesar, Haverhill, MA, USA), L-cysteine ($C_3H_7NO_2S$, AR, Sinopharm), ethanol ($C_2H_5OH$, AR, Sinopharm, Shanghai, China), and glycerin ($C_3H_8O_3$, AR, Sinopharm, Shanghai, China) used in this paper were analytical-grade materials without further purification. High-purity $CO_2$ gas (99.999%) was obtained from Changchun Juyang Gas Co., Ltd. (Changchun, China).

### *3.2. Synthesis of Catalysts*

#### 3.2.1. Fabrication of $WO_3$ Nanosheets ($WO_3$ NSs)

Two-dimensional $WO_3$ nanosheets were synthesized through an acid-mediated hydrothermal process. (a) Specifically, 0.3 g of APT was added into 30 mL of deionized water under magnetic stirring for 30 min. The solution pH was adjusted to 3.0 ± 0.1 using a 3 M HCl solution delivered through a calibrated burette. Subsequently, 0.6 g of OA was introduced as a structure-directing agent. The mixture was heated in a water bath to 90 °C under vigorous stirring for 3 h using a PTFE-coated magnetic bar. The resulting yellow precursor (P-$WO_3$) was obtained via the centrifugation processes and then washed with water and alcohol three times, respectively. (b) Final crystallization was achieved by calcining P-$WO_3$ in a muffle furnace under static air at 500 °C for 2 h (5 °C/min).

#### 3.2.2. Construction of $WO_3/In_2S_3$ Heterojunctions

The $WO_3/In_2S_3$ composites were synthesized through a modified hydrothermal reaction. (c) A precursor solution was prepared by dissolving stoichiometric $In_2S_3{\cdot}4H_2O$ and $C_3H_7NO_2S$ in 60 mL of deionized water, adding 15 mL of ethanol and 5.5 mL of glycerol to the deionized water. Four composite samples with varying theoretical mass ratios ($In_2S_3$:$WO_3$ = 0.2, 0.4, 0.6, 0.8) were prepared by dispersing 200 mg of pre-synthesized $WO_3$ NSs into respective solutions via 30 min ultrasonication, followed by magnetic stirring. The homogeneous suspensions were transferred to 100 mL of Teflon-lined autoclaves and hydrothermally treated at 160 °C for 12 h. After that, the products (denoted as WI-20, WI-40, WI-60, and WI-80 based on $In_2S_3$ content) were washed via centrifugation with ethanol/water and dried at 60 °C under vacuum for 12 h. Pure $In_2S_3$ control samples were synthesized identically without $WO_3$ addition. A schematic representation of the synthesis process is provided in Scheme 1.

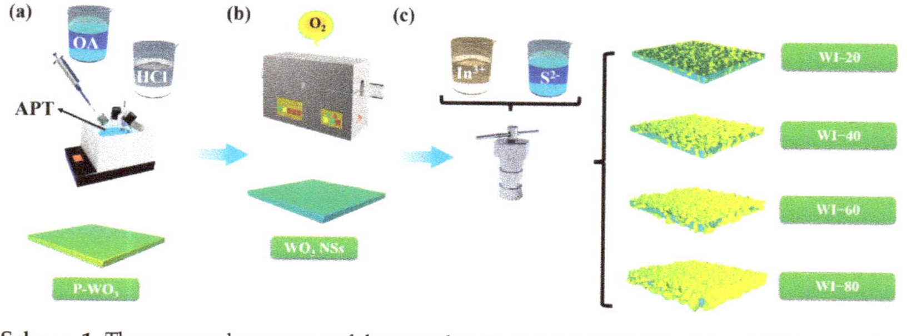


**Scheme 1.** The prepared processes of the samples, P-$WO_3$ (**a**); $WO_3$ NSs (**b**) and WI composites (**c**).

### 3.3. Characterization

X-ray diffraction (XRD) patterns of samples were tested using a powder XRD (MAC science, Yokohama, Japan, Ni-filtrated Cu-Kα radiation). The nanostructure property of prepared catalysts was analyzed using scanning electron microscopy (FE-SEM, JEOL (Tokyo, Japan), 7800F) and transmission electron microscopy (TEM, FEI Tecnai (Hillsboro, OR, USA) G2F20 instrument). The elemental composition and chemical states of the catalyst were studied using X-ray photoelectron spectroscopy (XPS) (PerkinElmer PHI 5300, a monochromatic Mg Kα source, Waltham, MA, USA), and the C 1s peak position at 284.6 eV was used as the standard for calibrating the XPS data. The photo-absorption ability of the prepared samples was researched using a UV–Vis spectrophotometer (UV-3600P; Shimazu, Tokyo, Japan). Photocurrent experiments were conducted on an electrochemical system (CHI 660B, Shanghai, China) with a standard three-electrode quartz cell. An AgCl and a Pt wire served as reference and counter electrodes, and samples served as the working electrodes. A 0.5 M $Na_2SO_4$ solution was the electrolyte. Electrochemical impedance spectroscopy (EIS) was performed in a 0.5 mM $Fe(CN)_6{}^{3-}/{}^{4-}$ and KCl solution at 10 mV sine amplitude and 0.1 Hz–100 kHz frequency on the same system. The photoluminescence (PL) spectra were analyzed by a FLS1000 photoluminescence spectrometer (Edinburgh Instrument Company, Edinburgh, UK). $N_2$ adsorption and desorption/$CO_2$ adsorption curves were measured by a specific surface area analyzer (Micromeritics, ASAP2460, Norcross, GA, USA). $CO_2$ photoreduction process: a Xe lamp (Education AU-light, Beijing, China) was selected as the light source. The Xe lamp was positioned approximately 15 cm away from the liquid surface, resulting in a total optical power of about 50.1 mW/cm$^2$ incident on the mixed liquid. The reaction was conducted in a 200 mL, fully transparent quartz reactor. The experimental procedure was as follows: 0.02 g of the catalyst was dispersed in 50 mL of deionized water in the reactor. Subsequently, $CO_2$ gas was purged into the reactor for 10 min. The suspension was then continuously stirred at a speed of 500 rpm and irradiated for 4 h before analysis. After the photoreduction process, the gas product was detected and analyzed through an off-line gas chromatography system (GC-8670M, Nanjing Dongcun Scientific Instrument Co., Ltd., Nanjing, China).

## 4. Conclusions

In this paper, serious $WO_3/In_2S_3$ S-scheme heterojunction photocatalysts were successfully synthesized via a facile hydrothermal method for $CO_2$ photoreduction. The Fermi level difference between components triggered interfacial electron transfer, establishing an IEF directed from $In_2S_3$ to $WO_3$. This IEF facilitated directional charge migration and substantially enhanced the separation efficiency of photoinduced carriers. UV-Vis DRS analysis and photoelectrochemical characterization revealed that the optimized WI-60 composite exhibited an extended photo-absorption range and effective $CO_2$ activation on $In_2S_3$ surfaces, synergistically contributing to the superior photocatalytic performance. Without requiring molecular co-catalysts or sacrificial agents, the WI-60 heterostructure achieved a CO evolution rate of 55.2 $\mu mol \cdot g^{-1}$ under 4 h of UV-Vis irradiation, representing a 3.9-fold enhancement compared to pristine $In_2S_3$ (14.1 $\mu mol \cdot g^{-1}$). Cyclic tests demonstrated exceptional stability, with the composite retaining 89.3% of initial activity after four consecutive runs. Mechanistic investigations combining XPS analysis and energy band alignment studies conclusively validated the S-scheme charge transfer pathway, wherein high-energy electrons in $WO_3$ and holes in $In_2S_3$ were preserved for redox reactions while low-energy carriers underwent recombination. This work provides fundamental insights into designing IEF-driven S-scheme systems for solar fuel production.

**Author Contributions:** Conceptualization: A.X. and Y.W.; data curation: A.X., Y.W., Z.Y. and B.Z.; investigation: A.X. and Y.W.; validation/visualization: A.X. and Y.W.; writing—original draft preparation: Y.W.; writing—review and editing, X.L. (Xin Li) and M.W.; supervision: J.C., X.L. (Xuefei Li), and K.Z.; project administration: J.L. and K.C.; funding acquisition: J.L., X.L. (Xin Li), M.W., X.L. (Xuefei Li), B.Z. and K.C. All authors have read and agreed to the published version of the manuscript.

**Funding:** We gratefully acknowledge the financial support of the program for the National Natural Science Foundation of China (22378158), the Development of Science and Technology of Jilin province (YDZJ202301ZYTS246, 20220402031GH, 20230508040RC, 20240601047RC, YDZJ202401470ZYTS), the program for the Science and Technology of the Education Department of Jilin Province (grant no. JJKH20250941KJ), Training Program of Innovation and Entrepreneurship for College Students (202410203037), Young Teachers Research Ability Cultivation and Promotion Program of Jilin Normal University (0521426), and the Doctoral Research Project of Jilin Normal University (0420355).

**Data Availability Statement:** The datasets used and/or analyzed during the current study are available from the corresponding author upon reasonable request.

**Conflicts of Interest:** The authors declare no conflicts of interest.

## References


1. Zhu, Z.; Xing, X.; Qi, Q.; Li, H.; Han, D.; Song, X.; Tang, X.; Ng, H.; Huo, P. Regulation CN reduction of $CO_2$ products selectivity by adjusting the number of V sites and mechanism exploration. *Fuel* **2025**, *388*, 134509. [CrossRef]
2. Wang, L.; Liu, Y.; Perumal, S.; Wang, Y.; Ko, H.; Hwang, Y.; Wang, H.; Yang, T.; Zhao, S.; Kim, Y.D.; et al. Enhancing photocatalytic $CO_2$ reduction to butanol by facet-dependent interfacial engineering of $CeO_2/Cu_2O$. *Appl. Catal. B. Environ. Energy* **2025**, *368*, 125122. [CrossRef]
3. Wang, X.; Pan, Y.; Wang, X.; Guo, Y.; Ni, C.; Wu, J.; Hao, C. High performance hybrid supercapacitors assembled with multi-cavity nickel cobalt sulfide hollow microspheres as cathode and porous typha-derived carbon as anode. *Ind. Crops Prod.* **2022**, *189*, 115863. [CrossRef]
4. Yang, W.; Gao, M.; Zhang, Y.; Dai, Y.; Peng, W.; Ji, S.; Ji, Y. Self -driven photoelectrochemical sensor based on Z-type perovskite heterojunction for profenofos detection in milk and cabbage. *J. Food Compos. Anal.* **2024**, *136*, 106738. [CrossRef]
5. Liu, X.; Zhang, T.; Li, Y.; Zhang., J.; Du, Y. Construction of core-shell $ZnS@In_2S_3$ rhombic dodecahedron Z-scheme heterojunction structure: Enhanced photocatalytic activity and mechanism insight. *Chem. Eng. J.* **2021**, *423*, 130138. [CrossRef]
6. Huang, L.; Mo, S.; Zhao, X.; Zhou, J.; Zhou, X.; Zhang, Y.; Fan, Y.; Xie, Q.; Li, B.; Li, J. Constructing Co and Zn atomic pairs incore-shell $Co_3S_4/NC@ZnS/NC$ derived from MOF-on-MOF nanostructures for enhanced photocatalytic $CO_2$ reduction to $C_2H_4$. *Appl. Catal. B Environ. Energy* **2024**, *352*, 124019. [CrossRef]
7. Zhang, T.; Li, T.; Gao, M.; Lu, W.; Chen, Z.; Ong, W.L.; Wong, A.S.W.; Yang, L.; Kawi, S.; Ho, G.W. Ligand mediated assembly of CdS colloids in 3D porous metal–organic framework derived scaffold with multi-sites heterojunctions for efficient $CO_2$ photoreduction. *Adv. Energy Mater.* **2024**, *14*, 2400388. [CrossRef]
8. Wang, F.; Zhu, Y.; Qian, L.; Yin, Y.; Yuan, Z.; Dai, Y.; Zhang, T.; Yang, D.; Qiu, F. Lamellar $Ti_3C_2$ MXene composite decorated with platinum-doped $MoS_2$ nanosheets as electrochemical sensing functional platform for highly sensitive analysis of organophosphorus pesticides. *Food. Chem.* **2024**, *459*, 140379. [CrossRef]
9. Xu, Q.; Zhang, L.; Cheng, B.; Fan, J.; Yu, J. S-Scheme Heterojunction Photocatalyst. *Chem* **2020**, *6*, 1543–1559. [CrossRef]
10. Khan, K.; Tang, Y.; Cheng, P.; Song, Y.; Li, X.; Lou, J.; Iqbal, B.; Zhao, X.; Hameed, R.; Li, G.; et al. Effects of degradable and non-degradable microplastics and oxytetracycline co-exposure on soil $N_2O$ and $CO_2$ emissions. *Appl. Soil Ecol.* **2024**, *197*, 105331. [CrossRef]
11. Hu, Y.; Yu, C.; Wang, S.; Wang, Q.; Reinhard, M.; Zhang, G.; Zhan, F.; Wang, H.; Skoien, D.; Kroll, T.; et al. Identifying a highly efficient molecular photocatalytic $CO_2$ reduction system via descriptor-based high-throughput screening. *Nat. Catal.* **2025**, *8*, 126–136. [CrossRef]
12. Zeng, P.; Liu, H.; Jia, H.; Cai, J.; Deng, X.; Peng, T. In-situ synthesis of single-atom CoN clusters-decorated $TiO_2$ for highly efficient charge separation and $CO_2$ photoreduction. *Appl. Catal. B Environ.* **2024**, *340*, 123268. [CrossRef]
13. Zheng, X.; Liu, T.; Wen, J.; Liu, X. Flower-like $Bi_2S_3$–$In_2S_3$ heterojunction for efficient solar light induced photoreduction of Cr (VI). *Chem* **2021**, *278*, 130422. [CrossRef] [PubMed]
14. Li, X.; Wei, Y.; Ma, C.; Jiang, H.; Gao, M.; Zhang, S.; Liu, W.; Huo, P.; Wang, H.; Wang, L. Multichannel electron transmission and fluorescence resonance energy transfer in $In_2S_3/Au/rGO$ Composite for $CO_2$ Photoreduction. *ACS Appl. Mater. Interfaces* **2021**, *13*, 11755–11764. [CrossRef]

15. Xu, A.; Zhang, Y.; Fan, H.; Liu, X.; Wang, F.; Qu, X.; Yang, L.; Li, X.; Cao, J.; Wei, M. $WO_3$ Nanosheet/$ZnIn_2S_4$ S-Scheme Heterojunctions for Enhanced $CO_2$ Photoreduction. *ACS Appl. Nano Mater.* **2024**, *7*, 3488–3498. [CrossRef]
16. Zhang, M.; Zhang, X.; Zhang, Z.; Zhang, L.; Liao, J.; Zhang, X.; Ge, C.; Zhou, W. Lead-free perovskite $Cs_3Bi_2Br_9$ quantum dots (QDs)/ultra-thin $W_{18}O_{49}$ nanobelts S-scheme heterojunction toward optimized photocatalytic $CO_2$ reduction coupled with toluene oxidation. *Chem. Eng. J.* **2025**, *505*, 159635. [CrossRef]
17. Ebri, G.; Alhashmi, E.; Baghdadi, Y.; Daboczi, M.; Eslava, S.; Hellgardt, K. Simultaneous photocatalytic $CO_2$ reduction and C-C coupling of benzyl alcohol under high pressure and supercritical conditions. *Chem. Eng. J.* **2025**, *505*, 158356. [CrossRef]
18. Gouda, A.; Hannouche, K.; Mohan, A.; Mao, C.; Nikbin, E.; Carrière, A.; Ye, J.; Howe, J.Y.; Sain, M.; Hmadeh, M.; et al. In-situ restructuring of Ni-based metal organic frameworks for photocatalytic $CO_2$ hydrogenation. *Nat. Commun.* **2025**, *16*, 695. [CrossRef]
19. Yang, R.; Li, Q.; Ma, Z.; Liu, S.; Tian, D.; Li, D.; Jiang, D. Improving charge separation in poly (heptazine imide) via synergy of covalent S-scheme heterojunction and Schottky junction for efficient photocatalytic $CO_2$ reduction. *Chem. Eng. J.* **2025**, *506*, 160043. [CrossRef]
20. Ji, Q.; Yu, X.; Chen, L.; Naa, O.; Zhou, C. Facile preparation of sugarcane bagasse-derived carbon supported $MoS_2$ nanosheets for hydrogen evolution reaction. *Ind. Crops Prod.* **2021**, *172*, 114064. [CrossRef]
21. Chen, Q.; Jiang, J.; Wu, Y.; Fang, J.; Lu, N.; Wang, G.; Mao, J. Enhancing local $CO_2$ availability via amorphous $Bi_2O_3$ enable efficient and stable photocatalytic $C_2H_6$ production. *Chem. Eng. J.* **2025**, *504*, 158854. [CrossRef]
22. Zhang, H.; Yohannes, A.; Zhao, H.; Li, Z.; Xiao, Y.; Cheng, X.; Wang, H.; Li, Z.; Siahrostami, S.; Kibria, M.G.; et al. Photocatalytic asymmetric C-C coupling for $CO_2$ reduction on dynamically reconstructed $Ru^{\delta+}$-O/$Ru^0$-O sites. *Nat. Commun.* **2025**, *16*, 534. [CrossRef] [PubMed]
23. Wang, J.; Wang, W.; Deng, Y.; Zhang, Z.; Wang, H.; Wu, Y. Directed Inward Migration of S-Vacancy in $Bi_2S_3$ QDs for Selective Photocatalytic $CO_2$ to $CH_3OH$. *Nat. Commun.* **2025**, *12*, e2406925. [CrossRef] [PubMed]
24. Zhou, M.; Ding, D.; Shi, Y.; Wang, K.; Wang, Q.; Chen, X.; Shen, H.; Zhang, T.; Yang, Y.; Xia, J.; et al. Hydrogen-Bonded organic framework Containing stacked $Cu^{2+}$ for photocatalytic reduction of $CO_2$ to $C_2H_4$. *Chem. Eng. J.* **2025**, *503*, 158501. [CrossRef]
25. Sun, Y.; Lai, K.; Shi, X.; Li, N.; Gao, Y.; Ge, L. Regulating metal cation Cu vacancies on $ZnIn_2S_4$/$Cu_{1.81}S$ to achieve high selectivity for the photocatalytic reduction of $CO_2$ to $CH_4$. *Appl. Catal. B Environ. Energy* **2025**, *365*, 124907. [CrossRef]
26. Xiao, Y.; Li, H.; Yao, B.; Wang, Y. Hollow core-shell B-g-$C_3N_4$-x@$Bi_2S_3$/$In_2S_3$ dual S-scheme heterojunction photothermal nanoreactor: Boosting photothermal catalytic activity in confined space. *Chem. Eng. J.* **2024**, *484*, 149399. [CrossRef]
27. Chen, Z.; Sun, Y.; Shi, J.; Zhang, W.; Zhang, X.; Huang, X.; Zou, X.; Li, Z.; Wei, R. Facile synthesis of Au@Ag core-shell nanorod with bimetallic synergistic effect for SERS detection of thiabendazole in fruit juice. *Food Chem.* **2022**, *370*, 131276. [CrossRef]
28. Gong, E.; Ali, S.; Hiragond, C.B.; Kim, H.S.; Powar, N.S.; Kim, D.; Kim, H.; In, S.-I. Solar fuels: Research and development strategies to accelerate photocatalytic $CO_2$ conversion into hydrocarbon fuels. *Energy Environ. Sci.* **2022**, *15*, 880–937. [CrossRef]
29. Chen, X.; Cao, Y.; Li, F.; Tian, Y.; Song, H. Enzyme-assisted microbial electrosynthesis of Poly(3-hydroxybutyrate) via $CO_2$ bio-reduction by engineered *Ralstonia eutropha*. *ACS Catal.* **2018**, *8*, 4429–4437. [CrossRef]
30. Lu, N.; Jiang, X.; Zhu, Y.; Yu, L.; Du, S.; Huang, J.; Zhang, Z. Single-Atom-Layer Metallization of Plasmonic Semiconductor Surface for Selectively Enhancing IR-Driven Photocatalytic Reduction of $CO_2$ into $CH_4$. *Adv. Mater.* **2024**, *37*, e2413931. [CrossRef]
31. Cai, W.; Qian, Z.; Hu, C.; Zheng, W.; Luo, L.; Zhao, Y. Systematic investigation of MoS2-metal sulfides [Metal = In, Sn, Cu, Cd] heterostructure via metal-sulfur bond for photocatalytic $CO_2$ reduction. *Chem. Eng. J.* **2024**, *479*, 147718. [CrossRef]
32. Wang, K.; Hu, Y.; Liu, X.; Li, J.; Liu, B. Visible-light-driven $CO_2$ photoreduction over atomically strained indium sites in ambient air. *Nat. Commun.* **2025**, *16*, 2094. [CrossRef] [PubMed]
33. Liu, Z.; Chang, X.; Shi, Y.; Zhao, X. Synergistic effects of $Cu_7S_4$@$Cu_2O$ core-shell photocatalyst and S-scheme heterojunction in photocatalytic $CO_2$ reduction. *J. Alloys Compd.* **2025**, *1010*, 177572. [CrossRef]
34. Song, M.; Song, X.; Liu, X.; Zhou, W.; Huo, P. Enhancing photocatalytic $CO_2$ reduction activity of $ZnIn_2S_4$/MOF-808 microsphere with S-scheme heterojunction by in situ synthesis method. *Chin. J. Catal.* **2023**, *51*, 180–192. [CrossRef]
35. Liu, C.; Zhang, Y.; Shi, T.; Liang, Q.; Chen, Z. Hierarchical hollow-microsphere cadmium sulfide-carbon dots composites with enhancing charge transfer efficiency for photocatalytic $CO_2$ reduction. *J. Alloys Compd.* **2023**, *936*, 168286. [CrossRef]
36. Xiong, R.; Sun, Y.; Li, J.; Chen, K.; Liu, F.; Xiao, Y.; Cheng, B.; Lei, S. $MgCr_2O_4$/$MgIn_2S_4$ Spinel/Spinel S-Scheme Heterojunction: A Robust Catalyst for Photothermal-Assisted Photocatalytic $CO_2$ Reduction. *Inorg. Chem.* **2024**, *63*, 19309–19321. [CrossRef]
37. Yan, C.; Xu, M.; Li, J.; Wang, H.; Huo, P. Fabricated $ZnIn_2S_4$/$Cu_2S$ S-scheme heterojunction with snowflake structure for boosting photocatalytic $CO_2$ reduction in gas–solid reactor. *Fuel* **2024**, *378*, 132921. [CrossRef]
38. Cai, J.; Li, X.; Su, B.; Guo, B.; Lin, X.; Xing, W.; Lu, X.F.; Wang, S. Rational design and fabrication of S-scheme $NiTiO_3$/CdS heterostructures for photocatalytic $CO_2$ reduction. *J. Mater. Sci. Technol.* **2025**, *234*, 82–89. [CrossRef]
39. Chai, Y.; Kong, Y.; Lin, M.; Lin, W.; Shen, J.; Long, J.; Yuan, R.; Dai, W.; Wang, X.; Zhang, Z. Metal to non-metal sites of metallic sulfides switching products from CO to $CH_4$ for photocatalytic $CO_2$ reduction. *Nat. Commun.* **2023**, *14*, 6168. [CrossRef]

40. Nasir, M.; Sheng, B.; Zhao, Y.; Ye, H.; Song, J.; Li, J.; Wang, P.; Wang, T.; Wang, X.; Huang, Z.; et al. An integrated photocatalytic redox architecture for simultaneous overall conversion of $CO_2$ and $H_2O$ toward $CH_4$ and $H_2O_2$. *Sci. Bull.* **2024**, *70*, 373–382. [CrossRef]
41. Vertepov, A.; Fedorova, A.; Batkin, A.; Knotko, A.; Maslakov, K.; Doljenko, V.; Vasiliev, A.; Kapustin, G.; Shatalova, T.; Sorokina, N.; et al. $CO_2$ hydrogenation to methanol on CuO-ZnO/$SiO_2$ and CuO-ZnO/$CeO^{2-}$ $SiO_2$ catalysts synthesized with β-Cyclodextrin template. *Catalysts* **2023**, *13*, 1231. [CrossRef]
42. Zhang, F.; Xiong, J.; Yu, X.; Wang, L.; Wu, T.; Yu, Z.; Tang, M.; Liu, H.; Chao, Y.; Zhu, W. Coral reef-like CdS/g-$C_3N_5$ heterojunction with enhanced $CO_2$ adsorption for efficient photocatalytic $CO_2$ reduction. *Catalysts* **2025**, *15*, 94. [CrossRef]
43. Vattikuti, S.; Sudhani, H.; Habila, M.; Rosaiah, P.; Shim, J. $SnO_2$ quantum dot-decorated g-$C_3N_4$ ultrathin nanosheets: A dual-function photocatalyst for pollutant degradation and hydrogen evolution. *Catalysts* **2024**, *14*, 824. [CrossRef]
44. Pei, J.; Li, H.; Yu, D.; Zhang, D. g-$C_3N_4$-based heterojunction for enhanced photocatalytic performance: A review of fabrications, applications, and perspectives. *Catalysts* **2024**, *14*, 825. [CrossRef]
45. Fang, J.; Wang, M.; Yang, X.; Sun, Q.; Yu, L. Design and preparation of $ZnIn_2S_4$/g-$C_3N_4$ Z-scheme heterojunction for enhanced photocatalytic $CO_2$ reduction. *Catalysts* **2025**, *15*, 95. [CrossRef]
46. Wang, X.; Yao, X.; Bai, H.; Zhang, Z. Oxygen vacancy-rich 2D GO/BiOCl composite materials for enhanced photocatalytic performance and semiconductor energy band theory research. *Environ. Res.* **2022**, *212*, 113442. [CrossRef]
47. Wu, Q.; Liang, J.; Huang, Y.; Cao, R. Thermo-, electro-, and photocatalytic $CO_2$ conversion to value-added products over porous metal/covalent organic frameworks. *Acc. Chem. Res.* **2022**, *55*, 2978–2997. [CrossRef]
48. Wang, Y.; Ding, J.; Yin, Q.; Zhang, C.; Zeng, Y.; Xu, S.; Liang, Q.; Zhou, M.; Li, Z. Reinforcing the efficiency of photocatalytic $CO_2$ conversion with $H_2O$ vapor through the integration of photothermal Cu/C@Bi/C supported on 3D g-$C_3N_4$ under full-spectrum solar irradiation. *J. Energy Chem.* **2024**, *12*, 113008. [CrossRef]
49. Sun, J.; Zhai, N.; Miao, J.; Sun, H. Can Green Finance Effectively Promote the Carbon Emission Reduction in "Local-Neighborhood" Areas? Empirical Evidence from China. Environ. *Sci. Pollut.* **2022**, *12*, 10.
50. Wu, Q.; Zhong, Y.; Chen, R.; Ling, G.; Wang, X. Cu-Ag-C@$Ni_3S_4$ with core shell structure and rose derived carbon electrode materials: An environmentally friendly supercapacitor with high energy and power density. *Food Chem.* **2024**, *222*, 119676. [CrossRef]
51. Chen, J.; Jia, Y.; Zhao, H.; Wu, D.; Du, Y. Ternary Heterojunction ZnS-ZnIn2S4-In2S3 Efficiently Catalyzes Triethylamine to Enhance Electrochemiluminescence Imaging Signals. *Adv. Funct. Mater.* **2024**, *35*, 2420714. [CrossRef]
52. Hanye, C.; Shengjie, G.; Guocheng, H.; Qiaoshan, C.; Yanxin, G. Built-in electric field mediated S-scheme charge migration in COF/$In_2S_3$ heterojunction for boosting $H_2O_2$ photosynthesis and sterilization. *Appl. Catal. B Environ.* **2024**, *343*, 123545.

MDPI AG
Grosspeteranlage 5
4052 Basel
Switzerland
Tel.: +41 61 683 77 34

*Catalysts* Editorial Office
E-mail: catalysts@mdpi.com
www.mdpi.com/journal/catalysts

www.ingramcontent.com/pod-product-compliance
Lightning Source LLC
Chambersburg PA
CBHW061554210926
51850CB00009B/2101
*9783725855872*